THE RISE OF THE FOURTH REICH

ALSO BY JIM MARRS

RULE BY SECRECY

ALIEN AGENDA

CROSSFIRE

THE
RISE
OF THE
FOURTH
REICH

THE SECRET SOCIETIES THAT
THREATEN TO TAKE OVER AMERICA

JIM
MARRS

WILLIAM MORROW
An Imprint of HarperCollins*Publishers*

HarperCollins books may be purchased for educational, business, or sales promotional use. For information please write: Special Markets Department, HarperCollins Publishers, 10 East 53rd Street, New York, NY 10022.

FIRST EDITION

Designed by Lovedog Studio

Library of Congress Cataloging-in-Publication Data has been applied for.

ISBN 978-0-06-124558-9

08   09   10   11   12   WBC/RRD   10   9   8   7   6   5   4   3   2   1

This book is dedicated to my father, my uncles, and all the
Allied soldiers who sacrificed so willingly to serve their
country in World War II. They deserve better.

Grateful acknowledgment is given to the following persons,
who have contributed significantly to this work through
their encouragement and assistance: Thomas Ruffner,
Jim Castle, Dan R. Foster, Shawn and Clay Pickering,
Nick Redfern, Larry Sells, Mark Taylor, my conscientious
editors Henry Ferris and Peter Hubbard,
and my forbearing wife, Carol.

CONTENTS

PART THREE
THE REICH ASCENDANT

THE RISE OF THE FOURTH REICH

INTRODUCTION

Adolf Hitler's Third Reich ended in Berlin on April 30, 1945.

Thunder reverberated from a storm of Russian artillery that was bombarding the ruined capital. The day before, along with the incoming shells, came particularly bad news for the fuehrer, who by this late date in World War II was confined to his underground bunker beneath the Reich chancellery. Hitler had learned that two days earlier his Axis partner, Italy's Fascist dictator Benito Mussolini, had been captured by paramilitary Italian resistance fighters. Mussolini and his mistress, Clara Petacci, were executed and their bodies were left hanging from lampposts in a Milan piazza. This news was especially worrisome to Hitler because only hours earlier he had married Eva Braun in a small civil ceremony inside the *Fuehrerbunker*.

Hitler had previously vowed never to be captured alive, and reiterated to his entourage that neither he nor his new bride would be made a "spectacle, presented by the Jews, to divert their hysterical masses." He made obvious preparations for the end of his reign. He handed out poison capsules to his remaining female secretaries and had Blondi, his favorite Alsatian dog, poisoned. Two other household dogs were shot.

Dictating a last will, he stated, "I myself and my wife—in order to escape the disgrace of deposition or capitulation—choose death." He ordered that their bodies be burned immediately. But Hitler, decorated World War I soldier and hardened political fighter, made it clear that he and his philosophies would not leave the world stage quietly. He added, "From the sacrifice of our soldiers and from my own unity with them unto death will in any case spring up in the history of Germany the seed of a radiant renaissance of the National Socialist movement and thus of the realization of a true community of nations."

Hitler then passed along a line of his entourage, mostly women, and shook their hands while mumbling inaudibly. Frau Traudl Junge, one of the secretaries present, recalled that Hitler's eyes "seemed to be looking far away, beyond the walls of the bunker."

At about three P.M. on April 30, members of Hitler's entourage heard a single shot from their leader's quarters. Some time later, Hitler's valet, SS Sturmbannfuehrer Heinz Linge, and an orderly emerged with a blanket-covered body. Martin Bormann, Hitler's private secretary, head of the Nazi Party and the most powerful man in the Reich after Hitler, followed with the body of a woman. The corpses were carried up to a garden area, placed in a shell crater, and burned with gasoline. However, these remains were never found, reportedly due to the constant shelling.

By evening, a Soviet flag was flying atop the Reichstag. It appeared that Hitler and his Third Reich were finished.

THE ESCAPE OF ADOLF HITLER

It was well known and publicly reported that Hitler often made use of doubles, men who closely resembled him, for use at certain public presentations. Pauline Koehler, a maid at Hitler's Berghof in Berchtesgaden, insisted that she knew of at least three men who doubled for Hitler.

Did Hitler make use of one final double in the bunker? After all, the few persons who testified that he was dead were ardent Nazis who were eager to please their captors—whether Russian, British, or American—with accounts of the leader's death. Was the strange execution of Eva Braun's brother-in-law, Hermann Fegelein, due to his knowledge of Hitler's escape plan with the use of a double? Fegelein had left the bunker but protested when captured by an SS search party that he planned to return. He was later shot by a firing squad in the chancellery garden for desertion. Yet, days earlier, Hitler had urged others in the bunker to flee. "Get out! Get out!" he cried. "Go to South Germany. I'll stay here. It is all over anyhow." Why make Fegelein the exception?

Evidence that Fegelein was privy to secret knowledge comes from Kristina Reiman, an actress who met with Fegelein in Berlin on April 27. She told author Glenn B. Infield, "He was very worried. We had several drinks together and he kept repeating that there were two Hitlers in Berlin. . . . I thought he was drunk. Just before he left me, however, he said that if the fuehrer ever discovered that he, Fegelein, knew his secret, Hitler would kill him."

To fake Hitler's death would have been simple. A Hitler double could have been secreted into the bunker any time prior to his reported suicide. After Hitler got Eva to take poison—or a dead duplicate Eva brought in—the double, dressed in the fuehrer's clothing, could have been shot, a poison capsule placed in his mouth, and left to be covered by Bormann and retrieved by the unsuspecting valet Linge.

Hitler could have then passed from the study through his living quarters to a small conference room containing a stairway to the garden above. Hitler had instructed Linge to wait "at least ten minutes before entering the room." While Linge and others from the entourage waited in the hallway outside Hitler's study, the fuehrer's party and an armed SS escort could have made their way to a secluded spot to await darkness.

Under the cover of night, Hitler could have moved along Hermann Goering Strasse, then cut across the Tiergarten to the Zoo Station near Adolf Hitler Platz. From there, they could have followed the rail lines to the Reichssportfeld and crossed the Scharndorfestrasse to the Piechelsdorf Bridge, a short walk to the Havel River, where a Ju-52 floatplane would have been waiting to fly the fuehrer out.

Indeed a Ju-52 pontoon plane had landed on the Havel the previous night, at the radioed request of someone in the *Fuehrerbunker*. It took off that same night. Author Infield has suspected this was a practice run for the following night.

Once away from Berlin, an airplane could have taken Hitler almost anywhere in territory not under direct control of the Allies—Switzerland, Spain, or any number of other friendly locations.

But did this happen?

Conventional history says that Hitler and Eva Braun committed suicide in the bunker—end of story, despite tantalizing tidbits of information that have surfaced since the war. On July 17, 1945, during the Potsdam Conference, Soviet leader Joseph Stalin reportedly told U.S. president Harry S. Truman that Hitler did not commit suicide but probably escaped. Years later, the Russians produced photos purporting to be of Hitler's dead body, which contradicted their earlier accounts that the bodies of Hitler and his mistress had been immediately burned.

Today, while Hitler's fate may be intriguing and undoubtedly will be argued for years, it is immaterial, a moot point. What is certain is that Hitler's legacy—National Socialism—lives on.

THE HISTORY OF how the Nazis, armed with advanced technology and the greatest hoard of treasure in history, were able to escape justice at the end of World War II is perhaps the greatest untold story of the twentieth century.

From the days of Lyndon B. Johnson to those of George W. Bush, there has been talk of "Amerika" turning "fascist." Most people, this author included, dismissed this as radical rhetoric. Unfortunately, as shall be seen, this might not be so far from the truth.

The Germans were defeated in World War II . . . but not the Nazis. They were simply forced to move. They scattered to the four corners of the world. Many of them came to the United States and penetrated what President Dwight D. Eisenhower termed "the military-industrial complex."

They escaped with the loot of Europe as well as rocket science and even more exotic technologies. Some of this technology was so advanced that it remains classified in U.S. government files even today.

Both Nazi science and ideology were brought to America in the aftermath of World War II with the aid and assistance of the very same self-styled globalists who created National Socialism in the first place. Their agenda matches that of the old Bavarian Illuminati, who were long thought to have perished soon after the time of George Washington. But if the order died, its credo lives on—power and control through wealth by any means possible. From the seeds of Nazism planted in America during the Cold War sprang a whole new nation, one that today has become the greatest superpower in history but has also incurred a growing hatred among the nations of the world as well as alienation and dissension among its own citizens.

At the beginning of the third millennium after Christ, by most criteria, the once-free constitutional republic of the United States had become a National Socialist nation, an empire of the creators of the Third Reich—a Fourth Reich. If this assessment seems harsh and unbelievable, read on. Be advised that this work has no political conviction to advocate, no conspiracy theory to press, and no hidden agenda. It is a collection of supportable facts that leads to certain conclusions, uncomfortable and unconventional as they may be.

But first, one must understand the definition of the terms under consideration.

A Definition of Terms

Everyone has heard of Hitler's Third Reich, but what were the First and Second Reichs?

The First Reich is known as the Holy Roman Empire, although it was neither holy, nor Roman, nor an empire. It was founded by the Frankish king Charles I, called Charlemagne or Charles the Great, who was crowned emperor in 800 A.D. by Pope Leo III after conquering and annexing most of Europe, including Germany, Switzerland, Austria, the Low Countries, and parts of France, Italy, and Czechoslovakia. This monarchial empire, modeled after the Roman Empire and ruled by kaisers, or caesars, existed until 1806, when Napoleon marched his troops into Berlin.

The Second Reich was created by Prince Otto von Bismarck, who as premier of Prussia defeated Napoleon III in 1871 and became the "Iron Chancellor" over about three hundred independent states. Bismarck's reich, toward the end headed by Kaiser Wilhelm II, lasted until 1918 and ended with the defeat of the Central Powers of Germany and Austria-Hungary in World War I.

With Adolf Hitler's ascension to power in 1933, he proclaimed Greater Germany as the Third Reich, *Reich* being the German word for "empire." Interestingly enough, when used with a lowercase "r," the word *reich* means "rich" or "wealthy." A *Reich*, therefore, could mean "an empire of the wealthy."

The term "Nazi" stems from the acronym of "National Socialism." This was derived by combining the first syllable of "NAtional" and the second syllable of "*soZIalist*" in the name Nationalosozialistiche Deutsche Arbeiterpartei, the National Socialist German Workers Party. This was the small radical political party Hitler built into a fascist system that threatened the entire world. Nazism is a philosophy. One recent dictionary defines a Nazi as a person "holding extreme racist or authoritarian views or behaving brutally" or anyone "belonging to any organization similar to the Nazis."

One edition of the *American Heritage Dictionary of the English Language* defined fascism as "a philosophy or system of government that advocates or exercises a dictatorship of the extreme right, typically through the merging of state and business leadership together with an ideology of belligerent nationalism." Always remember that a typical attribute of fascism is the merging of state and business leadership.

Italian dictator Benito Mussolini is credited with coining the word "fascism," a name taken from his fascist Black Shirts called *Fascisti*. This term derived from the ancient Roman symbol of the fasces, a bundle of rods with a protruding axe blade. It was the symbol of central authority. Under fascism, the individual is subordinate to the state, usually headed by a single leader.

However, even Mussolini pointed out that "The first stage of fascism should more appropriately be called corporatism, because it is the merger of state and corporate power." For the remainder of this work, fascism will be defined as the merger of state and corporate power.

In fascist Italy and Nazi Germany, the state gained control over the corporations. In modern America, corporations have gained control over the state.

The end result is the same.

Mussolini proclaimed, "The maxim that society exists only for the well-being and freedom of the individuals composing it does not seem to be in conformity with nature's plans. . . . If classical liberalism spells individualism, fascism spells government."

And Thomas J. DiLorenzo, professor of economics at Loyola College in Baltimore, Maryland, wrote, "[I]t is important to recognize that, as an economic system, fascism was widely accepted in the 1920s and 30s. The evil deeds of individual fascists were later condemned, but the practice of economic fascism never was."

Party politics, slogans, and social issues are employed to distract the masses. The world's elite deal in only one commodity—power. They seek to gain and maintain the controlling power that comes from great wealth, usually gained through the monopoly of ownership over basic resources. Politics and social issues matter little to the globalist ruling elite, who move smoothly between corporate business and government service. The desire for wealth with its attendant power and control drives their activities. It is this unswerving attention to commerce and banking that lies behind nearly all modern world events. It is the basis for a "New World Order" mentioned by both Hitler and former president George H. W. Bush.

In twenty-first-century America, many thoughtful persons have witnessed what appears to be a recycling of the events of pre–World War II Germany: the destruction of a prominent national structure; rushed emergency legislation; the rise of a secretive national security apparatus; attempts to register both firearms and people, coupled with preemptive wars of aggression propelled by fervent nationalism.

This may be simply a coincidence, some synchronistic cycle of history. But this also may be a covert plan being carried out by individuals following a definite agenda.

COMMUNISM VERSUS NATIONAL SOCIALISM

As documented in *Rule by Secrecy*, the same financial powers that built the United States into the world's foremost superpower also created communism. After an aborted revolution in 1905, thousands of Russian activists had been exiled, including the revolutionaries Leon Trotsky and Vladimir Lenin. After years of attempts at reform, the czar was forced to abdicate on March 15, 1917, following riots in Saint-Petersburg believed by many to have been instigated by British agents.

In January 1917, Leon Trotsky, a fervent follower of Karl Marx, was living rent-free on Standard Oil property in Bayonne, New Jersey. He worked in New York City as a reporter for *The New World*, a communist newspaper. Trotsky had escaped Russia in 1905 and fled to France, from where he was expelled for his revolutionary behavior. "He soon discovered that there were wealthy Wall Street bankers who were willing to finance a revolution in Russia," wrote journalist William T. Still.

One of these bankers was Jacob Schiff, whose family had lived with the Rothschild family in Frankfurt, Germany. According to the *New York Journal-American*, "[I]t is estimated by Jacob's grandson, John Schiff, that the old man sank about $20 million for the final triumph of Bolshevism in Russia." Schiff, a Rockefeller banker, had financed the Japanese in the 1904–05 Russo-Japanese War for control of Manchuria, and had sent his emissary George Kennan to Russia to promote revolution against the czar.

Another was Senator Elihu Root, attorney for Federal Reserve cofounder Paul Warburg's Kuhn, Loeb & Company. Root, an honorary president of the secretive Council on Foreign Relations and a former U.S. secretary of state, who moved smoothly between government positions and his law practice in New York City, contributed yet another $20 million, according to the congressional record of September 2, 1919.

Schiff and Root were not alone. Arsene de Goulevitch, who was present during the early days of the Bolsheviks, later wrote, "In private interviews, I have been told that over 21 million rubles were spent by Lord [Alfred] Milner in financing the Russian Revolution." Milner, a German-born British statesman, was the primary force behind Cecil Rhodes's Round Tables, a predecessor of the Council on Foreign Relations. The American

International Corporation (AIC), formed in 1915, also helped fund the Russian Revolution. AIC directors represented the interests of the Rockefellers, Rothschilds, Du Ponts, Kuhns, Loebs, Harrimans, and the Federal Reserve, as well as Federal Reserve cofounder Frank Vanderlip and George Herbert Walker, the maternal great-grandfather of President George W. Bush.

Trotsky left the United States by ship on March 27, 1917—just days before America entered the war—along with nearly three hundred revolutionaries and funds provided by Wall Street. Trotsky, whose real name was Lev Davidovich Bronstein, was being trailed by British agents who suspected him of working with German intelligence since his stay in prewar Vienna. In a speech before leaving New York, Trotsky stated, "I am going back to Russia to overthrow the provisional government and stop the war with Germany."

When the ship carrying Trotsky and his entourage stopped in Halifax, Nova Scotia, they and their funds were impounded by Canadian authorities, who rightly feared that a revolution in Russia might free German troops to fight their soldiers on the Western Front. But this well-grounded concern was overcome by President Woodrow Wilson's alter ego, Colonel Edward Mandell House, who told the chief of the British Secret Service, Sir William Wiseman, that Wilson wanted Trotsky released. On April 21, 1917, less than a month after the United States entered the war, the British Admiralty ordered the release of Trotsky, who, armed with an American passport authorized by Wilson, continued on his journey to Russia and history.

At this same time, Lenin also left exile. Aided by the Germans and accompanied by about 150 trained revolutionaries, "[he] was put on the infamous 'sealed train' in Switzerland along with at least $5 million," according to Still. The train passed through Germany unhindered, as arranged by German banker Max Warburg (the brother of Paul Warburg, who cofounded the Federal Reserve System and handled U.S. financing during World War I) and the German High Command. Lenin, like Trotsky, was labeled a German agent by the government of Alexandr Kerensky, the second provisional government created following the czar's abdication. By November 1917, Lenin and Trotsky, backed by Western funds, had instigated a successful revolt and seized the Russian government for the Bolsheviks.

But the communist grip on Russia was not secure. Internal strife between the "reds" and the "whites" lasted until 1922 and cost some 28 million Russian lives, many times the war loss. Lenin died in 1924 from a series of strokes after establishing the Third International, or Comintern, an organization formed to export communism worldwide. Trotsky fled Russia when Joseph Stalin took dictatorial control, and, in 1940, was murdered in Mexico by an agent of Stalin's.

Some conspiracy authors have seen a dual purpose to the funding of the Bolsheviks. It is clear that revolutionaries like Lenin and Trotsky were being used to get Russia out of the war, to the benefit of Germany. And communism was being supported by the globalists to advance their plan of creating tension between the capitalist West and socialist East.

A. K. Chesterson, a right-wing British journalist and politician, who in 1933 joined Oswald Moseley's British Union of Fascists, observed that to understand politics, one must make a study of power elites. "These elites, preferring to work in private, are rarely found posed for photographers, and their influence on events has therefore to be deduced from what is known of the agencies they employ." He once wrote in his magazine, *Candour,* "At times capitalism and communism would appear to be in conflict, but this writer is confident that their interests are in common and will eventually merge for one-world control."

Because of the warring factions in post-revolution Russia, sending an official delegation to Russia was problematic. Therefore, American financiers came in the form of the American Red Cross Mission. One head of this group was Raymond Robins, described as "the intermediary between the Bolsheviks and the American government" and "the only man whom Lenin was always willing to see and who even succeeded in imposing his own personality on the unemotional Bolshevik leader." Lenin apparently came to understand that he was being manipulated. "The state does not function as we desired," he once wrote. "A man is at the wheel and seems to lead it, but the car does not drive in the desired direction. It moves as another force wishes." This other "force" was the globalists behind the birth of communism itself, "monopoly finance capitalists," as Lenin described them.

"One of the greatest myths of contemporary history is that the Bolshevik Revolution in Russia was a popular uprising of the downtrodden masses against the ruling class of the Czars," wrote author G. Edward Griffin. ". . . however, the planning, the leadership, and especially the financing came entirely from outside Russia, mostly from financiers in Germany, Britain, and the United States."

The flight of the privileged elite from Russia in 1918 sent shockwaves through the capitals of Europe and America and prompted a backlash that lasted for decades. The cry "Workers of the world, unite!" struck fear into the capitalists of Western industry, banking, and commerce who were not in the know. This fear trickled through their political representatives, employees, and on into virtually every home.

Mystified conspiracy researchers were puzzled for years about why such high-level capitalists as the Morgans, Warburgs, Schiffs, and Rockefellers could condone, much less support, an ideology that overtly threatened their position and wealth. Author Gary Allen explained, "In the Bolshevik Revolution we have some of the world's richest and most powerful men financing a movement which claims its very existence is based on the concept of stripping of their wealth men like the Rothschilds, Rockefellers, Schiffs, Warburgs, Morgans, Harrimans and Milners. But obviously these men have no fear of international communism. It is only logical to assume that if they financed it and do not fear it, it must be because they control it. Can there be any other explanation that makes sense?"

The manufactured animosity between the democracies of the West and the communism of the East produced continuous tension from 1918 through the end of the twentieth century. But it threatened to get out of hand. Some researchers believe that the threat of worldwide communist socialism caused these globalists to turn to German nationalists. They funded the rise of National Socialism in Germany and saw an armed Greater Germany as a barrier to communism in Europe. National Socialism was a form of socialism almost indistinguishable from communism, only it was confined within national geographic boundaries. Under National Socialism, the globalists could pit the various nations against each other. But following Germany's military successes in Poland, the Low Countries, and France, these globalists realized they faced the same problem they had with the communists. A total German victory would result in a worldwide National Socialist system unable to produce the tensions and conflicts necessary for maximizing profit and control. They also may have feared that Stalin's Soviet Union was about to launch an attack on Western Europe. Only Hitler's Germany had the strength to prevent this.

At some point, the globalists determined that the Axis, after blocking Russia's invasion of Europe, should lose the war. They also began drawing up plans for the survival and renewal of a new form of National Socialism, one not dependent on racism and ethnicity. Working with the same financiers and capitalists that had helped create German Nazism, these globalists began laying the foundation for a Fourth Reich.

Conspiracy researchers have long suspected that one element of this German influence has been centered in the secretive Skull and Bones fraternity on the campus of Yale University. Known variously as Chapter 322, the Brotherhood of Death, the Order, or, more popularly, as Skull and Bones or simply Bones, the Order was brought from Germany to Yale in 1832 by General William Huntington Russell and Alphonso Taft.

(Russell's cousin, Samuel Russell, was an integral part of the British-inspired Opium Wars in China. Taft, secretary of war in 1876 and U.S. attorney general and an ambassador to Russia, was the father of William Howard Taft, the only person to serve as both president and chief justice of the United States. Another prominent Bones member was Averell Harriman, who has been described as "a man at the heart of the American ruling class," and played a prominent role in the establishment of the new American empire.)

A pamphlet detailing an 1876 investigation of Skull and Bones headquarters at Yale, known as "the Tomb" by a rival secret society, stated, ". . . its founder [Russell] was in Germany before Senior Year and formed a warm friendship with a leading member of a German society. He brought back with him to college authority to found a chapter here. Thus was Bones founded."

The secret German society may have been none other than the mysterious and infamous Illuminati. Ron Rosenbaum, a former Yale student and one of the few journalists to take a serious look at Skull and Bones, noted that the official skull-and-crossbones emblem of the Order was also the official crest of the Illuminati. In an investigative piece for *Esquire* magazine, Rosenbaum wrote, "I do seem to have come across definite, if skeletal, links between the origins of Bones rituals and those of the notorious Bavarian Illuminists . . . [who] did have a real historical existence. . . . From 1776 to 1785 they were an esoteric secret society with the more mystical freethinking lodges of German Freemasonry."

Other researchers agree that the Order is merely the Illuminati in disguise, since Masonic emblems, symbols, German slogans, even the layout of their initiation room, all are identical to those found in Masonic lodges in Germany associated with the Illuminati. The Tomb is decked out with engravings in German, such as *"Ob Arm, Ob Reich, im Tode gleich"*— "Whether poor or rich, all are equal in death." According to *U.S. News & World Report,* one of the Bonesmen's traditional songs is sung to the tune of *"Deutschland Über Alles."*

The Bavarian Illuminati was formed on May 1, 1776, by Adam Weishaupt, a professor of canon law at Ingolstadt University of Bavaria, Germany. His Illuminati were opposed to what they saw as the tyranny of the Catholic Church and the national governments it supported. "Man is not bad," Weishaupt wrote, "except as he is made so by arbitrary morality. He is bad because religion, the state, and bad examples pervert him. When at last reason becomes the religion of men, then will the problem be solved."

Weishaupt also evoked a philosophy that has been used with terrible results down through the years by Hitler and many other tyrants. "Behold our secret. Remember that the end justifies the means," he wrote, "and that the wise ought to take all the means to do good which the wicked take to do evil." Thus, for the enlightened—or "illuminated"—any means to gain their ends is acceptable, whether this includes deceit, theft, murder, or war.

The key to Illuminati control was secrecy. "The great strength of our Order lies in its concealment. Let it never appear in any place in its own name, but always covered by another name, and another occupation," stated Weishaupt. He not only deceived the public, but he reminded his top leaders they should hide their true intentions from their own initiates by "speaking sometimes in one way, sometimes in another, so that one's real purpose should remain impenetrable to one's inferiors."

In 1777, Weishaupt rolled his brand of Illuminism into Freemasonry after joining the Masonic Order's Lodge Theodore of Good Counsel in Munich. This lodge integrated with the Grand Orient Lodges, which, according to several researchers, were at the core of the Illuminati penetration into Freemasonry. By 1783, the Bavarian government saw the Illuminati as a direct threat to the established order and outlawed the organization, which prompted many members to flee Germany, only spreading their philosophies farther.

Many researchers today believe the Illuminati still exists and that the order's goals are nothing less than the abolition of all government, private property, inheritance, nationalism, the family unit, and organized religion. This belief partially comes from the intriguing notion that the much-denounced *Protocols of the Elders of Zion*—used widely since its publication in 1864 to justify anti-Semitism—was actually an Illuminati document with Jewish elements added later for disinformation purposes. "Even though the Illuminati faded from public view, the monolithic apparatus set in motion by Weishaupt may still exist today," commented author William T. Still. "Certainly, the goals and methods of operation still exist. Whether the name Illuminati still exists is really irrelevant."

No one can doubt that socialism, whether Illuminati-inspired or not, has come to the United States, and socialism is the cornerstone of Nazi philosophy. Beginning with seemingly innocuous programs like Social Security and continuing through a myriad of government programs such as Medicare, farm subsidies, food stamps, and student entitlements, it seems that nearly every aspect of life today involves the centralized federal government, which, since the attacks of 9/11, continues to draw ever more power unto itself. *USA Today* reported, "A sweeping expansion of social programs since 2000 has sparked a record increase in the number of Americans receiving federal government benefits such as college aid, food stamps and health care. A *USA Today* analysis of 25 major government programs found that enrollment increased an average of 17% in the programs from 2000 to 2005."

Socialism has come to America because the National Socialists of the New World Order recognize that any social program requires central authority. And they know full well that with their immense wealth and power, they can control any central authority. Over the years, they have masked this creep of socialism by distracting appeals to nationalism. Americans are constantly reminded that the United States is God's gift to the world, the epitome of freedom and democracy. Patriotism has been used to fan the flames of nationalism among Americans. Today, anyone who criticizes foreign policy, overseas military interventions, or even questions national policies opens themselves to charges of being unpatriotic.

It is possible that the United States is indeed becoming the Fourth Reich, the continuation of a philosophy of National Socialism thought to have been vanquished more than half a century ago. Such a concept may seem absurd to those who cannot see past the rose-colored spin, hype, and disinformation poured out daily by the corporate mass media, most of which is owned by the same families and corporations that supported the Nazis before World War II.

Many today describe what they see as "neo-Nazism," the movement to revive National Socialism. But this is a misnomer. There is nothing neo, or new, about this trend. National Socialism never died. The philosophies of fascism are alive and active in modern America. Unfortunately, younger generations cannot understand the nuances of differences between fascism, corporate power, democracy, and a democratic republic.

While the USA helped defeat the Germans in World War II, we failed to defeat the Nazis. Many thousands of ranking Nazis came to the United States under a previously classified program called Project Paperclip. Many other Nazis and war criminals set up shop in a variety of other nations, and many traveled on passports issued by the Vatican. They brought with them miraculous technology, such as the V-2 rockets, but they also brought with them Nazi ideology. This ideology, based on the Illuminati premise that the end justifies the means, includes unprovoked wars of aggression and curtailment of individual liberties, and has gained sway in "the land of the free and the home of the brave."

Ranking Nazis, along with their young and fanatical protégés, used the loot of Europe to create corporate front companies in many countries, including Spain, Portugal, Sweden, Turkey, and Argentina. More than two hundred fronts were created just in Switzerland, that banking hub that continued to handle Nazi money before, during, and after the war. Utilizing the stolen wealth of Europe, which may have included the legendary treasure of Solomon, men with both Nazi backgrounds and Nazi mentality wormed their way into corporate America, slowly buying up and consolidating companies into giant multinational conglomerates. They met little resistance from corporate leaders who had supported them in previous years and could not resist the temptation of obscene profits. Nor were they checked by others, who had grown fearful over the "communist threat." In time, they all became partners in a new version of America.

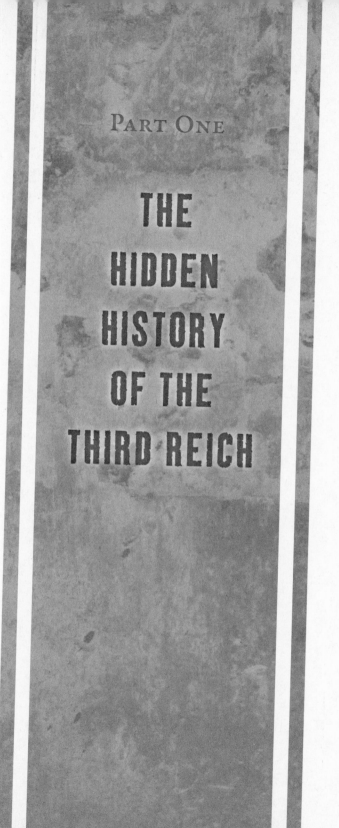

PART ONE

THE HIDDEN HISTORY OF THE THIRD REICH

CHAPTER 1

A NEW REICH BEGINS

HITLER'S SUPPORT GROUP

FOLLOWING THE ARMISTICE OF 1918, WHICH ENDED WORLD WAR I, German soldiers returned home, to a country economically devastated by the war. The Bavarian city of Munich was hit particularly hard, with jobless ex-soldiers wandering the streets and a number of splinter political parties vying for membership.

It was in this setting that Hitler, a twenty-nine-year-old veteran, came into contact with members of the Thule Gesellschaft, or Thule Society, ostensibly an innocent reading group dedicated to the study and promotion of older German literature. But the society, composed mostly of wealthy conservatives, ardent nationalists, and anti-Semites, actually delved into radical politics, race mysticism, and the occult under its emblem—a swastika superimposed over a sword.

The society also served as a front for the even more secretive Germanenorden, or German Order, a reincarnation of the old Teutonic Knights, which had branches throughout Germany patterned after Masonic lodges. It is believed that these lodges carried on the agenda of the outlawed Bavarian Illuminati, with its fundamental maxim that "the end justifies the means." In other words, members should pretend to be anything or anybody, adopt any philosophy, tell any lie, steal, cheat, even kill as long as it accomplishes the society's objectives.

Members of the Thule Society encouraged a Munich locksmith and toolmaker named Anton Drexler to bring workers into the political process. The unassuming Drexler founded the Deutsche Arbeiterpartei, or German Workers Party, which was guided to prominence by covert aid from conservative elements within industry and the military.

Hitler, unable to make a living as an artist, turned to earning extra money by serving as an army intelligence agent reporting to a Captain Karl Mayr. "One day I received orders from my headquarters to find out what was behind an apparently political society which, under the name of 'German Workers Party,' intended to hold a meeting. . . . I was to go there and look at the society and to report upon it," Hitler recalled in *Mein Kampf.* Arriving at the Sterneckerbräu beer hall, he was not overly impressed. "I met there about 20 to 25 people, chiefly from among the lower walks of life," wrote Hitler. However, the young military agent stood and "astonished" the small gathering by arguing against a proposal that Bavaria break ties with Prussia. Impressed with the nationalistic and anti-Semitic views of the fledgling party, military authorities allowed Hitler to join and began funding the party's work. He became the party's seventh registered member.

Hitler's work in the party was initially supported both by funds from Captain Mayr's army intelligence unit and the dedicated anticommunists and occultists of the Thule Society. Funding was passed through the publisher of occult literature, Dietrich Eckart, whom Hitler called the "spiritual founder of National Socialism." Eckart was soon introducing the new member to the right social circles in Munich and his intellectual friends in the Thule Society. The editors of Time-Life Books noted, "Dietrich Eckart took over as editor of the *Volkischer Beobachter,* the renamed *Munchener Beobachter,* which the party had purchased from the Thule Society with money supplied partly by Mayr's secret army account."

Author Joseph P. Farrell stated that the covert connections of Eckart and future deputy fuehrer Martin Bormann support the idea "that Hitler was deliberately manipulated and placed into power, and secretly manipulated behind the scenes by more powerful forces than even he wielded, and, when he had served his purpose, was deliberately sabotaged and cast aside." The forceful Hitler, armed with adequate funds, quickly gained control of the German Workers Party, which soon claimed three thousand members. In April 1920, Hitler changed the party's name to the Nationalsozialistiche Deutsche Arbeiterpartei, the National Socialist German Workers Party, abbreviated to Nazi.

Following an ill-fated attempt to take control of the government in 1923, known as the Beer-hall Putsch, Hitler and his lieutenants were imprisoned and the Nazi Party languished. Upon his release after only nine months, Hitler began to direct the Nazi Party into more effective, and legal, activities, which resulted in the Nazis becoming the largest political party in Germany by July 1932.

It was, in fact, wealthy businessmen in Western industrial and banking circles who guaranteed Hitler's success. After Hitler lost a popular election to Hindenburg in 1932, thirty-nine business leaders, with familiar names like Krupp, Siemens, Thyssen, and Bosch, signed a petition urging the aged president Paul von Hindenburg to name Hitler chancellor. In January 1933, through a compromise with German aristocrats, industrialists, and army officers, brokered by banker Baron Kurt Freiherr von Schroeder, Hitler was appointed chancellor of Germany. The deal to name Hitler chancellor of Germany was cut at von Schroeder's home on January 4, 1933. On hand were prominent industrialists, at least one director of the giant Deutsche Bank as well as I. G. Farben's Hermann Schmitz and Dr. Georg von Schnitzler representing Farben's board of directors. According to author Eustace Mullins, also attending this meeting were John Foster Dulles and Allen Dulles of the New York law firm Sullivan and Cromwell, which represented the Schroeder bank. This claim has been disputed by other researchers.

At that time, Germany was a free republic with one of the most educated and cultured populations in the world. The country was at peace and enjoying a blossoming of democratic freedom under a coalition government of the Weimar Republic.

Oddly, Hitler went against tradition by choosing not to work out of an office in the German Reichstag, or parliament building, and on February 27, 1933, the Reichstag was gutted by fire. In those slower, gentler times, this act was as great a shock to the German people as the destruction of the World Trade Center towers was to Americans in 2001. Hitler blamed the destruction on communist terrorists. Inside the building, police arrested an incoherent, half-naked retarded Dutch youth named Marinus van der Lubbe. They said he was carrying a Dutch Communist Party card. After some time in custody, the youth confessed to being the arsonist. However, later investigation found that one person could not have started the mammoth blaze and that incendiary devices had been carried into the building through a tunnel that led to the offices of Hitler's closest partner, Hermann Goering.

Despite misgivings in many quarters about the official explanation of the fire, it was announced that "the government is of the opinion that the situation is such that a danger to the state and nation existed and still exists." Law enforcement agencies quickly moved against not only the communists but also pacifists, liberals, and democrats. Less than a month later, on March 24, 1933, at Hitler's urging, a panicky German Parliament voted 441 to 94 to pass an "Enabling Act," which became the starting point for Hitler's dictatorship.

As a result of the Enabling Act, the Nazi government required national identity cards, racial profiling, the equivalent of a national homeland security chief (SS Reichsfuehrer Heinrich Himmler), gun confiscation, and, later, mass murders and incarcerations in concentration camps. "When Germany awoke," wrote British reporter Douglas Reed, "a man's home was no longer his castle. He could be seized by private individuals, could claim no protection from the police, could be indefinitely detained without preferment of charge; his property could be seized, his verbal and written communications overheard and perused; he no longer had the right to foregather with his fellow countrymen, and his newspapers might no longer freely express their opinions."

Hitler's financiers and especially Prussian military officers were becoming alarmed over Hitler's growing power, especially with some three million Sturmabteilung (SA) or Storm Detachment Brown Shirts under the command of Hitler's SA chief Ernst Roehm awaiting orders. The army proposed a deal—if the power of the SA was broken, the military would pledge loyalty to Hitler. Hitler agreed, and on June 30, 1934, trumped-up charges of plotting a revolution caused Roehm and hundreds of Brown Shirts to be fatally purged and the SA quietly faded away. The German military began pledging their allegiance not to their nation but to Hitler. With the German population firmly under control due to massive propaganda and fear of government retaliation, Hitler was free to launch invasions into former German territories as well as Poland.

It is instructive that Hitler did not invade Poland without pretext. A "false-flag" operation was accomplished first. SS men dressed in Polish uniforms fabricated an attack on a German radio station at Gleiwitz, which allowed Hitler to announce that a counterattack had been launched against Polish soldiers who had invaded German territory. Germany was simply making the world safe for National Socialism. World War II ensued.

WITH THE DEATH of the eighty-seven-year-old Hindenburg on August 2, 1934, Hitler merged the offices of president and chancellor and proclaimed himself commander in chief of the armed forces, the absolute leader—fuehrer—of all Germany.

He found a huge and powerful industrial base geared for war production already in place and awaiting his command. It had been put in place at the end of World War I thanks to an influx of Western capital investment. "This build-up for European war both before and after 1933 was in great part due to Wall Street financial assistance in the 1920s to create the German cartel system and to technical assistance from well-known American firms . . . to build the German Wehrmacht," noted financial investigator and author Antony C. Sutton, who added, "The contribution made by American capitalism to German war preparations before 1940 can only be described as phenomenal." (For example, in 1934 Germany produced only 300,000 tons of natural petroleum products and synthetic gasoline. In 1944, thanks to the transfer of hydrogenation technology from Standard Oil of New Jersey to I. G. Farben, Germany produced 6,500,000 tons of oil, 85 percent of which was synthetic.)

The intertwining of American capitalism with German corporations began following World War I, with two programs: the Dawes Plan (1924) and the Young Plan (1928). Both plans, engineered in America, virtually guaranteed success for the fledgling Nazi Party. The Dawes Plan, designed to restructure German war reparations, was named for chairman of the Allied Reparations Committee Charles G. Dawes and described by historian Carroll Quigley as "largely a J. P. Morgan production." This plan used American loans to create and consolidate the German steel and chemical giants, Vereinigte Stahlwerke and I. G. Farben, both major supporters of Hitler. It caused anger and frustration among the Germans, because it meant foreign control of Germany's finances—a fact constantly pointed out by Hitler in his speeches—and appeared open-ended, as no final reparation amount was ever announced. Its successor, the Young Plan, named for J. P. Morgan agent Owen D. Young, required burdensome monetary payments from Germany. It, too, led to support within Germany for Hitler and his Nazis.

Financing the rearmament of Germany in violation of the Versailles Treaty proved as profitable as it was dangerous to European peace. German steel magnate Fritz Thyssen, a major financial contributor to Hitler, stated, "I turned to the National Socialist Party only after I became convinced that the fight against the Young Plan was unavoidable if complete collapse of Germany was to be prevented."

Finance between Germany and the Allied nations was controlled by the Bank of International Settlements (BIS), headquartered in Basel, Switzerland. It was the brainchild of Hjalmar Horace Greeley Schacht, president of the Reichsbank (in 1930, he resigned in protest to the Young Plan but was reappointed by Hitler in 1933) and the financial genius behind Germany's economic revival. Although his father was an American citizen, Hjalmar was born in Germany during his mother's return there and named after the famous American editor and politician. It was Schacht who provided an ongoing link between Hitler and Germany's industrialists.

The BIS was administered by a multinational staff, which historian Quigley called the "apex of the system" of bankers, to secretly exchange information and plan for the coming war. One of the corporate giants created in post–World War I Germany with assistance from American capital was Internationale Gesellschaft Farbenindustrie A. G., better known in its shortened version as I. G. Farben. Created in 1926 by combining six existing chemical companies, it was the brainchild of Hermann Schmitz, who became the firm's president. Under his guidance, I. G. Farben became the largest chemical manufacturing enterprise in the world. It was so powerful during the Nazi regime that the firm became known as a "state within a state."

Farben had subsidiaries, offices, and representatives in ninety-three countries, including the United States. Paul Manning, a CBS news correspondent in Europe during World War II, explained Schmitz's connections by pointing out that the Farben chief once "held as much stock in Standard Oil of New Jersey as did the Rockefellers." By the time war began in 1939, I. G. Farben had doubled in size, gaining participation and managerial control over 380 other German companies as well as more than 500 foreign firms. This growth was made possible by bond sales in America, including one for $30 million offered by National City Bank, a forerunner of today's Citibank.

It was I. G. Farben's patented Zyklon-B, a prussic acid poison gas, that was used to kill victims in the "shower baths" of Auschwitz, Maidanek, and Treblinka. Previously, the firm had received a contract to produce carbon monoxide, used to gas the sick and mentally deficient under Germany's euthanasia program.

One example of the close business ties between the United States and Nazi Germany was Walter C. Teagle, chairman of Standard Oil of New Jersey, which was owned by Rockefeller's Chase Bank. Teagle also was a director of American I. G. Chemical Corporation, one of the subsidiaries of I. G. Farben, which changed its name to General Aniline and Film (GAF) in an effort to distance itself from its German owners.

Teagle, through Rockefeller banking and oil interests, made his superiors a handsome profit just prior to the war. "[Teagle] remained in partnership with Farben in the matter of tetraethyl lead, an additive used in aviation gasoline," wrote author Charles Higham. "[German Luftwaffe chief Hermann] Goering's air force couldn't fly without it. Only Standard, Du Pont and General Motors had the rights to it. Teagle helped organize a sale of the precious substance to [Farben president] Schmitz, who in 1938 traveled to London and 'borrowed' 500 tons from Ethyl, the British

Standard subsidiary. Next year, Schmitz and his partners returned to London and obtained $15 million worth. The result was that Hitler's air force was rendered capable of bombing London, the city that had provided the supplies. Also, by supplying Japan with tetraethyl, Teagle helped make it possible for the Japanese to wage World War II."

Following negative publicity regarding these tetraethyl transactions in 1938, Teagle resigned from the board of GAF, to be replaced by future Secretary of Defense James V. Forrestal. Curiously, it was this same Walter Teagle who helped create the National Recovery Administration, one of President Roosevelt's New Deal agencies designed to regulate American business. This was an odd choice if the captains of American industry were as opposed to socialism as they publicly claimed.

By the mid-1930s, with the government, military, and the German cartels now firmly in hand, Hitler knew it was time to strengthen his influence over international bankers and businessmen. Despite his declared intentions to nationalize German businesses and curtail the power of international business and finance, Hitler initially had little trouble getting funds from corporate sponsors who saw his National Socialism as a necessary alternative to worldwide communism.

"[H]is appeal to the common people offered a chance to win the working class away from communism," noted James Pool, author of *Who Financed Hitler*.

In America, efforts were under way to market Nazism to the public while concurrently practicing economic espionage. Teagle, together with Farben chief Schmitz, hired famed New York publicist Ivy Lee to pass proprietary information on American companies to Germany and to spin news stories so as to gloss over the darker side of Nazism. By the late 1930s, Lee was being paid $25,000 a year for disseminating pro-Nazi propaganda in America. Payments to Lee came from a Farben U.S. subsidiary, American I. G., and moved through Lee's company account with Chase Bank and his personal account at New York Trust Company. "They were American funds earned in the U.S. and under control of American directors, although used for Nazi propaganda in the United States," stated author Sutton.

Another solid conduit for Nazi propaganda and intelligence activities was the Hamburg-Amerika shipping line. Max Warburg, a leader of Deutsche Bank, sat on the board of Hamburg-Amerika Steamship Line along with Prescott Bush, father and grandfather of two future U.S. presidents. Max Warburg was the brother of Paul Warburg, America's first chairman of the Federal Reserve System and the man in charge of U.S. finances in World War I.

American I. G. Chemical Corporation was more than just a source of funds. It provided important intelligence to the Nazis throughout the war as noted by I. G. Farben director Max Ilgner, the nephew of Farben's chairman Schmitz. He wrote, "Extensive information which we receive continuously from [American I. G.] is indispensable for our observations of American conditions . . . [and] is, since the beginning of the war, an important source of information for governmental, economic and military offices."

"The full story of I. G. Farben and its worldwide activities before World War II can never be known," noted author Sutton, "as key German records were destroyed in 1945 in anticipation of Allied victory."

The banking industry, including foreign financial houses, provided Hitler and his Nazis with the funds to both consolidate and spread their National Socialist doctrine. Throughout World War II, the bank of Baron von Schroeder acted as financial agents for Germany in both Britain and the United States. Antony C. Sutton described how John Foster Dulles handled Schroeder bank loans in the USA. In fact, Dulles, in addition to providing legal services to a joint Rockefeller-Schroeder investment firm, the Schroeder-Rockefeller Company, also sat on the board of directors of General Aniline and Film (GAF) from 1927 to 1934. GAF, as it was known during the war, remained a subsidiary of I. G. Farben. Schroeder, the powerful head of the J. H. Stein & Company banking house of Cologne, had long provided financial support to the Nazis in hopes they would counteract the spread of communism. Hitler had given his word to von Schroeder that "National Socialism would engage in no foolish economic experiments"—in other words, he would not attack banking practices except in rhetoric.

This closeness between Hitler and the banking industry reached back to the earliest days of the Nazi Party. "On New Year's Day 1924, the financial fate of Germany was settled in London at a meeting between Hjalmar Schacht, the new Reich Commissioner for National Currency, and Montagu Norman, governor of the Bank of England," noted author John Toland. "Schacht, who had already abolished emergency money, began with a frank disclosure of Germany's desperate financial situation." He then proposed to open a German credit bank second to the Reichsbank, but one that would issue notes in pound sterling. Schacht asked Norman to provide half the capital for this new bank. "Within 48 hours Norman not only formally approved the loan at the exceptionally low interest of a flat five percent but convinced a group of London bankers to accept bills far exceeding the loan . . . ," Toland wrote.

A year after the meeting at Schroeder's home that launched Hitler into power, Nazi official and ideologist Alfred Rosenberg met with Schroeder Bank of London managing director T. C. Tiarks, who also was a director of the Bank of England.

Montagu Norman, governor of the Bank of England, in early 1934 informed a select group of City of London financiers that Hitler's regime was a system with a good future. With no opposition, it was decided to provide covert financial help to Hitler until Norman could persuade the British government to abandon its pro-French policy to one more favorable to Germany.

Thus a curious relationship developed between these powerful banks of two supposedly belligerent nations, which continued throughout the war. Of course, it only appears curious to those who do not understand the economic cooperation among the world's ruling elite.

In December 1938, Schacht came to England as a guest at Montagu Norman's home. The visit was declared to be purely personal. One month later, the hospitality was returned when Norman stopped over in Berlin for a visit with Schacht on his way to Switzerland. Although there was no public announcement as to what the two men discussed, it was rumored that they were attempting to create a common policy of settling Germany's foreign debts and expanding its markets. There were also rumors that Britain's bankers might extend to Germany some $375 million in export credits. The importance of such financial cooperation was explained by Sutton: "In the 1920s and 1930s, the New York Federal Reserve System, the Bank of England, the Reichsbank in Germany and the Banque de France also more or less influenced the political apparatus of their respective countries indirectly through control of the money supply and creation of the monetary environment. More direct influence was realized by supplying political funds to, or withdrawing support from, politicians and political parties."

Funds for this financial control were channeled through the Bank of International Settlements that, according to the bank's charter and with the agreement of the respective governments, was immune from seizure, closure, or censure, even if its owners were at war. These owners included the Morgan-affiliated First National Bank of New York (among whose directors were Harold S. Vanderbilt and Wendell Wilkie), the Bank of England, the Reichsbank, the Bank of Italy, the Bank of France, and other central banks. "The bank [BIS] soon turned out to be . . . a money funnel for American and British funds to flow into Hitler's coffers and to help Hitler build up his war machine," wrote Higham. By the start of World War II, the BIS was under Nazi control with the bank's directors including Schmitz, Schroeder, Dr. Walter Funk, and Emil Puhl of the Reichsbank.

In 1939, when the Nazis moved into Czechoslovakia, officials of the Czech National Bank removed $48 million in gold reserves to the Bank of England for safekeeping. Under pressure from the Nazis, Bank of England governor Montagu Norman unhesitatingly agreed to move the gold to Switzerland, where it went into Nazi accounts to purchase essential war materials for Germany.

Despite the obvious rearmament in Germany in the late 1930s, the Nazis continued to find support in Britain, even within the Rothschild-dominated Bank of England. This pro-Nazi proclivity by Bank of England officials will assume even more relevance in the events described in the next section.

Illustrating further interconnecting business associations of this time was International Telephone and Telegraph (ITT) German chairman Gerhardt Westrick, a close associate of John Foster Dulles, who was a partner to Dr. Heinrich Albert, head of Ford Motor Company in Germany until 1945. Two ITT directors were German banker Schroeder and Walter

Schellenberg, head of counterintelligence for the Nazi Gestapo—the Nazi *Geheime Staatspolizei* or Secret State Police. America's International Telephone and Telegraph Corporation sold Germany communication and war material, including as many as fifty thousand artillery fuses per month, more than three years after Pearl Harbor.

More interconnecting business ties between America and the Nazis can be seen by studying the Rockefeller-owned Chase National Bank, now JPMorgan Chase Bank, the largest corporate bank in the United States. Charles Higham explained how the Rockefellers owned Standard Oil of New Jersey, the German accounts of which were siphoned through their own bank, the Chase, as well as through the independent National City Bank of New York (NCB), which also handled Standard, Sterling Products, General Aniline and Film (part of the I. G. Farben combine), Swedish Enskilda Bank (SKF), and ITT, whose chief, Sosthenes Behn, was a director of NCB. Two executives of Standard Oil's German subsidiary were Karl Lindemann and Emil Helfferich, prominent figures in Himmler's Circle of Friends of the Gestapo—its chief financiers—and close friends and colleagues of the BIS's Baron von Schroeder.

Further banking connections, noted by author William Bramley, involved German banker Max Warburg and his brother Paul Warburg, who had been instrumental in establishing the Federal Reserve System in the United States. Both were directors of I. G. Farben. H. A. Metz of I. G. Farben was a director of the Warburg Bank of Manhattan, which later became part of the Rockefeller Chase Manhattan Bank. Standard Oil of New Jersey had been a cartel partner with I. G. Farben prior to the war. One American I. G. Farben director was C. E. Mitchell, who was also director of the Federal Reserve Bank of New York and of Warburg's National City Bank. I. G. Farben president Hermann Schmitz served on the boards of Deutsche Bank and the Bank for International Settlements. In 1929, Schmitz was voted president of the board of National City Bank, now Citibank.

In the 1930s, many people in both Britain and America were in agreement with Nazi ideology.

Automobile-maker Henry Ford became a guiding light to Hitler, especially in the realm of anti-Semitism. In 1920, Ford published an anti-Jewish book titled *The International Jew*. As Hitler worked on his book, *Mein Kampf*, in 1924, he copied liberally from Ford's writing and even referred to Ford as "one great man." Ford became an admirer of Hitler, provided funds for the Nazis, and, in 1938, became the first American to receive the highest honor possible for a non-German: the Grand Cross of the Supreme Order of the German Eagle.

Ford's son, Edsel, sat on the board of American I. G. Farben and GAF. In July 1940, at a meeting in Dearborn, Michigan, between ITT's Westrick and the Fords, it was decided that rather than build aircraft engines for beleaguered Britain, the Ford company would build five-ton military trucks for Germany, the "backbone of German Army transportation." And the Fords were not alone in providing Nazi Germany with the means to wage war. Bradford Snell told the *Washington Post* in 1998 that Nazi armaments minister Albert Speer once told him that Hitler would never have considered invading Poland without the synthetic fuel technology provided by General Motors.

One future American corporate giant provided the Nazis with the means of registering, correlating, and assigning shipment schedules to the millions of Jews and others that were rounded up and sent to their deaths in concentration camps. According to author Edwin Black, Hitler's desire to tabulate then eliminate these people was "greatly enhanced and energized by the ingenuity and craving for profit of a single American company and its legendary, autocratic chairman. That company was International Business Machines [IBM], and its chairman was Thomas J. Watson." Using recently discovered Nazi documents and the testimony of former Polish workers, Black found that IBM technology was passed not only through the company's German subsidiary, Deutsche Hollerith Maschinen Gesellschaft (Dehomag), but to a great extent through a subsidiary in Poland, Watson Business Machines in Warsaw, which reported directly to the IBM New York headquarters.

Watson kept in close contact with his German subordinates, traveling to Berlin at least twice a year from 1933 to 1939. Watson never sold IBM machines to the Nazis. They all were merely leased. This meant that all machines were dependent on IBM punch cards, parts, and servicing. Interestingly, IBM punch cards of that time were not standardized. Each batch sent to Nazi Germany was custom-designed by IBM engineers. "Railroad cars, which could take two weeks to locate and route, could be swiftly dispatched in just 48 hours by means of a vast network of punch-card machines. Indeed, IBM services coursed through the entire German infrastructure in Europe," noted Black.

After America's entry into the war, Nazi Hermann Fellinger was appointed as German enemy-property custodian. Fellinger maintained Watson Business Machines, keeping the original staff and ensuring continued profits for IBM. This subsidiary continued to send royalties and reports to the New York home office through IBM's Geneva office.

Watson, a well-connected Freemason, proclaimed "World Peace Through World Trade" in 1937, while in Berlin to be named president of the International Chamber of Commerce. In that same year, President Franklin D. Roosevelt named Watson U.S. commissioner general to the International Exposition in Paris, and Hitler created a special medal for Watson, called the Merit Cross of the German Eagle with Star, to "honor foreign nationals who made themselves deserving of the German Reich." "It ranked second in prestige only to Hitler's German Grand Cross," noted Edwin Black.

"Since the war, IBM . . . has obstructed, or refused to cooperate with, virtually every major independent author writing about its history, according to numerous published introductions, prefaces, and acknowledgments," he added.

Along with aviation hero Charles Lindbergh and newspaper magnate William Randolph Hearst, another American supporter of Hitler was Joseph P. Kennedy, father of the future president. Kennedy was appointed U.S. ambassador to Britain in 1939 but was recalled in November 1940 for voicing his sympathies for Hitler. Roosevelt had been advised by FBI director J. Edgar Hoover that "Joseph P. Kennedy, the former Ambassador to England, and Ben Smith, the Wall Street operator, some time in the past had a meeting with [Nazi Luftwaffe chief Hermann] Goering in Vichy, France, and that thereafter Kennedy and Smith had donated a considerable amount of money to the German cause. They are both described as being very anti-British and pro-German."

One important example of a prewar effort to install a fascist dictatorship within the United States is the attempted overthrow of Roosevelt early in his presidency.

Only a year after Hitler came to power in Germany, many wealthy Americans looked with favor on a fascist system to counteract international communism. Many were disgruntled with President Roosevelt's social policies and felt he was secretly a communist. Irénée Du Pont and General Motors president William S. Knudsen in early 1934 planned to finance a coup d'etat that would overthrow the president with the aid of a $3 million–funded army of terrorists, modeled on the fascist movement in Paris known as the Croix de Feu.

The undoing of this scheme was retired Marine Corps major general Smedley Butler, the most decorated marine in U.S. history, who was approached by the plotters and urged to head the new military government. Butler, who had openly attacked Roosevelt's New Deal programs, however, proved to be a loyal citizen and immediately informed Roosevelt of the treasonous conspiracy. "Roosevelt . . . knew that if he were to arrest the leaders of the houses of Morgan and Du Pont, it would create an unthinkable national crisis in the midst of a depression and perhaps another Wall Street crash. Not for the first or last time in his career, he was aware that there were powers greater than he in the United States," noted Higham. Roosevelt decided to leak the story to the press, which generally discounted it as a "ridiculous" rumor. Nevertheless, some of the primary plotters skipped the country until the furor died down. But the story did prompt Congress to appoint a special committee to look into the matter. Yielding to the powerful interests involved, the committee dragged its feet for four years before finally publishing a report marked for "restricted circulation." Although downplaying the significance of this attempted coup, the committee's report did state that "certain persons made an attempt to establish a fascist organization in this country" and that the committee "was able to verify all the pertinent statements made by General Butler."

Some researchers have speculated that this move against Roosevelt was merely a ploy orchestrated by the same elite families that put him into power. It was a scheme to paint FDR as an opponent of Wall Street and gain public support for his policies. If this plot was legitimate, it was the last overt move against an American president by powerful business interests until 1963. Ploy or not, this incident provides not only an example of hidden U.S. history but also the lengths to which powerful persons will go to subvert the principles of the United States.

Even at the time, some astute Americans could clearly see the connections between powerful national business leaders and Nazi Germany. U.S. ambassador to Germany William E. Dodd told reporters upon his arrival back home in 1937, "A clique of U.S. industrialists is hell-bent to bring a fascist state to supplant our democratic government and is working closely with the fascist regime in Germany and Italy. I have had plenty of opportunity in my post in Berlin to witness how close some of our American ruling families are to the Nazi regime." It was not just the ruling families that looked favorably on National Socialism and a fascist government. Prior to World War II, right-wing demagogues like Father Charles Coughlin and Gerald K. Smith, an ordained minister, drew thousands of supporters from Christian American workers into their America First and Union Party with their message of nationalism and fears of a "Jewish conspiracy." Smith's planned religious theme park in the 1960s was never completed, but his "Christ of the Ozarks" draws tourists to Eureka Springs, Arkansas, to this day. Prior to the war, no one paid serious attention to warnings against the spread of fascism, just as few people seem willing to consider the possibility of a fascist takeover of the USA today.

A successful political movement requires money, lots of it. There is no question that Hitler's rise to power rested heavily on the support of the major German banks—Schroeder's Cologne banking firm, the Deutsche Bank, Deutsche Kredit Gesellschaft, and the huge insurance firm Allianz—all with many interconnecting ties to foreign banks and companies. There were also close ties to prominent U.S. banks. Higham described how, in 1936, the J. Henry Schroeder Bank of New York had entered into a partnership with the Rockefellers. Named Schroeder, Rockefeller and Company, Investment Bankers, the firm became what *Time* magazine called the economic booster of "the Rome-Berlin Axis." "Avery Rockefeller owned 42 percent of Schroeder," Higham reported. "Their lawyers were John Foster Dulles and Allen Dulles of Sullivan and Cromwell. Allen Dulles (later of the Office of Strategic Services) was on the board of Schroeder." One Deutsche Bank executive outlined a few of the bank's wartime loans: 150 million reichsmarks to the aircraft industry; 22 million to Bavarian Motor Works (BMW); 10 million to Daimler-Benz (Mercedes) in 1943 alone. Similar amounts were loaned again in 1944.

But all the tightest business connections came to naught, for by 1941, the international order had turned against Hitler. Germany's blitzkrieg had shocked the ruling elite, as first Poland, then the rest of Europe, came under Nazi control. Britain was helpless to stop Hitler, who was already making preparations for a preemptive attack on the Soviet Union. In short, Hitler was getting out of hand.

CHAPTER 2

THE STRANGE CASE OF RUDOLF HESS

ALTHOUGH RELEGATED TO A MINOR FOOTNOTE IN HISTORY, THE strange case of Nazi deputy fuehrer Rudolf Hess in 1941 provides a rare glimpse of the elitist control over events during World War II.

The bushy-eyebrowed Hess flew alone to England in May 1941, in an effort to make peace. The conventional view of the Hess flight is that of an increasingly marginalized member of Hitler's inner circle who sought to regain favor with his fuehrer by making an unauthorized visit to Britain in the hope of personally negotiating an end to the war and even enlisting England's aid in the fight against Soviet expansionism. Hitler disavowed Hess as insane, while British prime minister Winston Churchill more kindly described Hess's attempt at negotiation as a "frantic deed of lunatic benevolence."

At the Nuremberg trials, Hess was found guilty of "crimes against peace" and spent the rest of his life a prisoner in Berlin's Spandau Prison. In August 1987, British military authorities announced that Hess had committed suicide, a judgment that continues to be disputed. Several recent studies of the Hess incident show there was much deeper meaning to this intriguing story, which was only magnified by his sudden and mysterious death just as his release from captivity seemed imminent.

RUDOLF HESS WAS born in Egypt in 1894, the son of a German importer. He was well schooled and well traveled by the time he joined the German Army during World War I, serving in the same regiment as Corporal Adolf Hitler. He was wounded twice and later became a fighter pilot, but the war ended before he could experience much combat.

Returning to Munich after the war, Hess helped other ex-servicemen in the paramilitary Freikorps to oust a short-lived Communist local government. After helping to break the Communist coup, Hess joined the Thule Society and enrolled as a student at the University of Munich, where he met his future wife and the man who was to prove a major influence on both Hitler and himself: Professor General Karl Haushofer.

According to author William Bramley, Professor Haushofer was a member of the Vril, another secret society based on a book by British Rosicrucian Lord Bulward Litton, about the visit of an Aryan "super race" to earth in the distant past. A mentor to both Hess and Hitler, Haushofer had traveled extensively in the Far East before becoming a general in the kaiser's army of World War I. "His early associations with influential Japanese businessmen and statesmen were crucial in forming the German-Japanese alliance of World War II," wrote author Peter Levenda. Haushofer became the first ranking Nazi to form relationships with South American governments in anticipation of a war with America. These relationships would prove instrumental in the later escape of war criminals from Europe.

Haushofer, as a professor at the University of Munich, worked out Hitler's policy of *Lebensraum,* "living space" for a hemmed-in Germany. Although he gained a reputation as the "man behind Hitler," Haushofer's views on geopolitics were largely accepted by Hitler, but only after they came from the mouth of Hess. "I was only able to influence [Hitler] through Hess," he told his American captors in 1945.

Both Hess and Haushofer first met Hitler at one of the beer hall meetings of the German Workers Party. During the abortive Beer-hall Putsch of 1923, when the new Nazi Party tried to seize power in Bavaria, Hess was at Hitler's side. When the coup failed, Hess drove off to Austria, where he was sheltered by members of a paramilitary wing of the Thule Society.

Voluntarily returning to Germany, Hess joined Hitler in Landsberg Prison after being convicted of conspiracy to commit treason. Due to the political climate at the time, both men were released within a year. During their months of imprisonment, Hess became a close confidant to Hitler and helped produce Hitler's book, *Mein Kampf.* Hess edited, rewrote, and organized the book so extensively that some researchers believe he should have been credited as coauthor. "As far as I know, Hess actually dictated many chapters of that book," Haushofer told interrogators in 1945.

Following the reorganization of the Nazi Party in 1925, Hess became Hitler's private secretary. He moved upward through other major party positions until 1933, shortly after Hitler became chancellor of Germany, when he was appointed deputy fuehrer. It was Hess who initiated the *"Heil Hitler!"* salute and was the first to call Hitler *"mein Fuehrer."*

Furthermore, as a member of the Geheimer Kabinettsrat—the Nazi Secret Cabinet Council—and the Ministerial Council for the defense of the Reich, Hess was well aware of the secret work to develop a German atomic bomb. Proof of this knowledge came during an interview with Britain's home secretary Sir John Simon, following his flight to England. "[O]ne day sooner or later this weapon will be in our hand and . . . I can only say that it will be more terrible than anything that has gone before," Hess revealed.

It is clear that Hess was much more powerful and well connected than is generally reported. He was the person closest to Hitler, one who shared his aspirations and beliefs. On the eve of war in 1939, Hess was even named the successor to Hitler after Reichsmarschall Hermann Goering.

THE EXCEPTIONAL POWER and position of Rudolf Hess demands close scrutiny of his ill-fated flight to England and its consequences. Just such a study was undertaken in 2001 by three British authors—Lynn Picknett, Clive Prince, and Stephen Prior. "It soon becomes apparent that the whole Hess affair, from 1941 onward, is riddled with so many contradictions and anomalies that it is obvious that the British authorities were desperate to conceal something," they concluded. "Judging by the fact that they are still desperate to conceal it, common sense dictates that they deem this secret to be unsuitable for public consumption, even after sixty years."

A detailed study of Hess's flight clearly indicates that it was not just a sudden whim of an unstable individual. There is evidence of foreknowledge in Germany. Hess prepared for the flight meticulously over a period of months, even having famed aircraft designer Willy Messerschmitt modify a twin-engine Messerschmitt-110. Hess also received special flight training from Messerschmitt's chief test pilot, as well as Hitler's personal pilot, Hans Baur—evidence that Hitler had knowledge of Hess's plans. On his flight, Hess carried the visiting cards of both Haushofer and his son, Albrecht Haushofer, yet another indication of his intent as a peace mission, since the elder Haushofer had long been an advocate of maintaining friendly relations with Britain as a cornerstone of German politics.

According to the French scholars Michel Bertrand and Jean Angelini (writing under the name of Jean-Michel Angebert), Haushofer passed along to Hess the names of members of the Order of the Golden Dawn, an occult society in England, as well as names of supporters of a peace initiative, such as the duke of Hamilton, the duke of Bedford, and Sir Ivone Kirkpatrick. The Golden Dawn, most popularly connected to England's foremost occultist, Aleister "the Beast" Crowley, was an outgrowth of the Theosophical Society, from which much Nazi mysticism was derived, and had close ties with the Thule Society.

According to some theories, British Intelligence manipulated Hess's belief in the occult to provoke his flight to England. Oddly enough, this scheme involved Crowley as well as British Intelligence agent Ian Fleming, who would later write the popular James Bond novels. "Via a Swiss astrologer known to Fleming, astrological advice was passed along to Hess (again, via the Haushofers and by Dr. Ernst Schulte-Strathaus, an astrological adviser and occultist on Hess's staff since 1935) advocating a peace mission to England," wrote Levenda. "May 10, 1941, was selected as the appropriate date, since an unusual conjunction of six planets in Taurus (that had the soothsayers humming for months previous) would take place at that time." Once in England, Hess was to be debriefed by fellow occultist Crowley.

One clue that such an outrageous plan may have been put into operation was somewhat supported by Nazi armaments minister Albert Speer, who wrote in later years, "[I]n Spandau Prison, Hess assured me in all seriousness that the idea had been inspired in him in a dream by supernatural forces." But whatever Hess's motivations, it is clear that foreknowledge of his flight existed in Britain as well as Germany. In fact, both Haushofer's son and Hess wrote to the duke of Hamilton, whom Hess had met briefly during the 1936 Berlin Olympics, in hopes of initiating peace talks between the two men, perhaps even opening a direct link to King George VI.

On orders of the government, Hamilton did not reply. However, one letter to Hamilton from the younger Haushofer was shown to British foreign secretary Lord Halifax and air minister Sir Archibald Sinclair. "Both of these ministers were supportive of the peace initiative, which had to be kept officially secret and distance [*sic*] from Churchill," wrote Picknett, Prince, and Prior. "So, while it is true to say that Hamilton showed the letter to his superiors, what is omitted is the fact that *they* kept quiet about it [emphasis in the original]."

It should be recalled that the Windsor family have always been sensitive about their German extraction. Peace with their relatives would have been very desirable during the war years. In 2000, senior British government sources confirmed that private letters between the Queen Mother and Lord Halifax showed hostility toward Churchill and even a willingness to submit to Nazi occupation if the monarchy was preserved. Even Churchill, who was tightly connected to the empire-builders of Britain, made it clear that the object of the war was to stop Germany—not the Nazis. "You must understand that this war is not against Hitler or National Socialism," Churchill once stated, "but against the strength of the German people, which is to be smashed once and for all, regardless whether it is in the hands of Hitler or a Jesuit priest."

In a letter to Lord Robert Boothby, Churchill explained that "Germany's unforgivable crime before the second world war was her attempt to extricate her economic power from the world's trading system and to create her own exchange mechanism which would deny world finance its opportunity to profit."

It also must be recalled that in 1941, despite the successful Battle of

Britain, England was economically strangled and near defeat. At the time Hitler seemed unstoppable and it was quite easy to envision a Nazi victory. The aristocracy, industrialists, bankers, and even the royal family were eager for peace. "Hess did not imagine a peace group," concluded Picknett, Prince, and Prior, "nor was it invented by MI6, but its existence at such a level [as the royals] would explain why so much about the Hess affair was—and continues to be—hushed up."

As for Hitler, Germany was preparing to strike Russia, and he did not want a two-front war, the very situation that caused Germany's defeat in World War I. Hitler wanted England as an ally against communism. "With England alone [as an ally], one's back being covered, could one begin the new Germanic invasion [of Russia]," Hitler wrote in *Mein Kampf.* In other words, Hitler needed peace with Britain before undertaking an attack on Russia.

Securing peace on the Western Front may have become an urgent priority for Hitler. According to former Soviet military intelligence officer Vladimir Rezun (writing under the pen name Viktor Suvorov), Hitler was forced to launch a preemptive assault against the Soviet Union in June 1941, to forestall an attack on Western Europe by Stalin in July.

Suvorov's work has been published in eighty-seven editions in eighteen languages, yet has received virtually no mention in the U.S. corporate mass media, despite the fact that his assertions turn conventional history upside down. Most people have been taught that Stalin naively trusted Hitler and was totally surprised by Hitler's attack.

Admiral N. G. Kuznetsov, who in 1941 was the Soviet Navy minister and a member of the Central Committee of the Soviet Communist Party, was quoted by Suvorov as stating in his postwar memoirs, "For me there is one thing beyond all argument—J. V. Stalin not only did not exclude the possibility of war with Hitler's Germany, on the contrary, he considered such a war . . . inevitable. . . . J. V. Stalin made preparations for war . . . wide and varied preparations—beginning on dates . . . which he himself had selected. Hitler upset his calculations." While Suvorov's conclusions grate against the conventional view of Hitler's attack on Russia, he has provided a compelling argument. Suvorov pointed out that by June 1941, Stalin had massed vast numbers of troops and equipment along Russia's European frontier, not to defend the Motherland but in preparation for an attack westward. Stalin's motive was to bring communism to Europe by force, a plan he expressed in a 1939 speech. "The experience of the last twenty years has shown that in peacetime the Communist movement is never strong enough to seize power. The dictatorship of such a party will only become possible as the result of a major war," stated Stalin.

Noting that when the German attack began on June 22, 1941, they could field a mere 3,350 tanks, mostly lightly armored and gunned, as compared to the Russians 24,000 tanks, many of superior armor and armament, retired U.S. Department of Defense official Daniel W. Michaels wrote, "Stalin elected to strike at a time and place of his choosing. To this end, Soviet development of the most advanced offensive weapons systems, primarily tanks, aircraft, and airborne forces, had already begun in the early 1930s. . . . The German 'Barbarossa' attack shattered Stalin's well-laid plan to 'liberate' all of Europe."

Suvorov supported his contention by pointing to the fact that Russian troops were prepared to attack, not defend, which led to the early German victories; that Russian troops had been issued maps only of Eastern European cities, not for the defense of Russia; that Russian troops had been issued Russian-German phrase books with such expressions as "Stop transmitting or I'll shoot"; and that none of Stalin's top commanders were ever held accountable for the Barbarossa debacle, since they had all merely followed Stalin's orders.

Suvorov concludes, "Stalin became the absolute ruler of a vast empire hostile to the West, which had been created with the help of the West. For all that, Stalin was able to preserve his reputation as naive and trusting, while Hitler went down in history as the ultimate aggressor. A multitude of books have been published in the West based on the idea that Stalin was not ready for war while Hitler was."

He also said the resources of Stalin's war machine have been underestimated. "Despite its grievous losses, it had enough strength to withdraw and gather new strength to reach Berlin. How far would it have gone had it not sustained that massive blow on 22 June, if hundreds of aircraft and thousands of tanks had not been lost, had it been the Red Army and not the Wehrmacht which struck the first blow? Did the German Army have the territorial expanse behind it for withdrawal? Did it have the inexhaustible human resources, and the time, to restore its army after the first Soviet surprise attack?"

Perhaps the best support for Suvorov's claims came from Hitler himself. "Already in 1940 it became increasingly clear from month to month that the plans of the men in the Kremlin were aimed at the domination, and thus the destruction, of all of Europe. I have already told the nation of the build-up of Soviet Russian military power in the East during a period when Germany had only a few divisions in the provinces bordering Soviet Russia. Only a blind person could fail to see that a military build-up of unique world-historical dimensions was being carried out. And this was not in order to protect something that was being threatened, but rather only to attack that which seemed incapable of defense. . . . I may say this today: if the wave of more than twenty thousand tanks, hundreds of divisions, tens of thousands of artillery pieces, along with more than ten thousand airplanes, had not been kept from being set into motion against the Reich, Europe would have been lost," the fuehrer stated in his speech on December 11, 1941, when he declared war against the United States.

Of course, the victors always write history, so whether Hitler's attack on the Soviet Union was sheer aggression or a necessary preemptive strike will probably be argued for many years. But, if it proves true that Hitler was merely forestalling an imminent attack by the Soviet Union, it places the history of World War II in an entirely different context. It would certainly go far in explaining Hitler's otherwise inexplicable actions in starting a two-front war, the very situation he had warned against in *Mein Kampf*. It also would help explain why Franklin Roosevelt, at the bidding of the globalists, was arming the Soviet Union in blatant violation of the Neutrality Acts of 1935, 1936, and 1937. By the end of 1940, with all Europe under German control and Britain threatened, they may have determined to stop Hitler.

Hitler clearly indicated what he saw as the machinations undertaken to prevent any negotiated end to hostilities in 1941. In a speech to the Reichstag less than a week before Hess's arrival in Scotland, he declared, "All my endeavors to come to an understanding with Britain were wrecked by the determination of a small clique, which, whether from motives of hate or for the sake of material gain, rejected every German proposal for an understanding due to their resolve, which they never concealed, to resort to war, whatever happened."

Picknett, Prince, and Prior even argue that Seelowe—or Sea Lion, the code name for the proposed German invasion of England—was a "sham right from the beginning," an effort by Hitler to distract Stalin by feinting west when he actually planned to strike to the east. It was merely a cover for the mobilization of men and equipment needed for the invasion of the Soviet Union. One clue that this tactic was in play can be seen in the fact that Hitler, who was known for constantly interfering with his generals on the smallest of details, never showed any real interest in the plans for an invasion of England, according to German military historian Egbert Kieser. These authors, along with other historians, explain that Hitler's strange order to halt the German advance at Dunkirk allowed the British Army to escape the continent. Hitler wanted his future ally intact.

And the prelude to such an alliance was Hess's peace initiative. On May 10, 1941, when Hess's ME-110 arrived over Scotland, he was to have landed at an airstrip near the Hamilton ancestral home, negotiate peace terms with the anti-Churchill faction, and then be flown to Sweden as the first leg of a return trip home. This faction was prepared to oust Churchill and agree to a ceasefire with Germany.

This proposition may not be as absurd as it first sounds. Picknett, Prince, and Prior noted, "The extravagant postwar mythologizing of Churchill has obscured the fact that he remained in a very insecure position politically for at least the first two years of his premiership, largely because it was well known that he did not—to put it mildly—enjoy the support and confidence of the king."

The notion of an internal coup against Churchill was even broached to President Roosevelt by FBI director J. Edgar Hoover. In a memorandum written only a week before the Hess flight, Hoover informed Roosevelt, "[I]t was reported that the Duke of Windsor entered into an agreement which in substance was to the effect that if Germany was victorious in the war, Hermann Goering through his control of the army would overthrow Hitler and would thereafter install the Duke of Windsor as the King of England."

Interestingly, Hess's flight brought him through the weakest section of the British coastal radar net, plus he overflew a Royal Air Force base twice without provoking any response—clues that orders had been given somewhere along the chain of command to facilitate his arrival. But he missed his landing spot and, low on fuel, finally was forced to bail out over a farm just south of Glasgow. Unarmed, his ankle broken from the jump, he was captured by a farmer with a pitchfork. The Home Guard quickly became involved and the whole secret operation was blown. Although Hess initially claimed to be a Luftwaffe pilot named Alfred Horn, this subterfuge quickly failed. "Horn" kept asking to be taken to the duke of Hamilton.

The whole scheme was a massive embarrassment to all concerned. Everyone, including Hitler, had to disavow any connection with the plot. After all, for Hitler to admit that he was preparing to make peace with Britain would have tipped off Stalin that a German attack on Russia was imminent.

British intelligence found itself conflicted over the Hess affair. "MI6 was supportive of a negotiated peace with Germany, as it saw Communist Russia as the real enemy," noted the three authors, "whereas SOE [Special Operations Executive] was in favor of an alliance with Stalin against Hitler. As the Prime Minister's creation, SOE was naturally pro-Churchill." This rivalry led to strange occurrences along Hess's path to prison. He was moved to a variety of locations, some recollections of his whereabouts conflicting with others. There was every opportunity for pulling a switch before Hess was finally locked up in the Tower of London.

Picknett, Prince, and Prior introduced yet another mystery—the death of George, the Duke of Kent, King George VI's youngest brother and the first member of the royal family to die while on active military service since the fifteenth century. The duke, like others in the royal family, was an admirer of the Nazis and was likely to have joined a peace group in a negotiated peace. He also served as an unofficial intelligence officer to his brother, the king. (It may be noteworthy that in 1939, King George installed the duke as the grand master of English Freemasonry at a ceremony at Olympia in West London.)

Superficially, the duke's death on August 25, 1942, was the result of a routine wartime air accident. It was reported that the Sunderland flying boat in which he was a passenger crashed into a low hill called Eagles Rock, in Caithness, Scotland. The official account of the accident said the seaplane was taking the duke on a morale-boosting mission to Iceland when the pilot changed course "for reasons unknown," descended through clouds without making sure he was above water, and crashed into a hillside. However, the complete file containing the details of the crash, made by a court of inquiry, has disappeared and anomalies abound.

The duke's plane clearly was in the process of ascending when it crashed, indicating it may have lifted off from nearby Loch More near Braemore Lodge, where accounts place the captive Hess. Second, when statements from witnesses and the one survivor are compared with the official rosters, it becomes clear that there was an unaccounted-for passenger on the craft.

These factors, coupled with much other evidence—both hard and circumstantial—support the conjecture that the anti-Churchill peace group waited until mid-1942, a low ebb in Britain's war fortunes, before attempting to fly the duke of Kent and Rudolf Hess to Sweden to announce a peace plan that would topple the Churchill government. Of course, this never happened, due to the plane crash. Whether this was sabotage or an accident has not been clearly established.

If Hess died in the duke's plane, it would have presented a thorny problem for Churchill—how to explain the mangled corpse of a man who was supposed to be their prized prisoner. Any investigation would have revealed the involvement of ranking members of British society, even the royals, in the peace initiative.

Here the story takes an even more bizarre twist. Evidence gathered for their book by Picknett, Prince, and Prior—including Hess being reported as seen in different locations at the same time, and inconsistencies in official reports—indicated that a duplicate Hess may have been prepared prior to the plane crash. "We are convinced that in the summer of 1942 there were two Hesses, one in Scotland and one at Maindiff Court, Abergavenny, Wales," they wrote. The real Hess died in the crash and the double lived to stand trial at Nuremberg and serve his sentence at Spandau.

But even these astute authors acknowledged a huge problem with such a scenario. "Even though it seems to fit the evidence perfectly, it has to be admitted that the mind skids on the thought that any man would allow himself to be tried and sentenced in Hess's name, not to mention continuing with the deception for the rest of a very long life in the harshest and most hopeless of conditions," they remarked.

The idea of a Rudolf Hess double is not new, and various theories have been advanced. One suggested that the look-alike was forced to play Hess out of fear for his family. Another was that the Hess double was a German—whoever the man was, German was his first language—and an ardent Nazi, who was convinced it was necessary to the party that he maintain the subterfuge, especially since he might become the founder of a Fourth Reich.

But the most provocative explanation comes from Picknett, Prince, and Prior, who learned that former CIA director Allen Dulles, a founder of the Council on Foreign Relations and high commissioner of Germany after the war, had dispatched Dr. Donald Ewen Cameron to Nuremberg to examine Hess. Dulles expressed to Cameron his belief that the Hess being held in Germany was an impostor and that the real Hess had been secretly executed on orders from Churchill. Knowing of Hess's war wounds, Dulles wanted Cameron to especially note if there were any scars on the prisoner's chest. Interestingly enough, British military authorities in Nuremberg refused to allow such an examination.

But the story grows stranger. Dr. Cameron was a Scot who pioneered brainwashing techniques before the end of the war, at the Allen Memorial Institute at McGill University, funded by the Rockefeller Foundation. He went on to become president of the American Psychiatric Association as well as the first president of the World Psychiatric Association. He also became part of the CIA's notorious MKULTRA mind-control program. Various researchers have wondered if Dulles's choice of Dr. Cameron to study Hess might have grown from the knowledge or suspicion that the man posing as Hess had been brainwashed into actually believing he was the Nazi deputy fuehrer. Mind-control experimentation was much further along—particularly in Europe, as shall be seen—than most people realize. Why else should Dulles have chosen a brainwashing expert to study Hess when any competent physician could have checked for scar tissue?

This subterfuge could account for Hess's eccentric behavior at the

Nuremberg trials, during which he repeatedly claimed he had lost his memory, a convenience for someone who had not lived Hess's life.

Once the peace plan went awry, all the usual methods of cover-up came into play—documents disappeared or were locked away from public scrutiny, witnesses were coerced into silence, and multiple "theories" from authoritative sources were spread.

One clue that a geopolitical game was being played out in the Hess affair is that the last person to dine with the duke of Kent prior to the fatal crash that killed him and perhaps the real Hess was a foreign exile, Prince Bernhard of the Netherlands. The dinner represented an unusual gathering of the British royals at Balmoral Castle in Scotland, which, in addition to the duke and Prince Bernhard, included King George VI and Queen Elizabeth.

But it is Bernhard's presence that has caught the interest of researchers. Prince Bernhard originated meetings of the Bilderberg Group, a collection of world movers and shakers so secretive they have no proper name. Bernhard was a former member of the Nazi SS and an employee of Germany's I. G. Farben in Paris. In 1937, he married Princess Juliana of the Netherlands and became a major shareholder and officer in Dutch Shell Oil, along with Britain's Lord Victor Rothschild.

After the Germans invaded Holland, the royal couple moved to London. It was here, after the war, that Rothschild and the founder of the European Movement for Unity, Polish socialist Dr. Joseph Hieronim Retinger, encouraged Prince Bernhard to create the Bilderberg Group. The prince personally chaired the group until 1976, when he resigned following revelations that he had accepted large payoffs from Lockheed to promote the sale of its aircraft in Holland.

It is impossible to know for certain whether Prince Bernhard sided with the British royal family and the peace initiative or was monitoring their activities for the prowar Churchill clique. But it is an indication of the machinations of the global elite. The peace initiative was stopped and the globalists' decision to stop National Socialism at all costs proceeded.

There can be little doubt that the failure of Hess's peace mission to Britain on the eve of the attack on Russia created the unwanted two-front war that cost Hitler the victory. After the failure of Hess's ill-fated flight, his place in the Nazi hierarchy was taken by Martin Bormann, a man who will be discussed later. Some Nazi leaders, including Himmler and Bormann, became uncertain of victory and began laying plans for their survival. They also turned to science for new *Wunderwaffen,* or wonder weapons, that might turn the tide of war in their favor.

CHAPTER 3

NAZI WONDER WEAPONS

JUST SIX DAYS AFTER THE D-DAY INVASION OF EUROPE, ON JUNE 12, 1944, the residents of London were startled to hear a droning buzz in the skies over their city. They were more startled when the sound suddenly stopped and moments later a huge explosion rocked the East London neighborhood of Mile End, killing eight civilians.

It was the first of the V-1 Buzz bombs—a forerunner of today's cruise missiles.

The V-1 and the later V-2 rockets that terrorized London are two of the more famous examples of German war technology. These *Vergeltungs-waffe,* or retaliation weapons, were developed at the secret German rocket facility Peenemunde and put into operation just after the D-Day landings in Normandy, France. From June 12, 1944, until August 20, more than eight thousand of the V-1 rockets (each carrying a ton of explosives) rained down on London, inflicting 45,479 casualties and destroying 75,000 buildings. The less numerous V-2 rockets—which, unlike the V-1, could not be seen, heard, or intercepted in flight—nevertheless produced more than 10,000 casualties in the British capital.

In addition to the vengeance weapons, the Germans produced a number of scientific breakthroughs in their quest for weapons technology during World War II. German ingenuity and efficiency appeared capable of overcoming almost any obstacle. One clear example may be found simply by comparing figures from its armaments industry. Despite constant bombing by the Allies, overall production of tanks, small arms, ships, and aircraft was higher at the beginning of 1945 than in 1941, when Germany was victorious on all fronts and America had not yet entered the war.

Technological advances were seen in almost every area. The rate and quality was astounding. Plastics, which only came into general use in the United States during the 1950s, were developed in Nazi Germany. Bakelite, polystyrene (under the name Trolitul), Plexiglas, polyethylene (forerunner of today's plastic Baggies and syringes), polyamide (nylon), and aldols (a derivative of polyvinyl) were all produced during wartime. The various forms of plastic were produced under a consortium of companies but led by I. G. Farben, which also in 1941 synthesized the opiate methadone and Demerol under the name "pethidine."

Television, which most Americans did not get to see until the early 1950s, was highly developed in Nazi Germany. More than 150,000 persons in twenty-eight public viewing rooms in Berlin saw clear television broadcasts of the 1936 Olympics. They watched screens equipped with Fernseh 180-line cathode ray tube projectors that presented a picture about forty-eight by forty-two inches. In 1939, the German firm Fernseh began developing a miniaturized TV system that allowed pilots to guide both bombs and missiles after launching. This system was used in the anti-aircraft rocket Wasserfall, or waterfall. "Many of these tests failed," noted author Joseph P. Farrell. "But by the war's end, a successful test of the television-guided 'Tonne' missile was conducted by German scientists for the Allies in Berlin, with the target being a photograph of a little girl's face. The test was successful, much to the impressed, and doubtless shocked, Allied observers."

Tanks, which began the war as little more than armor-plated bulldozers designed to support infantry, were developed into independent, thickly armored machines powered by gas turbines, with guns stabilized while moving, hydrokinetic power transmissions, and defenses against chemical and biological attacks. Some German tanks were so far ahead of their time, they were still being utilized in other nations as late as battles in the 1970s. To counter the threat of modern tanks, the Germans developed simple, but very effective, portable rocket launchers armed with a hollow charge such as the Panzerschreck bazooka and the easily produced Panzerfaust, a forerunner of today's hand-carried rocket-propelled grenade (RPG). The innovative 9-mm German MP-40 Schmeisser machine pistol saw extensive use during the war, and its successors, the MP-43 and the MP-44 assault rifles, became the forerunners of today's ubiquitous AK-47. Late in the war, some MP-44s carried an early but effective night-vision light and scope called the Vampyr, or Vampire.

At the end of the war in 1945, American military intelligence officers were shocked by the technology they found as Allied forces overran German research facilities. Supersonic rockets, nerve gas, jet aircraft, guided missiles, stealth technology, hardened armor—even flying saucers—were just some of the groundbreaking technologies being developed in Nazi laboratories, workshops, and factories. To give some idea of the aspirations of Nazi scientists, the huge ME-264 was dubbed the "America Bomber," while a three-stage rocket was named the "Mars Rocket."

As respected British historian Barrie Pitt noted, "[T]he Nazi war machine swung into action utilizing as much as it could of the most up-to-date scientific knowledge available, and as the war developed, the list of further achievements grew to staggering proportions. From guns firing 'shells' of air to detailed discussions of flying saucers; from beams of sound that were fatal to a man at 50 yards, to guns that fired around corners and others that could 'see in the dark'—the list is awe-inspiring in its variety." Pitt stated that while some German technology was less developed than imagined at the time, "some were dangerously near to a completion stage which could have reversed the war's outcome."

Former Polish military journalist Igor Witkowski described German wartime research as "the greatest technological leap in the history of our civilization." He said the Germans ignored Einstein and developed an approach to science based on quantum theories. "Don't forget that Einsteinian physics, relativity physics, with its big-picture view of the universe, represented Jewish science to the Nazis. Germany was where quantum mechanics was born. The Germans were looking at gravity [and other matters] from a different perspective to everyone else. Maybe it gave them answers to things that pro-relativity scientists hadn't even thought of," explained Witkowski, who had unprecedented access to German wartime documents that only recently because available, due to the collapse of communism.

Consider that at the beginning of the war, aircraft were made of canvas stretched over a wooden frame. By 1945, Germany had become the first nation in the world to put into service an all-metal, jet-propelled jet fighter—the Messerschmitt-262. They also produced the world's first operational helicopter and vertical takeoff and landing aircraft.

As German scientists worked feverishly to perfect the V-2 rockets and other, more secret weapons, SS chief Heinrich Himmler was taking steps to separate his SS from normal party and state control. "In the spring of 1944 Hitler approved Himmler's proposal to build an SS-owned industrial concern in order to make the SS permanently independent of the state budget," wrote Nazi armaments minister Albert Speer. Employing methods later used by the CIA, SS leaders created a number of business fronts and other organizations—many using concentration camp labor—with an eye toward producing revenue to support SS activities. These highly compartmentalized groups headed by young, ambitious SS officers neither required nor desired any connection with Germany's high-profile leaders. Their purpose was to create an economic base that could continue pursuing Nazi goals long after the defeat of Germany.

Armaments minister Speer conceded that there were weapons development programs that he knew nothing about. He admitted that an SS scheme in 1944 to construct a secret weapons plant requiring 3,500 concentration camp workers had been concealed from him. Speer even hinted at the possibility of secret weapons that "were secretly produced by the SS toward the end of the war and concealed from me."

While the V-2 rocket program began under the aegis of the German Army, and the ME-262 jet fighter under the Luftwaffe, they were ultimately transferred to SS control. "In short, anything that had shown any real promise as a weapon system—in particular, anything that appeared to represent a quantum leap over the then-state-of-the-art—had ended up under the oversight of the SS," noted Nick Cook, an aviation editor and aerospace consultant to *Jane's Defence Weekly*. With secret projects in the hands of hardcore SS fanatics, and with factories and research facilities scattered over—and under—the countryside, it is entirely conceivable that weapons far in advance of the V rockets could have been developed without the knowledge of anyone except Himmler and his top lieutenants.

Other notable secret Nazi weapons nearing completion in 1945 included the Messerschmitt-163 Komet and the vertically launched Natter rocket fighters, the jet-powered flying wing Horten Ho-IX and the delta-winged Lippisch DM-1. It has been noted that some of top-secret Nazi weaponry development was moved outside Germany, to such places as Blizna, Poland—the same area where Allied aircrews first encountered the infamous "foo-fighters," small glowing balls of light that shadowed Allied bombers. The "foo-fighters" soon caught the attention of the American news media. The *New York Times,* on December 13, 1944, reported news authorized by the Supreme Headquarters of the Allied Expeditionary Force. "Floating Mystery Ball Is New Nazi Air Weapon," read the headline. The story stated:

> Airmen of the American Air Force report that they are encountering silver-colored spheres in the air over German territory. The spheres are encountered either singly or in clusters. Sometimes they are semi-translucent.
>
> The new device, apparently an air defense weapon, resembles the huge glass balls that adorn Christmas trees. There was no information available as to what holds them up like stars in the sky, what is in them or what their purpose is supposed to be.

According to author Renato Vesco, the "foo-fighters" were actually the Feuerball, or fire ball, which was "a highly original flying machine . . . circular and armored, more or less resembling the shell of a tortoise, and was powered by a special turbojet engine, also flat and circular, whose principles of operation . . . generated a great halo of luminous flames. . . . Radio-controlled at the moment of takeoff, it then automatically followed enemy aircraft, attracted by their exhaust flames, and approached close enough without collision to wreck their radar gear." Vesco claimed that the basic principles of the Feuerball were later applied to a "symmetrical circular aircraft" known as the Kugelblitz, or ball lightning, automatic fighter that became an "authentic antecedent of the present-day flying saucers." He said this innovative craft was destroyed after a "single lucky wartime mission" by retreating SS troops.

Even though the public has been conditioned for more than sixty years to dismiss any notion of flying saucers, or UFOs, the accumulation of evidence available today makes it impossible to reject the reality of such craft out of hand. Obviously, the Nazis were experimenting with new and exotic energy technology. The extraordinary development of the Feuerball may have provided the first public glimpse into the heart of Nazi superscience.

Several writers have produced articles about the Nazi development of flying saucers. British author W. A. Harbinson claimed that he got his ideas after discovering postwar German articles mentioning a former Luftwaffe engineer, Flugkapitan Rudolph Schriever. According to information gleaned by Harbinson from articles in *Der Spiegel, Bild am Sonntag, Luftfahrt International,* and other German publications, Schriever claimed to have designed a "flying top" prototype in 1941, which was actually test-flown in June 1942. In 1944, Schriever said he constructed a larger, jet version of his circular craft, with the help of scientists Klaus Habermohl, Otto Miethe, and an Italian, Dr. Giuseppe Belluzzo. Drawings of this saucer were published in the 1959 British book *German Secret Weapons of the Second World War and Their Later Development,* by Major Rudolph Lusar, an engineer who worked in the German Reichs-Patent Office and had access to many original plans and documents. Lusar described the saucer as a ring of separate disks carrying adjustable jets rotating around a fixed cockpit. The entire craft had a height of 105 feet and could fly vertically or horizontally, depending on the positioning of the jets.

Schriever later said the Allied advance into Germany put an end to his "flying disc" experiments, with all equipment and designs lost or destroyed. However, a Georg Klein told the postwar German press that he had witnessed the Schriever disc, or something like it, test-flown in February 1945.

Schriever reportedly died in the late 1950s and, according to a 1975 issue of *Luftfahrt International,* notes and sketches related to a large flying saucer were found in his effects. The periodical also stated that Schriever maintained until his death that his original saucer concept must have been made operational prior to war's end. This possibility is acknowledged by British author Brian Ford, who wrote, "There are supposed to have been 'flying saucers' too, which were near the final stages of development, and indeed it may be that some progress was made toward the construction of small, disc-like aircraft, but the results were destroyed, apparently before they fell into enemy hands."

These accounts would seem to be corroborated by a CIA report dated May 27, 1954. As reported in Nick Redfern's 1998 book, *The FBI Files: The FBI's UFO Top Secrets Exposed,* the document stated, "A German newspaper (not further identified) recently published an interview with Georg Klein, famous German engineer and aircraft expert, describing the experimental construction of 'flying saucers' carried out by him from 1941 to 1945. Klein stated he was present when, in 1945, the first piloted 'flying saucer' took off and reached a speed of 1,300 miles per hour within three minutes. The experiments resulted in three designs—one designed by Miethe was a disk-shaped aircraft, 135 feet in diameter, which did not rotate; another, designed by Habermohl and Schriever, consisted of a large rotating ring, in the center of which was a round, stationary cabin for the crew. When the Soviets occupied Prague, the Germans destroyed every trace of the 'flying saucer' project and nothing more was heard of Habermohl and his assistants. Schriever recently died in Bremen, where he had been living. In Breslau, the Soviets managed to capture one of the saucers built by Miethe, who escaped to France. He is reportedly in the USA at present."

Another candidate for inventor of a German UFO is the Austrian scientist Victor Schauberger, who, after being kidnapped by the Nazis, reportedly designed a number of "flying discs" in 1940, using a flameless and smokeless form of electromagnetic propulsion called "diamagnetism." Schauberger reportedly worked for the U.S. government for a short time after the war before dying of natural causes. Prior to his death, he was quoted as saying, "They took everything from me. Everything." No one knows for certain if he meant the Nazis or the Allies.

That someone was flying highly unconventional disc-shaped objects shortly after World War II was made plain by the now-public comments of U.S. Army Lieutenant General Nathan Twining, then in charge of the Army Air Forces' Air Material Command (AMC).

In mid-1947, two years after the war ended, "flying saucers" were being reported both in Europe and America. General Twining wrote that the "phenomenon reported is something real and not visionary or fictitious." He went on to describe attributes of such discs as having "extreme rates of climb, maneuverability (particularly in roll), and action which must be considered evasive when sighted or contacted by friendly aircraft and radar, lend[ing] belief to the possibility that some of the objects are controlled either manually, automatically or remotely."

Allowing a small glimpse into the reality of such radical technology, Twining concluded, "It is possible within the present U.S. knowledge—provided extensive detailed development is undertaken—to construct a piloted aircraft which has the general description of the object [described above] which would be capable of an approximate range of 7,000 miles at subsonic speeds."

If technical knowledge in the 1940s was advanced enough to construct a workable flying saucer, the public was never to hear about it. Beginning in the late 1940s, a national security "lid" was placed on the subject.

But it is fascinating to recall that one of the first and best documented cases of mysterious abductions took place in September 1961, when Betty and Barney Hill under hypnosis recalled being taken aboard a circular craft manned by men in black uniforms. Barney Hill described the leader as a "German Nazi" wearing a shiny black jacket, scarf, and cap.

Before anyone rushes out to proclaim that all UFOs are really secret Nazi technology, serious attention should be given to the wealth of public literature that clearly indicates that while some saucers, especially in the years following World War II, may indeed have been Nazi test vehicles, any objective review of the material suggests the presence of some unconventional source as well.

Another amazing—and chilling—aspect of Nazi technology involved their development of nuclear weapons. Researcher and author Farrell concluded from new material released from the former East Germany that the Nazis were much closer to developing an atomic bomb than previously accepted by postwar writers. He characterized the idea that the

Germans had neither the talent nor the capability to construct an operational atomic bomb—recall the well-known story of the destruction of the heavy-water plant in Norway by commandos—an "Allied Legend" designed to distract the public from a horrible reality. "[A]ll the evidence points to the conclusion that there was a large, very well-funded, and very secret German isotope-enrichment program during the war, a program successfully disguised during the war by the Nazis and covered up after the war by the Allied Legend," wrote Farrell, after concluding that the conventional story that "the German failure to obtain the atom bomb because they never had a functioning reactor is simply utter scientific nonsense because a reactor is needed *only* if one wants to produce *plutonium*. It is an unneeded, and expensive, development, if one only wants to make a *uranium* A-bomb [emphasis in the original]."

Plus, there is the cryptic remark made by Kurt Diebner, a physicist involved with the Nazi atomic bomb project. Surreptitiously recorded by British intelligence during postwar internment at Farm Hall, England, Diebner mentioned a "photochemical process" to enrich uranium bypassing the need for a centrifuge. Since no modern researcher understands what process was referred to by Diebner, this may mean that the Nazis discovered a method of isotope separation and uranium enrichment that even now remains classified.

Adding to the idea that the Nazis already had perfected a method of enriching uranium are the words of nuclear scientist Karl Wirtz, who was also secretly taped at Farm Hall. Upon learning of the atomic bomb dropped on Hiroshima, Otto Hahn, who discovered atomic fission, commented, "They can only have done that if they have uranium isotope separation." To which Wirtz agreed by responding, "They have it too," a clear indication that he knew of a German separation process. Farrell noted, "Thus, there is sufficient reason, due to the science of bomb-making and the political and military realities of the war after America's entry, that the Germans took the decision to develop only a uranium bomb, since that afforded the best, most direct, and technologically least complicated route to acquisition of a bomb."

Based on his research, Farrell wrote, "American progress in the plutonium bomb, from the moment [physicist Enrico] Fermi successfully completed and tested a functioning reactor in the squash court at the University of Chicago, appeared to be running fairly smoothly, until fairly late in the war, when it was discovered that in order to make a bomb from plutonium, the critical mass would have to be assembled much faster than any existing Allied fuse technologies could accomplish. Moreover, there was so little margin of error, since the fuses in an implosion device would have to fire as close to simultaneously as possible, that Allied engineers began to despair of making a plutonium bomb work. . . . I believe a strong prima facie case has been outlined that Nazi Germany developed and successfully tested, and perhaps used, a uranium atom bomb before the end of World War II," Farrell concluded.

Farrell was not alone in this assessment. In 2005, Berlin historian Rainer Karlsch, in a book titled *Hitlers Bombe,* claimed that the Nazis indeed tested nuclear weapons on Rugen Island near Ohrdruf, Thuringia, site of a subsidiary concentration camp to the infamous Buchenwald. Reportedly, many prisoners were killed during these tests, which were conducted under the supervision of the SS. Karlsch's primary evidence consists of "vouchers" for "tests" and a patent for a plutonium weapon dated 1941. He also claimed to have found traces of radioactivity in soil from the site. However, in February 2006, the German government reported no abnormal radiation levels at the site, even after taking into account elevated levels due to the 1986 Chernobyl disaster in Russia.

Although Nazi armaments minister Speer was questioned about a mysterious blast at Ohrdruf during the Nuremberg war crimes trials, no significant information on a nuclear test was found, either because it never happened or because a postwar cover-up was quite successful.

Mainstream historians, at the mercy of carefully concocted cover stories in both Germany and the USA, have remained skeptical that Nazi scientists could have advanced their nuclear knowledge to the point of actual testing. However, evidence that the Nazis were planning a nuclear strike near the end of the war came from varied sources, including a news article in the *Washington Post* dated June 29, 1945, which reported on an amazing find by Allied troops in Norway:

R.A.F. [Royal Air Force] officers said today that the Germans had nearly completed preparations for bombing New York from a "colossal air field" near Oslo when the war ended.

Forty giant bombers with a 7,000-mile range were found on this base—"the largest Luftwaffe field I have ever seen," one officer said.

They were a new type bomber developed by Heinkel. They now are being dismantled for study. German ground crews said the planes were held in readiness for a mission to New York.

It should also be noted that the Nazis had two prototypes of the Junkers-390, a massive six-engine modification of the Junkers-290, known to have made flights to Japanese bases in Manchuria.

In late 1944, one JU-390 was flown from a base in Bordeaux, France, to within twelve miles of New York City, snapped photographs of the skyline, and returned—a nonstop flight of thirty-two hours.

What weapon was to be transported by these massive bombers? After the war, authorities discovered a feasibility study by the German Luftwaffe detailing the blast effects of an atomic bomb over New York's Manhattan Island. The Nazi study was based on an atomic bomb in the fifteen- to seventeen-kiloton range, approximately the same yield as the Little Boy uranium bomb dropped on Hiroshima.

If Nazi Germany had a nuclear weapon, they surely must have tested it, and a collection of disparate sources seems to indicate this was accomplished. Italian dictator Benito Mussolini, in a "Political Testament" written shortly before his death at the hands of partisans in April 1945, stated, "The wonder weapons are the hope. It is laughable and senseless for us to threaten at this moment, without a basis in reality for these threats. The well-known mass destruction bombs are nearly ready. In only a few days, with the utmost meticulous intelligence, Hitler will probably execute this fearful blow, because he will have full confidence. . . . It appears there are three bombs—and each has an astonishing operation. The construction of each is fearfully complex and of a lengthy time of completion."

Mussolini's mention of three bombs is intriguing because of a statement of a former Russian military translator who served on the staff of Marshal Rodion Malinovsky, the officer who took Japan's surrender to the Soviet Union in 1945. As reported by the German magazine *Der Spiegel* in 1992, Piotr Titarenko had written a letter to the Communist Party Central Committee, in which he stated that the three atomic bombs were dropped on Japan. One of these, dropped on Nagasaki prior to the blast of August 9, 1945, failed to detonate and subsequently was given to the Soviet Union by Japanese officials. If Titarenko's account is accurate, this would mean that America had three atomic bombs on hand in the summer of 1945. Yet, a report to Manhattan Project leader Robert Oppenheimer just days after President Roosevelt died on April 12, 1945, stated that not enough enriched uranium existed to create a viable critical mass for even one atomic bomb.

News stories in Britain point to a possible Nazi atomic bomb test in 1944. An August 11, 1945, article in London's *Daily Telegraph* reported, "Britain prepared for the possibility of an atomic bomb attack on this country by Germany in August 1944. It can now be disclosed that details of the expected effect of such a bomb were revealed in a highly secret memorandum which was sent that summer to the chiefs of Scotland Yard, chief constables of provincial forces and senior officials of the defense services." Another odd story also was published in England's *Daily Mail* on October 14, 1944, under the headline "Berlin Is 'Silent' 60 Hours, Still No Phones." The story, filed by a correspondent from Stockholm, stated that all telephone service in Berlin had been interrupted for three days with "no explanation for the hold-up, which has lasted longer than on any previous occasion." The story ended by saying, "It is pointed out, moreover, in responsible quarters that if the stoppage were purely the technical result of bomb damage, as the Germans claimed, it should have been repaired by now." A modern readership would know that such disruption can be caused by the electromagnetic pulse associated with a nuclear detonation.

Other intriguing hints of a German atomic test came in the form of three separate intelligence reports. A once-classified U.S. military intelligence report dated August 19, 1945, and titled "Investigations, Research, Developments and Practical Use of the German Atomic Bomb" details the experience of a German pilot named Hans Zinsser, a Flak rocket expert, while piloting a Heinkel bomber over northern Germany. Note that his experience coincides with the dates of the Berlin telephone blackout. Zinsser reported:

> At the beginning of October 1944, I flew from Ludwigslust (south of Luebeck) about 12 to 15 km from an atomic bomb test station, when I noticed a strong, bright illumination of the whole atmosphere, lasting about 2 seconds.
>
> The clearly visible pressure wave escaped the approaching and following cloud formed by the explosion. This wave had a diameter of about 1 km when it became visible and the color of the cloud changed frequently.... Personal observations of the colors of the explosion cloud found an almost blue-violet shade. During this manifestation reddish-colored rims were to be seen, changing to a dirty-like shade in very rapid succession. The combustion was lightly felt from my observation plane in the form of pulling and pushing.... About an hour later...I passed through the almost complete overcast (between 3,000 and 4,000 meter altitude). A cloud shaped like a mushroom with turbulent billowing sections (at about 7,000 meter altitude) stood, without any seeming connections, over the spot where the explosion took place. Strong electrical disturbances and the impossibility to continue radio communications turned up. Because of the P-38s operating in the area Wittenberg-Mersburg I had to turn to the north but observed a better visibility at the bottom of the cloud where the explosion occured [sic]. Note: It does not seem very clear to me why these experiments took place in such crowded areas.

Then there was the report of an Italian officer, Luigi Romersa, who claimed to have been present at the testing of a "disintegration bomb" on the night of October 11–12, 1944. Romersa was granted a special pass from Oberkommando Der Wehrmacht, or German High Command, to visit the test site on the island of Rugen. Romersa was a special envoy from Mussolini, who had wanted more information since Hitler had mentioned to him "a bomb with a force which will surprise the whole world."

According to Romersa, he and others were told the "disintegration bomb" was "the most powerful explosive that has yet been developed" and that "nothing can withstand it." They were sent to a bunker about a mile from the actual test site. He also was warned against radioactivity. "Around 4 P.M., in the twilight, shadows appeared, running toward our bunker," recalled Romersa. "They were soldiers and they had on a strange type of 'diving suit.' They entered and quickly shut the door. 'Everything is *kaput*,' one of them said as he removed his protective clothing. We also eventually had to put on white, coarse, fibrous cloaks. I cannot say what the material was made of, but I had the impression that it could have been asbestos. The headgear has a piece of *Glimmerglas* [mica glass?] in front of the eyes."

After making their way to the test site proper, Romersa stated, "The houses that I had seen only an hour earlier had disappeared, broken into little pebbles of debris. As we drew nearer [to the point of explosion], the more fearsome was the devastation. The grass had the same color as leather. The few trees that still stood upright had no more leaves."

Romersa's credibility is supported by the fact that he eventually came to the United States, where he was granted a high-security clearance.

A third report dated December 14, 1944, but only declassified by the National Security Agency in 1978, is titled "Reports on the Atom-splitting Bomb." This purports to be a decoded intercept of a message from the Japanese embassy in Stockholm to headquarters in Tokyo. It reads:

This bomb is revolutionary in its results, and will completely upset all ordinary precepts of warfare hitherto established. I am sending you, in one group, all those reports on what is called the atom-splitting bomb. It is a fact that in June of 1943, the German Army tried out an utterly new type of weapon against the Russians at a location 150 kilometers southeast of Kursk. Although it was the entire 19th Infantry Regiment of the Russians which was thus attacked, only a few bombs (each round up to 5 kilograms) sufficed to utterly wipe them out to the last man.

The following is according to a statement by Lieutenant Colonel . . . Kenji, adviser to the attaché in Hungary and formerly . . . in this country, who by chance saw the actual scene immediately after the above took place: "All the men and the horses [within radius of] the explosion of the shells were charred black and even their ammunition had all been detonated. Moreover, it is a fact that the same type of war material was tried out in the Crimea too. At that time the Russians claimed that this was poison gas, and protested that if Germany were ever again to use it, Russia, too, would use poison gas." . . . Recently the British authorities warned their people of the possibility that they might undergo attack by German atom-splitting bombs. The American authorities have likewise warned that the American east coast might be the area chosen for a blind attack by some sort of flying bomb. . . .

The Japanese report then goes into a remarkably accurate description of the splitting of the atom, ending with the statement, "[T]he German atom-splitting device is the Neuman disintegrator. Enormous energy is directed into the central part of the atom and this generates an atomic pressure of several tens of thousands of tons per square inch. This device can split the relatively unstable atoms of such elements as uranium. Moreover, it brings into being a store of explosive atomic energy. . . . That is, a bomb deriving its force from the release of atomic energy."

Some elements of the Japanese report were obviously in error, such as the confusion over descriptions of a fission versus a fusion bomb and the date of the Kursk offensive, which did not begin until July 5, 1943. Mistakes notwithstanding, it is clear that Japanese intelligence was firmly convinced that the Germans had used a revolutionary type of weapon on the Eastern Front.

But if the Nazis had deployed a tactical nuke or other exotic weapon on the Eastern Front, why would the Soviets have kept such an attack secret? Farrell pointed out that had Nazi Germany used such a weapon, it would most likely have been against the Russians, whom the Nazis considered "subhuman," in Nazi ideology. Fully one-half of the 50 million casualties of the war occurred in Russia, and several massive explosions, such as the one that destroyed a section of Sevastopol, have never been fully explained. It was announced that a hundred-foot below-ground ammunition bunker was destroyed after being struck by a lucky shot from Dora, a 31½-inch German railway gun considered the largest in the world.

Such attacks were never reported by Soviet leader Joseph Stalin, due to the fear of losing control over a panicked and war-weary Russian population. The use of a super-weapon on the Eastern Front also might explain why more is not known about this issue. Accurate war news from Russia was extremely hard to come by during the war and grew more so during the Cold War. To make public the use of a nuclear or unconventional weapon "would have been a propaganda disaster for Stalin's government," noted Farrell. "Faced with an enemy of superior tactical and operational competence in conventional arms, the Red Army often had to resort to threats of execution against its own soldiers just to maintain order and discipline in its ranks and prevent mass desertion. Acknowledgment of the existence and use of such weapons by the mortal enemy of Communist Russia could conceivably have ruined Russian morale and cost Stalin the war, and perhaps even toppled his government."

IF THE NAZIS had operational atomic weapons, is it possible they were transferred to the United States? Documents exist showing that America's secret development of the atomic bomb, the Manhattan Project, could not have produced enough enriched uranium to make a bomb by mid-1945. Since only a plutonium bomb was tested at Alamogordo, New Mexico, on July 16, 1945, researchers have wondered where America acquired the uranium bombs dropped on Japan less than a month later. Some have speculated that the United States used a Nazi bomb or used Nazi enriched uranium to manufacture its bombs.

The Trinity bomb exploded near Alamogordo, New Mexico, on July 16, 1945, was a plutonium bomb. Why then would the United States first drop the Little Boy, an untested uranium bomb, on Japan on August 6, 1945? "A rational explanation is [that] 'Little Boy' was not tested by the Americans because . . . [t]he Americans did not need to test it, because its German designers already had," surmised Farrell. This idea is supported by the statement of German authors Edgar Meyer and Thomas Mehner that J. Robert Oppenheimer, the "father of the atomic bomb," maintained that the bomb dropped on Japan was of "German provenance." Of course, this idea would fly in the face of the long-accepted Allied Legend that Germany simply couldn't manufacture an atomic bomb by the war's end.

Where could the Nazis have obtained enriched uranium for such a bomb? One potential source was the secure underground laboratory of Baron Manfred von Ardenne, built in Lichterfelde outside Berlin, which contained a 2-million-volt electrostatic generator and a cyclotron. In 1941, von Ardenne, along with Fritz Houtermans, had calculated the critical mass needed to create U-235. It should be noted that Hitler visited the laboratory toward the end of the war, at a time when he spoke enthusiastically of a new wonder weapon that would turn the tide in Germany's favor.

Some researchers contend that the Nazi development of a uranium bomb was kept secret because the work was not part of the German military- industrial system but hidden within the German Postal Service. According to Carter Plymton Hydrick, author of a well-documented book *Critical Mass: How Nazi Germany Surrendered Enriched Uranium for the United States' Atomic Bomb*, "[A]ll of Ardenne's facilities . . . were provided by and ongoing funding made available through, the patronage of one man, Reich minister of posts and a member of the Reich President's Research Council on Nuclear Affairs, Wilhelm Ohnesorge."

Reportedly, Hitler once remarked that while his party and military leadership worried about how to win the war, it was his postal minister who brought him the solution.

Farrell explained that the Reichspost was "awash with money, and could therefore have provided some of the massive funding necessary to the [uranium enrichment] project, a true 'black budget' operation in every sense."

Another source may have been a giant synthetic-rubber plant built by I. G. Farben next to Auschwitz, the notorious death camp. The site was chosen for its proximity to transportation hubs, both rail lines and rivers, as well as the nearby supply of slave labor found at the Auschwitz camp. This site probably was also selected with the idea that the Allies would not bomb a concentration camp, a supposition that proved correct. Yet, despite the facts—established during the Nuremberg trials—that more than $2 billion in today's dollars were spent; that 300,000 slave laborers had been used in both the construction and operation of this plant; and that it had consumed more electricity than Berlin, not one pound of buna, or synthetic rubber, was ever produced.

So, what was produced? "The facility has all of the characteristics of a uranium enrichment plant," noted Hydrick, adding, "the various components of the German atomic bomb efforts could have been implemented with a high degree of secrecy, even from other high-level Nazis, given Bormann's close-knit relationships with Ohnesorge; Schmitz, who was chief of I. G. Farben; [Rudolf] Hoess, the commandant of Auschwitz; and Heinrich Mueller, who, among his many other duties as head of the Gestapo, oversaw the supplying of forced laborers to Auschwitz."

A theory has been offered that, late in the war, certain Nazis arranged the transfer of enriched uranium to the United States in exchange for immunity from prosecution. At the heart of this transfer theory lies the saga of a Nazi submarine—the U-234.

Unterseeboot-234 was originally designed as a mine-layer but was converted to a cargo carrier prior to its only mission into enemy waters: the last German shipment to its ally, Japan. It sailed from Kiel in March 1945, with a most unusual cargo consisting of several high-level German officials, including Dr. Heinz Schlicke, the inventor of fuses for atomic bombs, and two Japanese officers—Air Force Colonel Genzo Shosi and Navy Captain Hideo Tomonaga. Also listed on the boat's manifest of 240 metric tons of cargo were two dismantled ME-262 jet fighters, ten gold-lined cylinders containing 560 kilograms of uranium oxide, wooden barrels of "water," and infrared proximity fuses.

On May 14, 1945, six days after the German surrender, the U-234 was intercepted by the USS *Sutton* and taken into captivity. Oddly enough, the sub had been overflown several times by Allied aircraft but never fired upon. The circumstances implied a preplanned meeting and surrender. Here the mystery began. Who issued the orders for this enemy sub to surrender, and why to the Americans? Upon arrival at Portsmouth, New Hampshire, it appeared that some of the boat's cargo was missing.

The two Japanese officers, after learning that the ship's captain planned to surrender, had committed suicide and were buried at sea with full honors. But, suspiciously, the two ME-262s were missing, as well as the uranium oxide. In fact, when the U.S. Navy prepared its own manifest for the U-234, there was no accounting for seventy tons of cargo.

Dr. Velma Hunt, a Colorado environmental scientist, said she uncovered information that the U-boat made a secret stop at South Portland, Maine, sometime between May 14 and May 17, 1945, where the cargo in question could have been unloaded. There has been controversy as to whether this uranium had been enriched enough for use as a weapon.

Cook noted that the gold-lined cylinders indicated the uranium was emitting gamma radiation, which meant the normally harmless uranium oxide had been brought to enrichment through the use of a working nuclear reactor. "And yet, officially, there had been no nuclear reactor in Germany capable of fulfilling this task," wrote Cook. "[At least] not in Speer's orbit of operations."

Farrell further explains, "The use of gold-lined cylinders is explainable by the fact that uranium, a highly corrosive metal, is easily contaminated if it comes into contact with other unstable elements. Gold, whose radioactive shielding properties are as great as lead, is also, unlike lead, a highly pure and stable element, and is therefore the element of choice when storing or shipping *highly enriched and pure* uranium for long periods of time, such as a voyage. Thus, the uranium oxide on board the U-234 *was highly enriched uranium*, and most likely, highly enriched U-235, the last stage, perhaps, before being reduced to weapons grade or to metalicization for a bomb (if it was already in weapons grade purity) [emphasis in the original]."

Adding weight to Farrell's deduction is an anecdote regarding the German crew of the U-234. Some crew members were amused when they saw the Japanese officers bring on board cargo marked "U-235." They apparently thought their Japanese guests couldn't even get the number of the boat correct. Some now believe the labels indicated the presence of uranium 235, the only isotope found in nature that has the ability to cause an expanding fission chain reaction—in other words, the element needed for a uranium fission bomb. Uranium that has undergone an extraction process to boost its U-235 proportion is known as enriched uranium.

Wolfgang Hirschfeld, radioman on the U-234, stated the submarine's

orders were "only to sail on the orders of the highest level. Fuehrer HQ."
He also revealed after the war that crew members believed Japan had succeeded in testing an atomic weapon before their departure from Germany
in March 1945. The U-234 met an inglorious end in November 1947,
when it was used as a torpedo target and sunk off Cape Cod.

Hydrick published copies of documents from the National Archives to
show a connection between the Manhattan Project and the U-234. One
such document is a secret cable from the commander of naval operations
directing that a three-man party take possession of the sub's cargo. In addition to two naval officers was the name of Major John E. Vance with the
Army Corps of Engineers, the department of the army under which the
Manhattan Project operated.

A few days after the visit by Vance, a manifest of the cargo indicated
the uranium was no longer in navy possession. Furthermore, telephone
transcripts between Manhattan intelligence officers about a week later
stated a captured shipment of uranium powder was being tested by a person identified only as "Vance." "That there could have been another
'Vance' who was working with uranium powder—especially 'captured'
uranium powder—is improbable," noted Hydrick.

But author Henry Stevens found an even more disturbing cover-up.
After receiving a statement from the National Archives denying that any
canisters containing fissionable material was onboard the U-234, Stevens,
recalling that the submarine had surrendered to the USS *Sutton,* wrote to
the Naval Historical Center at the Washington Navy Yard requesting a
cargo manifest from the U-234 in the files of the *Sutton.* For a $5 micro-
fiche charge, Stevens received the manifest that was identical to the one
from the National Archives except that the uranium oxide canisters were
listed. This discrepancy in the manifests can only be explained by someone altering the documents.

A plutonium bomb, such as the one Manhattan scientists were developing, required a critical mass to be achieved within 1/3000th of a second, a speed far exceeding the capabilities of fuses available at that time.
According to Farrell, there is evidence to support the idea that the necessary fuses were obtained from U-234 passenger Dr. Schlicke. A message
from the chief of Naval Operations to the authorities in Portsmouth,

where the U-234 was taken after its surrender, indicated that Dr. Schlicke along with his fuses were to be taken to Washington accompanied by naval officers. Once there, the doctor was scheduled to present a lecture on his fuses in the presence of a "Mr. Alvarez," apparently meaning Dr. Luis Walter Alvarez, the man who is credited with producing fuses for the plutonium bomb. Alvarez and his student Lawrence Johnson are credited with designing the exploding-bridgewire detonators for the spherical implosives used in the Trinity bomb test as well as the Nagasaki bomb.

On March 3, 1945, President Roosevelt received an ominous memo from Senator James F. Byrnes, a Democrat from South Carolina and a longtime confidant to the president. This "Memorandum for the President" stated, "I understand that the expenditures for the Manhattan Project are approaching 2 billion dollars with no definite assurance yet of production. . . . Even eminent scientists may continue a project rather than concede its failure." Byrnes, who went on to become a secretary of state and a Supreme Court justice, was voicing the concern of many that the atom bomb project was foundering and might even prove a failure. Byrnes may have been aware of a letter dated December 28, 1944, in which Eric Jette, chief metallurgist at Los Alamos, expressed reservations over the lack of sufficient amounts of uranium for the atomic bomb. He wrote, "A study of the shipment of [weapons grade uranium] for the past three months shows the following . . . : At present rate we will have 10 kilos by February 7 and 15 kilos about May 1." According to Hydrick, Edward Hammel, a metallurgist who worked at Los Alamos, where enriched uranium was made into material for the atomic bomb, reported that very little enriched uranium was received there until less than a month before the bomb was dropped on Hiroshima.

Little Boy, the bomb dropped on Hiroshima, carried 64.15 kilograms of enriched uranium, virtually the entire quantity that could have been produced since mid-1944 by the enrichment facilities at Oak Ridge, Tennessee, even working around the clock. One explanation for the lack of enriched uranium was that some of this fissionable material had been used to produce plutonium in Enrico Fermi's breeder reactors at Hanford, Washington.

The mounting pressure on Manhattan Project directors to produce a bomb before the planned invasion of the Japanese home islands must have been terrific. If the submarine's cargo did indeed include U-235 and Dr. Schlicke's fuses, its acquisition by the United States solved two pressing problems of the Manhattan atomic bomb project—a lack of sufficient amounts of uranium and adequate fuses. The American bomb-makers may have been greatly relieved that the two major problems facing the Manhattan Project were solved with the surrender of the U-234. "The fact that U-234 arrived on American soil carrying 560 kilograms of uranium that was enriched and went on to be used in the bombs that were dropped on Japan can scarcely be argued any longer except by those who refuse to consider the evidence," concluded Hydrick. While it may remain a controversy whether the acquisition of the U-234 was a fortuitous capture or the planned transfer of technology from Germany to the United States, the evidence strongly indicates the latter.

If additional uranium was obtained from the U-234, this would have provided more than could ever have been produced by the Manhattan Project, and the equivalent of about eight Hiroshima bombs. It also means the German nuclear program was much further advanced than believed by conventional historians. In late July 1945, atomic bomb components— and perhaps additional German uranium bombs—were delivered to Tinian Island in the Pacific following a secret and rushed voyage from California by the USS *Indianapolis*. After delivering its deadly cargo, this Portland-class heavy cruiser suffered the largest single at-sea loss of life in U.S. naval history and became the last American ship sunk in World War II after being torpedoed by a Japanese submarine in the Philippines.

Farrell voiced the suspicion that the *Indianapolis* may have delivered much more than America's atomic bomb—it may have carried a German bomb in addition to its cargo of uranium and fuses. He was supported by Stevens, who wrote that the "unexploded German atomic bombs fell into the hands of the Americans at the end of the war in Europe in May 1945, two months before the 'first' explosion of an atomic weapon in the New Mexico desert. What a present for the Americans! All they did was to put new tail fins on the bombs, repaint them, and drop them on Japan. Naturally, the American scientists involved with the Manhattan Project were given credit."

But, if the Nazis had developed a working atomic bomb, why was it not used as Allied armies closed in on Germany? One answer seems to be that they did not have a reliable delivery system in place. The Nazis' V-3, a smooth-bore 150-mm gun dubbed the Centipede, designed to launch large-finned shells into London, along with its multistage A-10 rocket, was still undependable. Witkowski voiced his suspicions that the fatal flight of Lieutenant Joseph Kennedy, older brother to the future president, might have been an ill-fated attempt to destroy the V-3 complex at Mimoyecques, France. The giant airfield in Norway, home to the massive six-engine bombers, had not yet been completed.

This idea was echoed by Stevens, who became convinced that the Third Reich produced an atomic bomb. "The Germans did make atomic bombs," he stated emphatically. "Not only did they make atomic bombs, they made uranium as well as plutonium bombs and other atomic weapons which remain somewhat of a mystery. What the Germans could not do, in these dying days of the Third Reich, was to match up one of these nuclear weapons with an effective delivery system. The reasons for this differ with each weapon, individually, and run the [gamut] from mistake to treachery to incompetence."

One thought that must have crossed the minds of Nazi leaders was the total destruction of Germany that would have resulted from the use of a nuclear weapon. The devastation of London or New York would not have materially altered the course of the war in the spring of 1945. And the retaliation of the Allies would have been unimaginable. Further, high-ranking Nazis, such as Hitler's secretary Martin Bormann, who by war's end had become the second most powerful man in Nazi Germany, realized the war was lost, and used advanced technology as a bargaining chip with the Western allies.

Hydrick proposed just that intriguing possibility: that the U-234 was purposely handed over to U.S. authorities on the order of Bormann in exchange for immunity as part of a covert plan for the continuation of Nazi research. Although there was criticism over Hydrick's technical descriptions of both the atomic bomb and its detonators, his mass of documentation concerning the transfer of nuclear technology from Germany to

America is compelling. Hydrick's claim is supported by Farrell, who wrote, "I have argued that most likely all of it [extra uranium and even atom bombs] came from Nazi Germany, courtesy of Nazi Party Reichsleiter Bormann and SS Obergruppenfuehrer Hans Kammler."

But Farrell had an even more horrifying thought about why the Nazis did not drop an atomic bomb. Considering Nazi research into quantum physics and energy manipulation, Farrell speculated that their atomic bombs "were being developed as detonators for something far more destructive." Since only a few scattered plans to Nazi super-science were recovered after the war, the question arises, "What became of their advanced technology?" There has never been a public answer.

HOWEVER, THE ANSWER to this question may be found by studying the man in charge of Germany's high-tech weapons programs, Dr. Engineer Hans Kammler.

Kammler, whose name has been largely lost to history, may have played a large role in developing and hiding away the technology secrets of Hitler's Third Reich. Kammler did not have higher purposes in mind when he set out to develop rockets and energy manipulation. He was searching for new weapons.

Born in 1901, Kammler completed engineering studies at a technical university and began working for the German Air Ministry. After joining the Nazi SS, he managed finances and construction for the SS until 1942, when he became chief of Group C under the Wirtschafts und Verwaltungshauptamt, or the Economic and Administrative Central Office (WVHA) of the SS, one of five key branches of the Black Shirts. This branch controlled all economic enterprises as well as all concentration and extermination camps. Beginning in 1943, Kammler took control of all "special tasks," which included "Kammler special construction"—the creation of secret underground facilities as well as exotic weapons programs. His official title was SS Obergruppenfuehrer, or lieutenant general, and he had worked his way up to command the Third Reich's most precious wartime secrets.

In mid-1943, SS chief Heinrich Himmler sent a letter to armaments minister Speer. "With this letter, I inform you that I, as SS Reichsfuehrer . . . do hereby take charge of the manufacture of the A-4 instrument," it read. The A-4 rocket was later designated by Hitler as the V-2. Himmler then placed Kammler in charge of the project, one of Germany's most secret high-tech weapons systems. Due to the devastation brought on by incessant Allied air raids, by the end of 1944, Kammler had taken control of weapons research as well as the construction of underground factories and concentration camps.

"Thus—just a few weeks before the end of the war—he had become commissioner general for all important weapons," wrote Speer, who later bemoaned the fact that Himmler's SS gradually assumed total control over Germany's weaponry, production, and research.

In connection with his new responsibilities, Kammler created an SS Sonderkommando, or special command, independent from the normal German military and bureaucracy. "What Kammler had established was a 'special projects office,' a forerunner of the entity that had been run by the bright young colonels of the USAF's stealth program in the 1970s and 1980s," noted Cook. It was "a place of vision, where imagination could run free, unfettered by the restraints of accountability. Exactly the kind of place, in fact, you'd expect to find anti-gravity technology, if such an impossible thing existed."

Kammler also had use of computer technology that was only dreamed of in American science fiction stories. "Dr. Kammler had the benefit of knowledge, hardware and software that was developed by the computing pioneer, Dr. Konrad Zuse," wrote Stevens. "In spite of everything churned out by the computer industry and 'history' as we know it, Dr. Zuse built the first digital computer in 1938 and the first programmable software language, Plankaikuel. He also was instrumental in developing magnetic tape as a computer storage medium. By 1944 the Germans were using computers, the Zuse-built Z-3, to plot the course of ballistic attack by the V-2 at Peenemunde and Nordhausen." Stevens, who spent more than fifteen years researching the Reich's most secret technology, including flying saucers, wrote, "By the end of the war a whole new research and production command and control structure had been set up which reduced or replaced the figures we normally think of as running the Third Reich, such as, for instance, Hermann Goering and Albert Speer." It was Kammler and his Sonderkommando that became the repository for the Reich's most advanced technology, going far beyond the rockets and flying discs.

But Kammler's immediate concern was the V-2 rocket program. Kammler worked closely with Wernher von Braun and his superior, Luftwaffe Major General Walter Dornberger. Von Braun, who had been a member of the SS since 1940, carried the rank of SS Sturmbannfuehrer, or major.

Alarmed by progress on the V-2 rockets, Britain's Bomber Command sent 597 bombers on the night of August 16–17, 1943, to raid Peenemunde—Germany's top-secret rocket facility built on an island at the mouth of the Oder River near the border of Germany and Poland. Because of a navigation "blunder," much of the underground and well-camouflaged Peenemunde site was left undamaged. Brian Ford described the results: "Even so, over 800 of the people on the island were killed. . . . After this, it was realized that some of the facility had better be dispersed throughout Germany; thus the theoretical development facility was moved to Garmisch-Partenkirchen, development went to Nordhausen and Bleicherode, and the main wind-tunnel and ancillary equipment went down to Kochel, some 24 miles south of Munich. This was christened Wasserbau Versuchsanstalt Kochelsee—experimental waterworks project—and gave rise to the most thorough research center for long-range rocket development that, at the time, could have been envisioned."

Mary Bennett and David S. Percy, authors of *Dark Moon: Apollo and the Whistleblowers,* speculated that the British air raid on Peenemunde was designed not to knock out the V-rocket site but to force it to move to safer environs, to ensure the safety of the rocket program. They showed how the raid bombed the site's northern peninsula rather than the main facility, due to misplaced target indicators. These authors noted that of the eight hundred personnel who died in the air raid, about half were mostly Russians from the prisoner labor force and the other half were technicians and their families. After this raid, the irreplaceable Hermann Oberth was transferred to the safety of the Reinsdorf works near Wittenberg, to continue his work.

"Instructions from the highest level, it seems, had been to target personnel and certainly *not the V-2 rocket production facilities*. It was clearly CRUCIAL that these rockets, plans and parts were spared," they stated (emphasis in original). Someone with high authority wanted this Nazi technology available to them after the war.

Nick Cook also saw the connection between such exotic technology and the mysterious Hans Kammler. "There was, via the Kammler trail, a mounting body of evidence that the Nazis, in their desperation to win the war, had been experimenting with a form of science the rest of the world have never remotely considered," he wrote. "And that somewhere in this cauldron of ideas, a new technology had been born; one that was so far ahead of its time it had been suppressed for more than half a century."

One clue to what this revolutionary technology might involve was found in the capture of physicist Walter Gerlach, one of the Nazi scientists brought to the United States after the war. Gerlach has been connected with the German attempts to build an atomic bomb, yet his background indicated even more esoteric knowledge.

In 1921, Gerlach received a Nobel Prize, not for nuclear research but for magnetic spin polarization, dealing with the momentum of electrons of atoms situated in a magnetic field. Such work had little to do with the atomic bomb but much to do with energy manipulation to include antigravity.

In 1931, a paper titled "About Gravitation, Vortices and Waves in Rotating Media" was published by O. C. Hilgenberg, a student of Gerlach, which indicated the focus of Gerlach's work. "And yet, after the war, Gerlach, who died in 1979, apparently never returned to these subject matters, nor did he make any references to them; almost as if he had been forbidden to do so," noted Cook. Interestingly, Gerlach's wartime work diaries were confiscated by U.S. authorities and remain classified today.

At the turn of the current century, both Cook and the Polish military journalist Witkowski tracked Kammler and his top-secret Nazi energy work to the Wenzeslaus Mine, located about 215 miles west of Warsaw in Lower Silesia, near the border with Czechoslovakia. This mine is in Ludwikowice Klodzkie, formerly Ludwigsdorf. The location was perfect for security purposes as it was outside Germany yet within the Greater Third Reich. Additionally, Kammler spoke fluent Czech.

During their journey, Witkowski revealed his access to a formerly classified Soviet document detailing the interrogation at the end of the war of a Rudolf Schuster, who had been a member of the Reichssicherheitshauptamt, or Reich Central Security Office, Nazi Germany's version of the Department of Homeland Security. Schuster revealed that in June of 1944, he was transferred to a special evacuation Kommando called General Plan 1945, formed by Martin Bormann to evacuate valuable science and technology from the Reich. Schuster, who was not privy to the plan's overall agenda, nevertheless located much of these evacuation activities in the area of the Wenzeslaus Mine.

Schuster's testimony, coupled with other information, convinced Cook that the Bormann evacuation plan had been one of the Nazis' greatest secrets. "There has never been any official acknowledgment of the existence of the special evacuation Kommando," he wrote. It was this unacknowledged evacuation operation that saved the Reich's most precious technology. Once at the mine site, Cook and Witkowski found remnants of what once had been a secret SS testing and production facility that may have even included a giant early superconductor.

In 1931, the Wenzeslaus Mine suffered an accident that caused bankruptcy and a takeover by the Polish government. With the occupation of Poland, the mine was reconditioned by the Nazis as a gigantic science center. "The whole area, in the center of which was located the main left shaft, proved to be the interior of a deep valley, which was accessible only through two 'mountain passes,'" noted Witkowski. "Since the remnants of watchtowers could be seen in them, it was obvious that the whole area had been closely guarded, and its configuration caused that in this way the whole valley was physically cut off [from] the outside world." This valley, about three hundred yards across, was bisected by rail lines, and lined with a variety of structures, concrete bunkers, and guard stations, many covered with dirt and trees to act as camouflage. Today the site is virtually ruins and overgrown with trees and vegetation.

Cook saw that "the Germans had gone to a great deal of trouble to ensure that the place looked pretty much as it had always looked since mining operations began here at the turn of the last century, a clear indication that whatever had happened here during the war had been deeply secret. . . . Almost everything that was known about the Wenzeslaus Mine had been handed down from [SS General Jakob] Sporrenberg [the officer appointed to command the 'northern route' of General Plan 1945's evacuation Kommando]. It had been run by the SS, had employed slave-labor and had been sealed from the outside world by a triple ring of check points and heavily armed guards." Sporrenberg's testimony and affidavits, the only known description of the strange experiments at the mine, were given during a postwar trial in Poland. He was found guilty of war crimes and executed.

In the closing days of the war, most of the local population was evacuated westward. In fleeing the Russians, many of these refugees died during the fighting or froze in one of the coldest winters on record. Today, most of the local residents are newcomers with no recollection of what transpired at the mine during the war.

A central shaft led downward to the original mine as well as a labyrinth of additional underground facilities dug by Germans. But what most intrigued Cook and Witkowski was a huge circular concrete structure. Green camouflage paint was still visible on the edges. The circular structure was formed by twelve thick columns supporting a dodecagon-shaped reinforcing concrete ring about ninety feet in diameter.

Initially, Witkowski thought this might be the remains of a cooling tower. He abandoned this idea once he saw cooling towers at a different location on photographs of the area, taken in 1934. Next he thought of the structure as a "fly trap," similar to those used to test helicopters and other hovering aircraft. Yet, this answer was not satisfactory either in that the researchers found a concrete duct containing thick electric cables leading to a power-generating station. Learning that high-voltage current cannot be used in mines with the potential for flammable gas—such as the Wenzeslaus Mine—Cook and Witkowski determined that the structure had nothing to do with mining but was used in connection with the strange experiments described to his captors by the SS officer Sporrenberg.

These experiments centered around a bell-shaped object—appropriately enough codenamed Die Glocke, or the Bell—which was housed in a concrete chamber hundreds of feet underground. According to the research of Witkowski and Cook, the Bell was made from hard, heavy metal and cylindrical in shape with a semicircular cap and hook or clamping device on top. Huge quantities of electricity were fed into it through thick cables dropping into the housing chamber from the outside. Inside the Bell was a thermos-like tube encased in lead and filled with a metallic liquid.

During operation, the Bell was covered by a ceramic material, apparently to act as insulation. Inside, two contra-rotating cylinders filled with a mercury-like and violet-colored substance spun a vortex of energy, which emitted a strange phosphorescent blue light and made such a buzzing sound that operators nicknamed it the Bienenstock, or beehive.

Due to the phosphorescent light and reports that operators suffered from nervous-system disruption, headaches, and a metallic taste, Witkowski concluded the Bell's operation involved iodizing radiation as well as a very strong magnetic field of energy. The scientists experimenting with the Bell would place various plants, animals, and animal tissue within its energy field. "In the initial test period from November to December 1944, almost all the samples were destroyed," noted Cook. "A crystalline substance formed within the tissues, destroying them from the inside; liquids, including blood, gelled and separated into clearly distilled fractions."

Very little is known for certain about the Bell. However, it was given the highest—and perhaps most unique—classification possible in the Third Reich. In a few captured documents, experimenters with the Bell were said to be working on something *Kriegsentscheidend,* or decisive for the war. Most top-secret German weapons, including the V rockets, were classified *Kriegswichtig,* or important to the war.

One major reason that so little is known about the Bell was the loss of the scientists involved in the project. "They were taken out and shot by the SS between the 28th of April and the 4th of May, 1945," explained Witkowski. "Records show that there were 62 of them, many of them Germans. There were no survivors, but then that's hardly surprising. . . . It's quite clear that someone had gone to great lengths to clean up."

The whole concept is a nightmare—Nazis tinkering with the building blocks of the universe. And it gets even worse.

TO TRY AND understand the purpose of the Bell requires a brief side trip into the amazing world of cutting-edge science and quantum physics. While discussions and articles about energy manipulation—whether termed cold fusion, antigravity, or free energy—have been generally discouraged as science fiction in mainstream America, many credible writers have dealt with the subject.

In his 2003 book *Winning the War: Advanced Weapons, Strategies, and Concepts for the Post-9/11 World*, Colonel John B. Alexander noted, "A potential link between superconductor quantum mechanics and gravity has been inferred from recent quantum gravity research. Another approach to modifying gravity involved the manipulation of the quantum vacuum ZPE [Zero Point Energy found in the vacuum of space] field. One proposed experiment to manipulate the ZPE involves the use of ultrahigh-intensity lasers to irradiate a magnetized vacuum. If any of these are successful it will change energy issues on Earth and our relationship with the universe by allowing deep space travel."

The idea of gaining mastery—and power—from the environment around us is nothing new. Such ideas were advanced by American physicist Thomas Townsend Brown, who, in the early 1920s, experimented with antigravity based on his understanding that a charged capacitor tended to move toward a positive plate when sufficiently energized in the hundred kilovolt and upward range. Brown contended that all matter is essentially an "electrical condition." "It fact, it might be said that the concrete body of the universe is nothing more than an assemblage of energy which, in itself, is quite intangible."

Brown's theories echoed those of U.S. electrical engineer Nicola Tesla, whose discovery in 1888 of the rotating magnetic field led to alternating-current (AC) electricity transmission. Tesla foresaw limitless free energy by simply tapping into the Earth's natural magnetic energy field.

In 1908, long before the idea of rotating magnetic fields was commonplace in science, Tesla stated:

Every ponderable atom is differentiated from a tenuous fluid, filling all space merely by spinning motion, as a whirl of water in a calm lake. By being set in motion this fluid, the ether, becomes gross matter. Its movement arrested, the primary substance reverts to its normal state [stillness].

It appears, then, possible for man through harnessed energy of the medium and suitable agencies for starting and stopping ether whirls to cause matter to form and disappear. At his command, almost without effort on his part, old worlds would vanish and new ones spring into being. He could alter the size of this planet, control its seasons, adjust its distance from the sun, guide it on its eternal journey along any path he might choose, through the depths of the universe. He could make planets collide and produce his [own] suns and stars, his heat and light, he could originate life in all its infinite forms. To cause at will the birth and death of matter would be man's grandest deed, which would make him the master of physical creation, make him fulfill his ultimate destiny.

The belief that antigravity or other exotic technologies were passed from the Nazis to the Allies has been further supported by sporadic periodical coverage of antigravity in the late 1940s and early 1950s. This was at a time before the total blackout of news concerning energy-manipulation experimentation was enforced as a "matter of national security."

In 1956, the Swiss aviation journal *Interavia Aerospace Review* published an article titled "Towards Flight Without Stress or Strain . . . or Weight." The article carried the dateline of Washington, D.C., and stated, "Electro-gravitics research, seeking the source of gravity and its control, has reached a stage where profound implications for the entire human race begin to emerge. Perhaps the most startling and immediate implications of all involve aircraft, guided missiles and free space flight of all kinds." The article added, "There are gravity research projects in every major country of the world. *A few are over 30 years old* [emphasis added]." It also mentioned that, over and above theoretical research, there was empirical research into the "study of matter in its super-cooled, super-conductive state, of jet electron streams, peculiar magnetic effects [and] the electrical mechanics of the atom's shell." The article stated that the weight of some materials utilized in this research had been reduced by as much as 30 percent by "energizing" them. But in a premonition of what was to come, it added, "Security prevents disclosure of what precisely is meant by 'energizing' or in which country this work is underway."

Proving the ability of superconductors to produce antigravity effects, researchers at Pacific National Laboratory, in the late 1980s, cooled a ceramic superconductor with liquid nitrogen and levitated a round magnet in midair.

Some of the companies involved in this cutting-edge research, according to the *Interavia Aerospace Review* article, included Lear, Inc., Glenn L. Martin Company, Sperry-Rand Corporation, Bell Aircraft, Clarke Electronics Laboratories, and the U.S. General Electric Company.

The names of these firms are especially noteworthy, because in his 2001 book on Zero Point energy, author Cook cited another 1956 magazine article naming aviation experts Lawrence D. Bell, George S. Trimble, and William P. Lear as stating that work was then under way with "nuclear fuels and equipment to cancel out gravity." This article, from an unnamed publication and titled "The G-Engines Are Coming!" may have let slip mention of an incredible new technology.

"All matter within the ship would be influenced by the ship's gravitation only," Lear was quoted as saying. "This way, no matter how fast you accelerated or changed course, your body would not feel it any more than it now feels the tremendous speed and acceleration of the earth."

During the 1960s and 1970s, public discussion of energy manipulation such as antigravity was virtually closed off, scorned as fantasy or conspiracy theory. Yet, it is clear that within government and military circles, work continued secretly in this area. Could it have been based on transferred Nazi super-science?

Bruce L. Cathie, a former New Zealand commercial pilot, theoretician, and an advocate of the existence of a worldwide energy grid, wrote in 1971, "Somewhere, I knew, [my proposed energy grid] system contained a clue to the truth of [Einstein's] Unified Field which, he had postulated, permeates all of existence. I didn't know at the time that this clue had already been found by scientists who were well ahead of me in the play. . . . for many years they have been carrying out full-scale research into the practical applications of the mathematical concept contained in that theory." Cathie speculated:

> The only way to traverse the vast distances of space is to possess the means of manipulating, or altering, the very structure of space itself; altering the space-time geometric matrix, which to us provides the illusion of form and distance.
>
> . . . for distance is an illusion. The only thing keeping places apart in space is time. If it were possible to move from one position to another in space in an infinitely small amount of time, or "zero time," then both the positions would co-exist, according to our awareness. By speeding up the geometric of time we will be able to bring distant places within close proximity. This is the secret of UFOs—they travel by means of altering the spatial dimensions around them and repositioning in space-time.

One hint that the U.S. government experimented with such technology came in December 1980, when Betty Cash, Vickie Landrum, and Landrum's seven-year-old grandson Colby encountered a large, glowing, diamond-shaped object hovering in the air near the small town of Huffman, Texas. The trio, supported by other witnesses in the area, said the object was surrounded by military CH-47 helicopters. Days later, the trio experienced painful swellings and skin blisters, along with headaches, nausea, and hair loss, all symptoms of intense electromagnetic radiation. In 1985, the three victims sought $20 million in damages from the U.S. government, but the following year, their suit was dismissed, based on denials by the government that any such craft existed in its inventory.

Yet another small public exposure to exotic energy manipulation may have come with the accidental discovery of single-atom (monatomic) elements in the 1970s by Phoenix-area cotton farmer David Hudson. His discovery was followed by several scientific papers exploring the mysteries of the atomic structure, nucleus deformation, and electromagnetism. Hudson himself obtained eleven worldwide patents on his "Orbitally Rearranged Monatomic Elements (ORME)."

Hudson found that the nuclei of such monatomic matter acted in an unusual manner. Under certain circumstances, they began spinning and creating strangely deformed shapes. Oddly, as these nuclei spun, they began to come apart on their own.

It was found, for example, that in the element rhodium 103, the nucleus became deformed in a ratio of two to one, which made it twice as long as it is wide, and entered a high-spin state. When all electrons are brought under the control of the nucleus of an atom, the nucleus attains a "highward," or high-spin, state. When reaching a state of reciprocal relationship, the electrons turn to pure white light and the individual atoms fall apart, producing a white monatomic powder.

Using thermo-gravimetric analysis, it was found that a sample of Hudson's monatomic matter lost 44 percent of its original weight when reduced to this white-powder state. By being either heated or cooled, it would gain or lose weight. "By repeated annealing we could make the material weigh less than the pan weighed it was sitting in," said Hudson, ". . . or we could make it weigh 300–400 times what its beginning weight was, depending on whether we were heating or cooling it. . . . [I]f you take this white powder and put it on a quartz boat and heat it up to the point where it fuses with the quartz, it becomes black and it regains all its weight again. This makes no sense, it's impossible, it can't happen. But there it is."

British author Laurence Gardner noted, "Hudson was then asked to reverse the process fully by turning the powder back into a piece of metallic gold. It was like asking someone to remake an apple from a pan of apple sauce—seemingly impossible! Early trial led to some disastrous results. . . . By late 1995, the difficulties had been overcome and the figurative apple had indeed been rebuilt from the apple sauce. From this, there was no doubt that it was possible (just as in ancient metallurgical lore) to manufacture gold from a seemingly non-gold base product. From a commencing sample which registered as iron, silica, and aluminum, emerged an ingot which analyzed as pure gold. After centuries of trial, error, frustration, and failure, the Philosopher's Stone of ancient times had at last been rediscovered."

Gardner amassed a wealth of material linking the white powder of gold to alchemists, the legendary Knights Templar, Solomon's treasure, the manna of the Israelites, Moses, and ancient Egypt. The significance of these connections will become apparent in the next section.

By the early 1990s, scientific papers were being published by the Niels Bohr Institute and Argonne National and Oak Ridge National Laboratories, substantiating the existence of these high-spin, monatomic elements and their power as superconductors.

Hudson also met with Dr. Hal Puthoff, director of the Institute for Advanced Studies in Austin, Texas. Puthoff performs cutting-edge research into zero-point energy and gravity as a zero-point fluctuation force. He and other scientists have theorized that enough energy exists in the space found in the atoms inside an empty coffee cup to boil all the oceans of the Earth if fully utilized.

Puthoff had also theorized that matter reacting in two dimensions should lose about 44 percent of its gravitational weight, exactly the weight loss found by Hudson. When it was found that Hudson's elements, when heated, could achieve a gravitational attraction of less than zero, Puthoff concluded the powder was "exotic matter" capable of bending time and space. The material's antigravitational properties were confirmed when it was shown that a weighing pan weighed less when the powder was placed in it than it did empty. The matter had passed its antigravitational properties to the pan.

Adding to their amazement, it was found that when the white powder was heated to a certain degree, not only did its weight disappear but the powder itself vanished from sight. When a spatula was used to stir around in the pan, there apparently was nothing there. Yet, as the material cooled, it reappeared in its original configuration. The material had not simply disappeared; it apparently had moved into another dimension.

Hudson also saw evidence of perpetual energy through the use of a superconductor. "You literally start the superconductor flowing by applying a magnetic field," he said. "It responds to this by flowing light inside and building a bigger Meissner Field [Walter Meissner in 1933 discovered that light flowing within a superconductor produces an electromagnetic energy field that excludes external magnetic fields] around it. You can put your magnet down and walk away. Come back a hundred years later and it is still flowing exactly as when you left. It will never slow down. There is absolutely no resistance; it is perpetual motion and will run forever."

This new technology dealt with the manipulation and control of basic energy. Some scientists believed that such control at the atomic and subatomic level might do much more than offer new propulsion technology. It might open the door to antigravity, limitless free energy, a cure for diseases such as AIDS and cancer, an end to the aging process, faster-than-light speeds, and much more, perhaps even inter-dimensional and time travel.

Since science is coming to the conclusion that gravity and time are interconnected aspects of energy, it is possible that the Bell was used for experimenting with time travel. This possibility is not as outrageous as it sounds, as many notable scientists and authors have written seriously about the possibility of time travel.

Astronomer and Pulitzer Prize winner Carl Sagan, director of the Laboratory for Planetary Studies at Cornell University at the time of his death in 1996, when asked about time travel, stated:

> Right now we're in one of those classic, wonderfully evocative moments in science when we don't know, when there are those on both sides of the debate, and when what is at stake is very mystifying and very profound.
>
> If we could travel into the past, it's mind-boggling what would be possible. For one thing, history would become an experimental science, which it certainly isn't today. The possible insights into our own past and nature and origins would be dazzling. For another, we would be facing the deep paradoxes of interfering with the scheme of causality that has led to our own time and ourselves. I have no idea whether it's possible, but it's certainly worth exploring.

Jenny Randles, a science-oriented British author, presented compelling examples of recent discoveries in her 2005 book *Breaking the Time Bar-*

rier, which indicate the very real possibility of time travel. She noted that "a race to build a time machine has been going on since at least the Second World War." After discussing "worm holes" in time and space, and other possible means of time travel, she pointed out, "... [F]rom our understanding of physics—if you travel faster than light, then you can overtake the flow of events that light happens to transmit. Since the passage of these events forms what we interpret as time, then by traveling faster than light you ought to travel through time. Spaceships that outstrip light speed are always going to moonlight as time machines." Today, more than one scientist has claimed to have broken the light barrier, though official acceptance has been lacking.

TIME-TRAVELING NAZIS. This horrendous idea sounds preposterous, but the science is there and the Bell did exist.

No wonder certain powerful persons would go to any lengths to obtain or conceal such knowledge. Just such attempts began in the closing days of World War II, as the victors sought to learn the secrets of Nazi super-science.

It is clear that certain members of the American military were keen to learn Nazi secrets, as shown by this portion of a 1945 letter from Major General Hugh J. Knerr to Lieutenant General Carl Spaatz, the commander of U.S. Strategic Air Forces in Europe: "Occupation of German scientific and industrial establishments has revealed the fact that we have been alarmingly backward in many fields of research. If we do not take this opportunity to seize the apparatus and the brains that developed it and put the combination back to work promptly, we will remain several years behind while we attempt to cover a field already exploited."

Consider the rush into Czechoslovakia by General George S. Patton's Third Army even as the European war wound to a close. "The madcap, and some would say, militarily and politically indefensible, Allied dash away from Berlin and to south-central Germany and Prague are consistent with American knowledge, at some very high level, of Kammler's SS Sonderkommando black projects and secret weapons empire," wrote Farrell.

Vernon Bowen, whose 1950s-era book on UFOs was classified by the

U.S. government, relates how one of Patton's officers, Colonel Charles H. Reed, organized the escape of the Lippizan horses from the Spanish Riding School at the end of the war, an event memorialized in the 1963 Disney film *Miracle of the White Stallions*. Bowen noted that Reed saved the horses "while on his mission of persuading the head of German intelligence to turn over to the U.S. the many truckloads of documents buried on the Czech-Austrian border—documents which are still secret today." Could these documents have been Kammler's technology files?

The Allies' rejection of SS chief Himmler's last-minute offer to surrender may not have been due to the "frantic attempts of a desperate mass murderer to avoid his inevitable fate," as described by mainstream historians, but instead because Himmler had lost real control over the exotic technology. After all, Himmler was too high-profile a person to be allowed to live on after the war. He reportedly committed suicide by taking a poison capsule on May 23, 1945, after being caught trying to sneak through British lines disguised as a German Army private.

Hans Kammler, on the other hand, was largely unknown to the public, though he undoubtedly was high on the list of wanted Nazi war criminals, considering his involvement in the construction of concentration camps and their gas chambers as well as his participation in the leveling of the Warsaw ghetto. "Unlike Himmler," noted Cook, "Kammler had something of value to deal—something tangible. By early April [1945], Hitler and Himmler had placed under his direct control every secret weapon system of any consequence within the Third Reich—weapons that had no counterpart in the inventories of the three powers that were now bearing down on central Germany from the east and the west."

"The deal had already probably been cut between Kammler's representatives and OSS [the U.S. Office of Strategic Services] station chief in Zurich, Allen Dulles, or via General Patton himself," Farrell surmised.

If such a deal was made with Patton, he did not live to see the results. On December 9, 1945, while riding in his 1939 Cadillac staff car, Patton suffered a head injury when his car was struck by a 2½-ton military truck that turned in front of them. Patton's driver and a passenger, his chief of staff Major General Hobart "Hap" Gay, were uninjured. Paralyzed from the neck down, Patton was taken to a military hospital in Heidelberg, Germany, where he died on December 21.

Since the war, there have been several conspiracy theories regarding Patton's death—one being that he was killed by his own government. Most have concentrated on his vocal assertions that the United States should have carried the war on into Russia and put an end to communism, plus his public advocacy of reinstating ranking Nazis to help rebuild Germany.

Noting that Patton, whose forces drove straight to the heart of Nazi research in Czechoslovakia, may well have been aware of Kammler and his Nazi superweapons, Farrell stated that if Patton was deliberately silenced, "then surely this [knowledge of Nazi super-science] is the most plausible motivation for the deed."

Did knowledge of the incredible ability to manipulate energy die with top Nazis at the end of the war? Consider the fate of Hans Kammler.

As the war drew to a close, Kammler made no secret that he intended to use both the V-2 scientists and rockets under his control as leverage for a deal with the Allies. On April 2, 1945, on Kammler's orders, a special train carried rockets and five hundred technicians and engineers escorted by a hundred SS troopers to an Alpine redoubt in Bavaria. According to von Braun and Dornberger, Kammler planned to "bargain with the Americans or one of the other Allies for his own life in exchange for the leading German rocket specialists."

"[Kammler] came to me in early April in order to say good-bye," recalled Nazi armaments minister Speer. "For the first time in our four-year association, Kammler did not display his usual dash. On the contrary, he seemed insecure and slippery with his vague, obscure hints about why I should transfer to Munich with him. He said efforts were being made in the SS to get rid of the fuehrer. He himself, however, was planning to contact the Americans. In exchange for their guaranty of freedom, he would offer them the entire technology of our jet planes, as well as the A-4 rocket and other important developments. . . ."

On April 4, 1945, when von Braun pressed Kammler for permission to resume rocket research, the SS officer quietly announced that he was about to disappear for "an indefinite length of time." He was true to his word: no one saw Kammler again. As everyone knows, von Braun and Dornberger, along with other scientists and many of the V-2 rockets, eventually made their way to the United States, becoming founding members of its modern space program with no help from Kammler.

Jean Michel, himself an inmate of concentration camp Dora, which provided slave labor for Kammler's rocket program, wrote of Kammler: "The chief of the SS secret weapon empire, the man in Himmler's confidence, disappeared without a trace. Even more disturbing is the fact that the architect of the concentration camps, builder of the gas chambers, executioner of Dora, overall chief of all the SS missiles has sunk into oblivion. There is the Bormann mystery, the Mengele enigma; as far as I know, no one, to this day, has taken much interest in the fate of SS Obergruppenfuehrer Hans Kammler." Michel, along with others, wondered, "Why had the 'cold and brutal calculator' described by Speer so abruptly discarded the trump cards he had so patiently accumulated?"

As the war drew to a close, Kammler "had the good fortune to inspect the Czechoslovakian stretch of the front," wrote Witkowski. "After this event, nobody knew what became of him. Perhaps he died, though it is unlikely that this would never have been recorded."

The reports of Kammler's death are varied and mutually exclusive. One version has him committing suicide in a forest between Prague and Pilsen two days after Germany surrendered, while another said he was shot by his own SS aide in Prague. Another version was that he died in a shootout with Czech partisans. The Red Cross initially reported Kammler as "missing," but this was later changed to "dead" upon the testimony of a relative. The one common denominator regarding Kammler's various death reports was that he was last seen in north central Czechoslovakia, in close proximity to the Wenzeslaus Mine—and the Bell.

Despite the lack of a body, no effort appeared to have been taken to establish the truth of Kammler's death and, unlike his superior Bormann, Kammler was not tried in absentia at Nuremberg.

Kammler was not alone in his escape. Dozens of high-ranking former SS or party members simply disappeared. Many of them were associated with advanced technology programs.

Did Kammler and his cohorts escape with weapons plans for the amazing Bell project? Whoever controlled such secret technology was certainly in a strong position to strike a deal with one of the Allied nations.

With secret projects in the hands of the fanatical SS and with factories and research facilities scattered over—and under—the countryside, it is entirely conceivable that saucers, uranium weapons, the Bell, and other exotic technologies could have been developed without the knowledge of anyone except Himmler, Bormann, and Kammler. The high-profile Himmler had been taken out of the loop as far back as 1943. The fates of Bormann and Kammler remain unproven. "[T]he evidence is strong enough to suggest collusion at the highest levels between the United States and Nazi Germany governments—and that collusion extends down to those within U-234, its officers, crew and passengers—and has been maintained by powerful parties with vested interests on both sides of the Atlantic ever since," stated Hydrick.

If the highest circle of America's ruling elite indeed obtained Nazi super-science in the wake of World War II, it came with a price—one these prewar, pro-Nazi sympathizers were willing to accept. When American authorities realized the alternative and nonlinear physics within Nazi science, they knew it was beyond the frame of reference of most U.S. scientists, which is why they recruited so many Germans and brought them to America.

"The trouble was," recounted one government insider, "when the Americans took it all home with them, they found, too late, that it came infected with a virus—you take the science on, you take on aspects of the ideology as well."

The intense interest of the Nazi leadership in occult or hidden subjects—from ancient artifacts to legends of prehistorical high-tech super-races—is well documented. Toward the end of the war, their acquisition of super-science may have been matched by the recovery of an amazing and precious treasure.

CHAPTER 4

A TREASURE TROVE

IN MARCH 1944, THE CRAGGY PEAKS SURROUNDING THE ANCIENT Cathar fortress of Montségur in southern France reverberated with the grinding gears and revving engines of military machines. The trucks and command cars belonged to a battalion of Nazi SS troopers led by Adolf Hitler's top commando, SS Standartenfuehrer Otto Skorzeny. Standing six feet and four inches, Skorzeny was larger than life among his comrades, and his exploits during World War II only enhanced this reputation. An old dueling scar creased his face from the left cheekbone to his chin, earning him the nickname Scar.

Born in Vienna in 1908, Skorzeny had joined the Nazi Party in 1930 while studying in Germany. By 1939, he had been accepted as a member of Hitler's personal bodyguards. Sent home from the Russian Front in 1942 due to wounds, Skorzeny soon was directing secret agents in other countries.

But worldwide attention became focused on Skorzeny in September 1943, when he led a glider assault by commandos on a mountaintop hotel where the dictator Benito Mussolini was being held captive following a coup in Italy. In a daring daylight operation, Skorzeny and his men liberated Mussolini, who had been contemplating suicide, and whisked him off to safety. Hitler declared Mussolini the rightful leader of Italy and the war there continued until Germany surrendered in May 1945.

Mussolini's ouster followed by Allied landings in Italy prompted Hitler to send his troops into what had until then been called Vichy, France, to ostensibly protect the "soft underbelly of Europe." The Nazis had gained freedom of movement in the historic Languedoc region, located in the foothills of the Pyrenees mountains, which separate France from Spain.

But in early March 1944, something more than military victory was on Skorzeny's mind as his troops entered the area encompassing Montsegur and the village of Rennes-le-Château, the site of a great mystery since the discovery of strange documents by a young priest in 1891.

SKORZENY'S THOUGHTS UNDOUBTEDLY were centered on the location of a fabulous treasure believed to have been located by a German author and occult researcher named Otto Rahn.

Little is known about Rahn's early life except that he was born February 18, 1904, in Michelstadt and educated in literature and philology at the University of Berlin. During his time in school, Rahn had become fascinated with legends of the Holy Grail as well as the little-understood Cathars—"pure ones," as they were called—who had opposed the Roman Church and thus suffered near annihilation in a papal military campaign in 1209, known as the Albigensian Crusade.

In the early 1930s, Rahn traveled widely in the Languedoc region of Southern France, even spelunking among the maze of cave systems in the foothills of the Pyrenees mountains. Here he gained firsthand knowledge about the Cathars and their descendants, many of whom became members of the fabled Knights Templar.

Rahn's book *Crusade Against the Grail* was published in 1933, the same year Hitler came to power. This and other books on the Cathars and Grail legends, as well as his many travel articles, brought him to the attention of SS chief Heinrich Himmler, who, along with many other top-ranking Nazis, had a keen interest in occult artifacts and knowledge. Rahn's knowledge of both the Cathars and the Templars apparently intrigued Himmler, as Rahn was inducted into the SS as a lieutenant in 1936.

Himmler and his cronies must have been entranced with Rahn, who had drawn connections between the Cathar fortress of Montségur and a fabulous cave housing the Holy Grail called Montsavat, mentioned in *Parzival* by Wolfram von Eschenbach in the thirteenth century. Rahn believed he had discovered the final resting place of a great treasure of antiquity, which included the Tables of Testimony, the Grail Cup known as Emerald Cup, and perhaps even the long-lost Ark of the Covenant.

And by March 1944, the Nazis were free to move troops into Languedoc in search of this ancient wealth, known as King Solomon's treasure.

It was much too late for Rahn. By 1939, Rahn had become disenchanted with his Nazi superiors, writing, "There is much sorrow in my country. [It is] impossible for a tolerant, liberal man like me to live in the nation that my native country has become." He resigned his commission in the SS in February 1939 and, barely a month later, reportedly died of exposure after having been caught in a snowstorm during a hiking expedition. Rumors circulated that he had been killed in a concentration camp. The National Socialist ideologue Alfred Rosenberg recorded that Rahn committed suicide by taking cyanide "for politico-mystical reasons as well as for personal ones." However he died, Rahn's knowledge was retained by Himmler.

THE FABLED TREASURE of King Solomon is the greatest cache of riches known to humankind. Its fascinating history serves as a timeline for the evolution of Western civilization as it can be traced from ancient Mesopotamia up to World War II.

Gold, silver, and precious gems—such as diamonds, pearls, emeralds, amber, amethyst, topaz, sapphires, rubies, turquoise, and others—comprised this priceless hoard of riches. But Solomon's treasure also contained riches of quite a different sort. It included ancient scrolls, texts, and tablets upon which was inscribed some of the world's most esoteric and occult knowledge. This knowledge had been handed down for thousands of years from the time of the world's first recorded civilization in ancient Sumer—present-day Iraq.

Thousands of translated Sumerian tablets along with their inscribed cylinder seals are now available and they tell of astonishing technology apparently in use prior to Noah's flood. With recent scientific advancements—such as powered flight, the space program, DNA manipulation, and cloning—many experts are beginning to rethink the idea that today's world is the apex of civilization's evolution. It is now possible to consider that a technically advanced civilization was on the Earth in the far distant past, possessing knowledge that mankind is only just now relearning.

Bits and pieces of ancient knowledge that survived the Great Flood formed the essence of the riches that were transported to Egypt by Abraham, the inheritor of the secrets of Enoch and the biblical patriarch of both Arabs and the Jews. Abraham, a native of Sumer, known early on as Abram, by some traditions was said to possess a tablet of symbols representing all of the knowledge of humankind handed down from the time of Noah. Known to the Sumerians as the Table of Destiny, it was this table of knowledge—known to the early Jews as the Book of Raziel—that reportedly provided King Solomon with his vast wisdom.

The Sumerian Table of Destiny is thought to be the same as the Tables of Testimony mentioned in Exodus 31:18. Other Bible verses—Exodus 24:12 and 25:16—make it clear that these tables are not the Ten Commandments.

British author Laurence Gardner believed this ancient archive was directly associated with the Emerald Table of Thoth-Hermes, and that its author was the biblical Ham. "He was the essential founder of the esoteric and arcane 'underground stream' which flowed through the ages," stated Gardner. This table of knowledge was passed from Egypt and Mesopotamia to Greek and Roman masters, such as Homer, Virgil, Pythagoras, Plato, and Ovid. In more recent times, it was passed through such secret societies as the Rosicrucians and Knights Templar and on to the Stuart Royal Society in England.

In Jewish history, the Cabala, also spelled as Kabbalah or Qabbalah, was supposed to contain hidden meanings. Such cleverly coded knowledge was thought to be found within the Torah and other old Hebraic texts, such as the Sefer Yezirah (Book of Creation) and the Sefer Ha-Zohar

(Book of Light). These books, which predate the Talmud, a compilation of early Jewish laws and traditions first written in the fifth century A.D., were produced centuries before the time of Jesus. According to the Book of Light, "mysteries of wisdom" were given to Adam by God while still in the fabled Garden of Eden, generally believed to have been located between the Tigris and Euphrates Rivers. These elder secrets were then passed on through Adam's sons to Noah and on to Abraham long before the Hebrews existed as a distinct people.

Much like our understanding of history and religion today, the information within the Cabala became both incomplete and garbled over the centuries through losses due to war and natural disasters as well as misinterpretations and foreign influences. But it was this ancient wisdom, taken from Egypt at the time of the Great Exodus, that formed the core of Cabalistic knowledge handed down through the centuries via several secret societies, some of which remain active among us even today.

But what of Solomon's treasure itself? What happened to the wealth—in both riches and knowledge—that found a resting place on the Temple Mount in Jerusalem with the construction of Solomon's Temple nearly a one thousand years before the birth of Jesus?

Much of this treasure fell into the hands of the Romans when they sacked Jerusalem following the Jewish Revolt of 66 A.D. By the time of this looting, Solomon's Temple had been built over to become the palace of King Herod. With the certain advance intelligence that the Romans would be sending troops to put down the Jewish rebellion, keepers of the knowledge buried away much of the treasure in the catacombs beneath Herod's palace. Alcoves and passages were closed off and sealed with earth.

Thus, when the Romans sacked Jerusalem following revolts in both 66 A.D. and 132 A.D., they found only a portion of the treasure. To have hidden all the treasure would have prompted a strenuous search by the Roman authorities. As it was, they were content to move what they found to Rome as war booty. The best part of the treasure, including both wealth and knowledge, was safely buried under the Temple Mount and all but forgotten, as most religious leaders were killed or taken to Rome as prisoners.

In 410 A.D., Alaric, who had been commander of Visigoth auxiliaries under Roman emperor Theodosius, sacked Rome. Alaric had been pro-

claimed king over the Visigoths with the death of Theodosius, and he began his march on Rome after invading Greece and Northern Italy. It was
the first successful attack on Rome in more than eight hundred years.

Alaric's troops took the portion of Solomon's treasure in Rome along
with other prizes of war. "When Alaric withdrew his forces, the treasures
of Solomon's Temple went with them," wrote Colonel Howard Buechner,
a former medical officer with the 45th Infantry Division.

This contention was supported by the work of Otto Rahn, who wrote in
1933 of four young men who discovered a casket in a Pyrenees cavern,
"Was this reliquary casket part of 'Solomon's treasure,' which was taken by
the Visigoth king Alaric from Rome to Carcassonne in A.D. 410? According to [the last major ancient Greek philosopher, Proclus], it was filled with
objects that once belonged to King Solomon, the king of the Hebrews."

The Visigoths secured their booty in the Pyrenees foothills located in
the Languedoc region of what was to become southern France. This area
encompasses the Cathar stronghold at Montségur as well as the small village of Rennes-le-Château. The secrets of this treasure were handed from
the Goths and the early Franks to the Cathars, the "pure ones" of southern France. They considered their Christian beliefs more pure than those
of the Church of Rome, probably because they had access to original
documents and were not dependent on the Church hierarchy to translate
and interpret the Bible. In fact, according to Otto Rahn, Cathar beliefs
were greatly influenced by Druids, priests, and soothsayers who had spread
from Mesopotamia through eastern and western Europe to the British
Isles. Catharism was an odd blending of ancient Earth worship, Eastern
mysticism, Gnosis, and basic Christianity.

The Cathar faith, sometimes described as "Western Buddhism," might
have spread to all the corners of Europe but for the blood and fire of the
Albigensian Crusade begun in 1209, which may well have been more of a
French civil war than a religious campaign. It was fought between the
tightly controlled and austere northerners and the cultured, freedom-
loving peoples of the south.

Rahn noted the exotic ethnic mixture of the Languedoc region. "In the
third century B.C., an immigration of peoples from the Caucasus to the
West took place: Phoenicians, Persians, Medeans, Getules (actually Berbers

of North Africa), Armenians, Chaldeans [Sumerians], and Iberians," he wrote. Prior to the Vatican-approved bloodletting, the provinces of southern France were virtually independent republics that allowed extraordinary freedom of education, culture, and diversity. Jews were accorded the same rights as the rest of the citizenry. Both agriculture and the arts were flourishing. Many of the Cathars, while Christians, nevertheless still worshipped the feminine goddesses—Isis and Athena—as had their Gothic and Frankish ancestors.

It is not known but strongly suspected by researchers that this hidden treasure guarded by the Cathars included a copper scroll similar to an etched scroll of copper found among the Dead Sea Scrolls in 1947 at Qumran, on the northwest shore of the Dead Sea. When translated in the mid-1950s at Manchester University, this scroll proved to be an inventory of a great treasure. It apparently was one of several copies. With its detailed directions to hidden Hebrew valuables, the Copper Scroll was literally a treasure map. Such an inventory in the hidden Visigoth cache would explain why certain French aristocrats, descendants of the Goths and the Cathars, who had access to the treasure, fomented the First Crusade, resulting in the capture of Jerusalem in 1099 A.D.

Less than twenty years after the crusaders took Jerusalem and placed King Baldwin II of Le Bourg in charge of the occupied territories, nine knights were granted a military order called the Poor Knights of Christ and of the Temple of Solomon. This title was soon shortened to the Knights of the Temple, or Knights Templar. They were allowed to be billeted in Herod's palace, the exact location of the hidden treasure as described in the Copper Scroll.

These knights were led by Hugh de Payens, a nobleman in the service of his cousin, Hughes, count of Champagne, and Andre de Montbard, the uncle of Bernard of Clairvaux, later known as the Cistercian Saint Bernard. Montbard also was a vassal of the count of Champagne. At least two of the original knights, Rosal and Gondemare, were Cistercian monks prior to their departure for Jerusalem. In fact, the entire group was closely related both by family ties and by connections to the Cistercian monks and Flemish royalty descended from the Cathars. They journeyed to the Holy Land with an agenda: to recover the remainder of the treasure.

Ostensibly, this order of knights was to protect the roads to Jerusalem, but their actions were of a very different nature. Rather than guard roads, the Templar knights spent years excavating under Herod's palace, the old Temple of Solomon. The digging was extensive. British Royal Engineers led by a Lieutenant Charles Wilson discovered evidence of the Templars while mapping vaults under Mount Moriah in 1894. They found vaulted passageways with keystone arches, typical of Templar handiwork. They also found artifacts consisting of a spur, parts of a sword and lance, and a small Templar cross, which are still on display in Scotland.

It was during their excavations, according to several accounts, that the Templars acquired material wealth as well as texts of hidden knowledge, most probably including some dealing with the life of Jesus and his associations with the Essenes and Gnostics. They also reportedly acquired the legendary Tables of Testimony given to Moses as well as other holy relics—perhaps even the legendary Ark of the Covenant and the Spear of Longinus—which could have been used to validate their later position as an alternative religious authority to the Roman Church.

When the Knights Templar transported the remainder of Solomon's treasure back to the Languedoc region of southern France, it was reunited with the portion the Goths had brought from Rome more than seven hundred years earlier. The material wealth—King Solomon's diamonds, precious gems, gold, and silver—formed the base of the Templars' legendary fortune. Much of this was transported to their temple in Paris. The esoteric treasure—scrolls and tablets of ancient knowledge—were kept hidden from the Roman Church in the elaborate cave systems of the Pyrenees.

On Friday, October 13, 1307, the greedy French king Philip, in debt to the Templars, moved against the Templars with the blessing of Pope Clement V. Like their Cathar forebears, the Templars were charged with all forms of heresy. Templars throughout Europe were hunted down, killed, and tortured. The last Templar grand master, Jacques de Molay, was burned at the stake in 1314. But many Templars simply cast off their distinctive surcoats, identifiable by the red Maltese cross, and blended into the local populations only to emerge in later years as Freemasons.

When authorities broke into the Paris temple, they found nothing. The treasure had been removed by the Templars, who apparently dispersed it to several different locations. Some went to Scotland, where Robert the Bruce provided the Templars sanctuary, some went to pre-Columbus America, and some returned to the caverns of Languedoc.

Centuries passed while the devout in southern France kept the secret of the hidden treasure from both church and state authorities. This secret briefly broke into public view in the late 1890s, when the young priest of the small village of Rennes-le-Château discovered some documents hidden in the alter of his church, which had been consecrated to Mary Magdalene in 1059 and stood on Visigoth ruins dating to the sixth century.

In 1891, Father Francois Berenger Sauniere discovered two genealogies dating from 1244 and 1644, along with two texts written in the 1780s by a former parish priest, Abbot Antoine Bigou. The Bigou texts were unusual and appeared to be written in different and indecipherable codes. Sauniere took his discovery to his superior, the bishop of nearby Carcassonne, who sent him on to Paris to meet with the director general of the Saint-Sulpice Seminary, reportedly a center for an unorthodox society called the Compagnie du Saint-Sacrement, reputed to be a front for the Priory of Sion. This priory is thought to include members committed to keeping secret the Templar treasure and knowledge.

Whatever was in the documents changed Sauniere's life. He journeyed to Paris, where he mingled with the Parisian cultural elite and soon came into great wealth. Before his sudden death in 1917, researchers estimated he had spent several million dollars on construction and renovations in the town. He also had the town's road and water supply upgraded, assembled a massive library, and built a zoological garden, a lavish country house named Villa Bethania, and a round tower named Tour Magdala, or Tower of Magdalene. Within the renovated church, Sauniere erected a strange statue of the demon Asmodeus—"custodian of secrets, guardian of hidden treasures, and, according to ancient Judaic legend, builder of Solomon's temple."

Sauniere began to exhibit a defiant independence toward his Church superiors, refusing to disclose the source of his newfound wealth or accept a transfer from Rennes-le-Château, where he and his housekeeper were seen digging incessantly in the graveyard around the church. Yet, when push came to shove, the Vatican supported Sauniere, a good indication of the significance of his discoveries.

On January 17, 1917, Sauniere suffered a sudden stroke. A nearby priest was called to administer last rites but, "visibly shaken," refused to do so after hearing Sauniere's confession, which has never been made public. His housekeeper and companion, Marie Denarnaud, kept her silence about Sauniere's activities, living quietly in the Villa Bethania. Toward the end of her life, she sold the villa to a man whom she promised she would tell a secret that would make him both wealthy and powerful. Unfortunately, she too died of a stroke before passing along this secret.

Thus began the mystery of Rennes-le-Château. "Speculation has varied over the years as to the true nature of Sauniere's discovery," wrote Lynn Picknett and Clive Prince; "most prosaically it has been suggested that he found a hoard of treasure, while others believe it was something considerably more stupendous, such as the Ark of the Covenant, the treasure of the Jerusalem Temple, the Holy Grail—or even the tomb of Christ.... The Priory claim that what Sauniere had discovered were parchments containing genealogical information that proves the survival of the [Franks] Merovingian dynasty." Whatever Sauniere found, it seems to have been linked to Solomon's treasure long hidden in the nearby cave systems by first the Goths and later the Templars.

In review of what is known about the Father Sauniere affair, it appears doubtful that the priest actually found the lost treasure. It is more likely that his find was some ancient genealogies inimical to the Catholic Church and perhaps some clues to the location of the treasure. Such clues were expanded upon by the work of Otto Rahn with further expeditions to Languedoc financed by SS chief Himmler. Rahn's work was getting him closer to the location of the treasure. "In a letter written to [Rahn's close friend Karl Maria Wiligut-Weisthor] in September 1935, Otto Rahn informed his friend that he was at a place where he had reason to believe the Grail might be found, and that Weisthor should keep the matter secret with the exception of mentioning it to Himmler," reported British authors David Wood and Ian Campbell.

By the start of the war, Rahn was dead but his knowledge was kept alive by Himmler. According to author Angebert, as early as June 1943, a group of German geologists, historians, and ethnologists camped near Montségur and began excavations that lasted into November. This expedition failed to produce the treasure.

BUT OTTO SKORZENY, dispatched by Himmler in early 1944, apparently had better luck. "The commando force reached Languedoc in early March 1944, and set up headquarters at the base of Montségur. They spent a few days exploring the Cathar fortress and in reconnaissance of the surrounding mountains. They discovered remnants of what had once been a 3,000-step stairway which led from the castle to an exit in the valley below," wrote Colonel Howard Buechner.

Skorzeny, disdaining intellectual study of the problem over the treasure's location, set about his work from the standpoint of a tactician. He quickly surmised that Rahn and the members of the 1943 expedition had looked in the obvious—and wrong—locations.

The Germans promptly found a secret path used as an escape route for the Cathars during the siege of Montségur, which ended in March 1244, exactly seven hundred years earlier. "Skorzeny and his men scouted along this path and soon discovered what appeared to be an ancient trail leading into the higher mountains," related Colonel Buechner. "At an undisclosed distance from Montségur they found a fortified entrance to a large grotto. Perhaps it was the grotto of Bouan, which was the last refuge of the Cathars after the fall of Montségur. Not far from this grotto was the mountain called La Peyre. Near the crest of this mountain was another grotto and in this cavern, it is said, they found the treasure."

On March 15, 1944, Skorzeny sent a one-word telegram to Berlin. It read: "*Ureka* [Eureka, or I have found it!]."

It was signed with Skorzeny's nickname, "Scar."

His message was soon answered with a cryptic note: "Well done. Congratulations. Watch the sky tomorrow at noon. Await our arrival."

This was signed "Reichsfuehrer SS."

According to Colonel Buechner, there followed an amazing coincidence of events. Each March 16, local descendants of the Cathars gathered at Montségur to pay homage to their ancestors who had died there seven hundred years earlier. In 1944, the local German military governor refused to grant permission, claiming Hitler's Third Reich had "historic rights" to Montségur. In defiance of this prohibition, a group of pilgrims traveled to Montségur anyway and there encountered Skorzeny and his men. The giant commander chief, who had a reputation for defying bureaucracy, granted their request, since he had control of the treasure.

The pilgrims placed special significance on the date March 16, 1944, because of an ancient prophecy that stated, "At the end of seven hundred years, the laurel will be green once more." Many assumed this meant the beginning of a revival of Catharism. That year's seven hundredth anniversary delegation of pilgrims was much larger than usual. "Thus it was that the worshippers were on top of the mountain [Montségur] at precisely the time when Skorzeny had been instructed to 'watch the sky,'" noted Colonel Buechner. Near noon, a Fieseler Storch, or Stork, light airplane bearing German markings approached and created a giant spectacle for the gathered crowd. The airplane, which may have carried either Himmler, Nazi ideologist Alfred Rosenberg, or both, used skywriting equipment to produce a huge Celtic cross across the sky over Montségur.

"The pilgrims on the mountaintop were awestruck and reacted as if a miracle had occurred," said Colonel Buechner. "They had no idea that the fabulous treasure of the Cathars had been discovered only a short time before and that the plane was saluting the victorious expedition."

The next afternoon, an official delegation arrived and congratulations and medals were handed out. This delegation was headed by Rosenberg and Oberst, or colonel, Wolfram Sievers, a ranking member of the Ahnenerbe SS, the organization dealing with esoteric and occult matters for Himmler's Black Shirts. According to Colonel Buechner's sources, the treasure was carried out of the Pyrenees by pack-mule train to the village of Lavelanet, where it was loaded onto trucks for the journey to a rail head. Guarded rail cars carried the treasure to the small town of Merkers, located about forty miles from Berlin, where it was catalogued by hand-picked members of the Ahnenerbe SS and then moved to other locations, including Hitler's redoubt at Berchtesgaden, where some of the treasure was carried into the extensive tunnel system, large parts of which remain inaccessible today.

"During its initial days at Merkers, the 'Treasure of the Ages' was intact for the last time," stated Colonel Buechner. The Nazis apparently had secured the world's greatest treasure trove—both of wealth and of lost secrets.

According to Colonel Buechner, the treasure consisted of:

◆ Thousands of gold coins, some of which dated back to the early days of the Roman Empire and earlier.
◆ Items believed to have come from the Temple of Solomon, which included gold plates and fragments of wood that provided strong evidence that the partially decomposed relic was the fabled Ark of the Covenant.
◆ Twelve stone tablets bearing pre-runic inscriptions, which none of the experts were able to read. These items comprised the stone grail of the Germans and of Otto Rahn.
◆ A beautiful silvery cup with an emerald-like base made of what appeared to be jasper. Three gold plaques on the cup were inscribed with cuneiform script in an ancient language.
◆ A large number of religious objects of various types, which were unidentifiable as to time and significance. However, there were many crosses from different periods, made of gold or silver and adorned with pearls and precious stones.
◆ An abundance of precious stones in all sizes and shapes.

By the time the Allies occupied Germany, much of the treasure had been melted down into bars and shipped out of the country. A vast amount of gold and silver, as well as pieces of art and religious artifacts, were taken into Allied hands in the town of Merkers, but the most rare and valuable items dropped from public view.

"When Martin Bormann's wife—Frau Gerda Buch Bormann—was captured at a small hotel in northern Italy, she had 2,200 antique gold coins in her possession," wrote Buechner. "These priceless coins were almost certainly a part of Hitler's personal share of the Treasure of Solomon. . . . Bormann himself sent gold coins to Argentina by submarine, where on arrival, his treasure was placed under the personal protection of

Evita Peron." Bormann's wife suffered from cancer and was released by the Allied authorities. She died of mercury poisoning on March 23, 1946.

Lest anyone consider Colonel Buechner's account of Otto Rahn and the taking of Solomon's treasure some personal fantasy, they would do well to consider his credentials. A native of New Orleans, Howard A. Buechner earned a bachelor's degree from Tulane University and a medical degree from Louisiana State University. During World War II, Dr. Buechner was a medical officer with the 3rd Battalion, 157th Infantry Regiment of the 45th Infantry Division, the unit that arrived first at Dachau concentration camp. Dr. Buechner was the first American physician to enter the camp upon its liberation. He was later promoted to colonel while serving in the postwar reserves. It was during his wartime experiences, on the scene, that Colonel Buechner first learned of the loss of Solomon's treasure. Buechner's awards included the Medical Combat Badge, the Bronze Star, three battle stars, the Army Commendation Medal, the War Cross, and the Distinguished Service Cross of Louisiana. He also became a professor of medicine at Tulane and served as emeritus professor of medicine at LSU, where an honorary professorship was established in his name. His papers on tuberculosis and other lung diseases made him an internationally recognized expert.

Colonel Buechner and other researchers have estimated the treasure trove recovered by Skorzeny in southern France in excess of $60 billion, based on the current price of gold. This, added to the other loot from Europe, gave the Nazis more than enough economic clout to continue their plans for world conquest long after the end of World War II. Such wealth made it possible for Bormann and other Nazis to misdirect West German investigations and silence foreign governments and news organizations. And it provided the means to infiltrate and buy out numerous companies and corporations, both outside the United States and within.

To understand how a shadowy Nazi empire was created, one must return to German business history and take note of Bormann's activities beginning in mid-1944.

CHAPTER 5

THE WRITING ON THE WALL

IN THE FALL OF 1942, THE GERMAN SIXTH ARMY WAS RAMPAGING virtually unhindered through the Ukraine in Russia. Its objectives were Baku and the rich Caucasian oil fields. With these oil reserves in hand, Hitler planned to turn south and capture the oil of the Middle East in a combined operation with Field Marshal Erwin Rommel's famed Afrika Korps' assault from North Africa. This scheme was thwarted by Rommel's defeat at El Alamein—made possible by the now-known decoding of German Enigma messages—and the eventual destruction of the German Sixth Army at Stalingrad, a city on the Volga River.

Stalingrad, which the Germans had entered in strength by late September of that year, soon turned into a cauldron of death and destruction. Even breathing became a chore due to the constant shelling and bombing. Though of dubious strategic value, both Hitler and Stalin insisted there be no withdrawal from the fiercely defended city, the namesake of the Soviet leader. Russian pincer attacks isolated the Sixth Army in late November, but organized resistance did not end until February 2, 1943, with the surrender of more than ninety thousand German soldiers—most of them reduced to skin and bones through lack of supplies. With the loss of the Sixth Army, ranking Nazis recognized that the war's momentum had turned against them on the Eastern Front. It was never to be regained.

Six months later, after the disastrous Battle of Kursk, in which the Nazi war machine lost nearly three-fourths of its entire mechanized force, it became clear that the defeat of Germany was more than a possibility, it was a probability. Top Nazis began to draw up plans for escape and the continuation of their goals.

Curt Reiss, a noted news correspondent of the time, who traveled extensively in Europe, wrote in detail about the Nazis' plans for survival, in his book *The Nazis Go Underground*. Astonishingly, this was published in the spring of 1944, prior to the Allied D-Day landings in France that June. Reiss wrote, "They had better means for preparing to go underground than any other potential underground movement in the entire previous history of the world. They had all the machinery of the well-organized Nazi state. And they had a great deal of time to prepare everything. They worked very hard, but they did nothing hastily, left nothing to chance. Everything was thought through logically and organized to the last detail. Himmler [along with Bormann] planned with the utmost coolness. He chose for the work only the best-qualified experts—the best qualified, that is, in matters of underground work."

Reiss pointed out that when the Nazi Party gained control in Germany, the apparatus of the party was simply transferred over to the apparatus of the state. "Now, when the party wished to go underground and still retain its organization, all it had to do was simply to act in reverse order; that is, to transfer—or, more accurately perhaps, retransfer—the apparatus of the state into the party apparatus—a not-too-difficult enterprise, since both apparatuses were still organized along parallel lines," he explained.

According to Reiss, some misgivings about the fate of Germany arose even before the defeat of the Sixth Army at Stalingrad. He reported on a private meeting on November 7, 1942, in Munich, between SS chief Heinrich Himmler and Hitler's top lieutenant Martin Bormann. This meeting occurred only two days after Allied armies had landed in North Africa. Himmler later confided the topic of discussion, telling his most trusted associates, "It is possible that Germany will be defeated on the military front. It is even possible that she may have to capitulate. But never must the National Socialist German Workers' Party capitulate. That is what we have to work for from now on."

In May 1943, in the wake of the defeat at Stalingrad, Reiss said German industrialists met in Chateau Huegel near Essen, home of the Krupps, and reviewed the situation of their nation. The decision was to distance German commerce from the Nazi regime, Reiss wrote, adding: "All future changes discussed at the meeting centered around the idea of divorcing German industry as far as possible from Nazism as such. Krupp [von Bohlen und Halbach] and [I. G. Farben Director Georg von] Schnitzler declared that it would be much easier for them to work after the war if the world were certain that German industry was not owned and run by the Nazis. He said that Goering as well as other influential party men saw eye to eye with him on this, and would consent to any arrangement that did not involve the prestige of the party."

Reiss explained why these captains of industry faked a divorce from Nazism rather than mounting genuine opposition—because they had prospered under Hitler. He had "liberated" them from the threat of worker unions and strikes, kept taxes much lower than other industrialized nations, and brought them unprecedented profits through his rearmament program. "But all these are only symptoms," wrote Reiss. "More important than these symptoms is the fact that the Nazis as a dynamic movement had assured German big businessmen of basic conditions far more favorable than those they enjoyed under the republic or even under the Kaiser. Could they wish for anything better than a world constantly on the brink of new wars?"

As noted previously, it was not only German businessmen who profited from the war. Their counterparts in England and America were all capitalizing on the worldwide conflict. Reiss pointed out that only days after the meeting of industrialists, Farben's von Schnitzler flew to Madrid and declared he had escaped Germany just ahead of the Gestapo.

"Spain scarcely seemed a logical asylum. Switzerland or Sweden would have been much healthier places to repair to," noted Reiss. "And anyway, why should Herr von Schnitzler have had to fear the Gestapo, since his son-in-law, Herbert Scholz, was one of its leading officials? No, there is no reason to believe a word of what Baron Schnitzler said in those first interviews." Reiss said Schnitzler's "flight" was nothing but an elaborate ruse, similar to that of Germany's steel magnate Fritz Thyssen, who moved to

France in 1940, reportedly to escape the Nazis, but ended the war in Germany's prestigious hotel Adion, where he remained in contact with his old friend, banker Kurt Freiherr von Schroeder.

By the end of 1943, another ranking Nazi had left the Fatherland. Reichsbank president Hjalmar Horace Greeley Schacht had left the Fatherland for Switzerland, ostensibly for health reasons. During his stay, accounts of his "Schlacht Plan" began to circulate. It was similar to that of Schnitzler—a collaboration between German and Allied corporate business with the major German banks acting as clearinghouses for such transactions. Naturally, Schacht was to direct this effort. Despite his activities as one of the Reich's principal money men, Schacht suffered no real penalties after the war. He was acquitted by the Nuremberg war crimes court, which stated that rearmament was not itself a criminal act. He was convicted in a German court and sentenced to eight years in prison, but this was overturned on appeal. Four more efforts to convict Schacht in court came to no avail.

By late August 1944, following the D-Day invasion of Europe and despite the advent of the V-1 wonder weapon, many in the Nazi leadership were beginning to see the writing on the wall. When the French town of Saint-Lô, center of the German defense line facing the Allied beachhead in Normandy, fell on July 18, opening all of southern France to Allied armor and infantry, they knew the end of the war was only a matter of time.

According to captured medical records, Hitler was on a roller-coaster ride of euphoria and depression due to large daily doses of amphetamines, and had increasingly lost contact with reality. However, the second most powerful man in the Reich, Hitler's deputy Martin Bormann, was not so incapacitated.

Bormann, a stocky, nondescript man with thinning brown hair, was born in 1900 in Halberstadt in central Germany. He was the son of a cavalry sergeant who later became a civil servant. Young Bormann dropped out of high school after one year and was later drafted into the army during World War I, where he served with the field artillery. Returning from the war, Bormann joined the right-wing Freikorps and served a year in prison in 1924 for his part in the murder of his former elementary school teacher, who had been accused of betraying a Nazi leader when the Ruhr was under French occupation. Following his release from a Leipzig prison, Bormann joined the Nazi Party and rose steadily through the ranks.

Shortly after Hitler became German chancellor in 1933, Bormann was appointed chief of staff to Deputy Fuehrer Rudolf Hess. After Hess's ill-fated flight to Scotland in 1941, Bormann assumed his duties as well as becoming secretary to Hitler. Nazi leaders dubbed Bormann the "brown eminence" and "the Machiavelli behind the office desk," as he soon became the most powerful man in Nazi Germany. No one got to Hitler but through Bormann.

In 1943, Bormann gained total control over both the Nazi Party and the German economy, including all top-secret technology. Already named to replace Hess as head of the Nazi Party, Bormann wrested economic and political control from Himmler by having Hitler prohibit the SS chief from issuing orders to the Gauleiters, or district leaders, through his SS commanders. According to Heinrich Hoffman, Hitler's personal photographer and the man who introduced him to his mistress Eva Braun, Hitler once said of Bormann, "I know he is brutal, but what he undertakes he finishes. I can rely absolutely on that. With his ruthlessness and brutality he always sees that my orders are carried out." Bormann reigned supreme.

On August 10, 1944, Bormann called top German business leaders and Nazi Party officials to the Hotel Maison Rouge in Strasbourg. According to captured transcripts of the meeting, its purpose was to see that "the economy of the Third Reich was projected onto a postwar profit-seeking track." This "track" came to be known as Aktion Adlerflug, or Operation Eagle Flight. It was nothing less than the perpetuation of National Socialism through the massive flight of money, gold, stocks, bonds, patents, copyrights, and even technical specialists from Germany.

An emissary for Bormann, SS Obergruppenfuehrer Dr. Scheid, a director of the industrial firm of Hermadorff & Schenburg Company, explained the purpose of the meeting to one attendee: "German industry must realize that the war cannot now be won, and must take steps to prepare for a postwar commercial campaign which will in time ensure the economic resurgence of Germany."

Scheid told attendees, "[A]fter the defeat of Germany, the Nazi Party recognizes that certain of its best-known leaders will be condemned as war criminals. However, in cooperation with the industrialists, it is arranging to place its less conspicuous but most important members with various German factories as technical experts or members of its research and designing offices." As part of this plan, Bormann, aided by the black-clad SS, the central Deutsche Bank, the steel empire of Fritz Thyssen, and the powerful I. G. Farben combine, created 750 foreign front corporations—58 in Portugal, 112 in Spain, 233 in Sweden, 214 in Switzerland, 35 in Turkey, and 98 in Argentina.

According to Paul Manning, a CBS Radio journalist during World War II and the author of *Martin Bormann: Nazi in Exile,* Bormann "dwelled" on control of the 750 corporations. He wrote: "[Bormann] utilized every known device to disguise their ownership and their patterns of operations: use of nominees, option agreements, pool agreements, endorsements in blank, escrow deposits, pledges, collateral loans, rights of first refusal, management contracts, service contracts, patent agreements, cartels, and withholding procedures." Copies of all transactions and even field reports were maintained and later shipped to Bormann's archives in South America.

Bormann followed strategies perfected by I. G. Farben chairman Hermann Schmitz. The names of various companies and corporations would be changed and interchanged to create confusion as to ownership. For example, I. G. Chemie became Societe Internationale pour Participations Industrielles et Commerciales SA, while in Switzerland, the same organization was known as International Industrie und Handelsbeteiligungen AG, or Interhandel.

Another tactic was to name a compliant citizen from each country as the nominal head of a given corporation. Meanwhile, the directors would be a blend of German administrators and bank officials. Officers at senior and management levels would be German scientists and technicians. The real ownership of the corporation would be Nazis holding bearer bonds as proof of stock ownership. These individuals, all part of the Bormann operation, would remain in the shadows. The targeted nations generally were appreciative of Bormann's scheme, as it meant increased employment and a more favorable balance of trade.

In 1941, 171 American corporations had more than $420 million invested in German companies. After war was declared, Bormann merely had operatives in neutral countries such as Switzerland and Argentina buy American stocks using foreign exchange funds in the Buenos Aires branch of Deutsche Bank and Swiss banks. Large demand deposits were also placed with major banks in New York City to include National City Bank (now Citibank), Chase (now JP Morgan Chase), Manufacturers and Hanover (now part of JP Morgan Chase), Morgan Guaranty, and Irving Trust (now part of the Bank of New York).

At the Strasbourg meeting, Scheid cited several prominent American companies that had been useful to Germany in the past. Due to patent obligations, United States Steel, American Steel and Wire, and National Tube had to work in conjunction with the Krupp empire. He also mentioned Zeiss Company, the Leica Company, and the Hamburg-Amerika line as firms that were especially effective in protecting Nazi interests.

Bormann's complex, yet well organized, flight capital operation confounded Orvis A. Schmidt, the U.S. Treasury Department's director of foreign funds control. In 1945, Schmidt stated, "The network of trade, industrial, and cartel organizations has been streamlined and intermeshed, not only organizationally but also by what has officially been described as 'personnel union.' Legal authority to operate this organizational machinery has been vested in the concerns that have majority capacity in the key industries, such as those producing iron and steel, coal and basic chemicals. These concerns have been deliberately welded together by exchanges of stock to the point where a handful of men can make policy and other decisions that affect us all."

AT THE HEART of this flight capital program lay the huge I. G. Farben conglomerate. The Farben complex already had produced many scientific breakthroughs for the Third Reich. "Its experts developed the noted Buna Process for the manufacture of synthetic rubber, freeing Germany from dependence on natural rubber," explained Paul Manning. "It developed the hydrogenation process for making motor fuels and lubricating oils from coal. Germany's shortage of bauxite, the raw material essential to manufacture aluminum, was surmounted by its developments in utilizing the element magnesium."

Schmidt said Treasury investigations discovered Farben documents that showed the firm maintained an interest in more than 700 companies around the world. This number did not include Farben's normal corporate structure, which covered ninety-three countries, nor the 750 corporations created under Bormann's flight capital program.

I. G. Farben also was at the hub of money transfers out of Nazi Germany. Even before the end of the war, for example, "I. G. Latin American firms all maintained, unrecorded, in their books, secret cash accounts in banks in the names of their top officials," wrote Manning. "These were used to receive and to disburse confidential payments; firms dealing with Farben wanted this business but certainly did not wish it known to British and United States economic authorities."

"The great German combines were the spearheads of economic penetration in the other American republics [South and Central American nations]," stated U.S. Treasury official Schmidt. "In the field of drugs and pharmaceuticals the Bayer, Merck, and Schering companies enjoyed a virtual monopoly. I. G. Farben subsidiaries had a firm hold on the dye and chemical market. German enterprises such as Tubos Mannesmann, Ferrostaal, AEG, and Siemens-Schuckert played a dominant role in the construction, electrical, and engineering fields. Shipping companies and, in some areas, German airlines, were well entrenched." The foundation for a multinational German business empire was in place.

AS IN THE 1930s, the largest banking enterprises provided the underlying financial foundation for the resurgence of National Socialism.

The chairman of Deutsche Bank, Dr. Hermann Josef Abs, was particularly important to the Nazi flight capital program. Abs was also a director of I. G. Farben, Daimler-Benz, and Siemens. Martin Bormann maintained a cordial relationship with the Berlin banker. Manning noted: "[Bormann] knew in 1943 . . . he had the means to ultimately take the reins of finance unto himself. . . . He could set a new Nazi state policy, when the time was ripe for the general transfer of capital, gold, stocks, and bearer bonds to safety in neutral countries." Deutsche Bank, Dresdner Bank, and Commerzbank constituted the three major German banks, but it was Abs's Deutsche Bank that took the lead in establishing economic authority over the banks and corporations of the occupied countries.

During the war, Deutsche Bank coordinated Nazi gold transactions, purchasing 4,446 kilograms of gold from the Reichsbank and selling it in Turkey. Much of this gold came from victims of Nazi persecution. It arrived at the Reichsbank in crates and suitcases, sometimes marked with their place of origin, such as Auschwitz or Lublin. This wealth was greatly expanded by the loot of occupied Europe. According to author Ladislas Farago, this included "millions in gold marks, pound sterling, dollars, and Swiss francs, 3,500 ounces of platinum, over 550,000 ounces of gold, and 4,639 carats in diamonds and other precious stones, as well as hundreds of pieces of works of art."

According to *The Guinness Book of World Records,* the "greatest unsolved bank robbery" in world history was the disappearance of the entire German treasury at the end of the war. But was it truly unsolved or merely covered up at the highest levels?

Both Abs and Schmitz taught Bormann how to protect his wealth by depositing it in Swiss banks. Bormann saw to it that while the Reich allowed occupied countries to continue printing their own currency, the major commercial banks of Germany dealt in gold. Manning wrote, "The gold, whatever its origin, would be stamped with Third Reich seals and periodically sold to leading Swiss banks, as well as to the Swiss National Bank. . . . The money [from the gold sales] was then left on deposit in various numbered accounts to be invested in Switzerland and in other neutral countries, and ultimately to maintain the Bormann party apparatus abroad."

Ironically, it was a 1934 law passed in Switzerland that preserved Nazi loot. The law that prohibited the disclosure of the owners of private bank accounts was initially meant to keep the Gestapo from locating the savings of German Jews. To this day, it has been used to hide Nazi wealth. Today many American corporations have followed Bormann's lead by depositing their money in Swiss banks.

Swiss officials claim that their policies toward the Allied and Axis powers were those of balanced neutrality, but the scales were heavily tipped in favor of the Nazis, at least on economic matters. "Declassified intelligence reports reveal that Swiss banks, particularly the Swiss National Bank, accepted gold looted from the national treasuries of Nazi-occupied countries and from dead Jews alike, gold they either bought outright or laundered for the Nazis before sending it on to other neutral countries," wrote Adam LeBor, author of *Hitler's Secret Bankers: The Myth of Swiss Neutrality During the Holocaust.* According to LeBor, "Swiss banks supplied the foreign currency that the Third Reich needed to buy vital war material. Swiss banks were the vital financial conduit that allowed Nazi economic officials to channel their loot to a safe haven in Switzerland. Swiss banks financed Nazi foreign intelligence operations by providing funds for German front companies in Spain and Portugal."

Bormann had a personal account at the Reichsbank under the fictitious name "Max Heiliger," into which he siphoned a substantial portion of the Reich's wealth. Utilizing both gold and "treasure," Bormann, through his chief of economics, Dr. Helmut von Hummel, sent these riches out of the country for later use.

Abs presents a classic example of the survivability of high-level bankers. He not only survived the war but was instrumental in Germany's postwar revival, becoming a financial adviser to West Germany's first chancellor, Konrad Adenauer. He maintained his positions on the boards of Deutsche Bank, Daimler-Benz, and Siemens. In 1978, Abs headed a West German consortium that managed to buy and return nearly $20 million worth of artwork taken from Germany in the 1930s by the Jewish Baron Robert von Hirsch. Later that same year, Abs addressed American business leaders at a meeting chaired by fellow banker John J. McCloy, one-time chairman of Chase Manhattan Bank, the Ford Foundation, and the Council on Foreign Relations. McCloy served as a member of President Lyndon B. Johnson's Warren Commission and was a legal adviser to the Rockefeller family.

It was Abs who had prevented two American banks in France—Morgan et Cie and Chase of New York—from being closed or controlled by the German occupation authorities. According to U.S. Treasury reports cited by Manning, this exemption came through an "unspoken understanding among international bankers that wars may come and go but the flux of wealth goes on forever."

This "understanding" was principally between Abs and Lord Hartley Shawcross, a leader in the City of London financial center and a board member of many international companies. Shawcross, a lawyer who was a special adviser to Morgan Guaranty Trust of New York as well as the two American banks in France, was later named chief prosecutor for Britain in the Nuremberg war crimes trials. In the 1950s, Shawcross, along with his friend Dr. Abs, formed the Society for the Protection of Foreign Investments of World War II, headquartered in Cologne, West Germany.

Bormann's Operation Eagle Flight was substantially helped by the close connections with foreign banks and businesses begun long before the war. According to former U.S. Department of Justice Nazi War Crimes prosecutor John Loftus, much of the wealth was passed out of Germany by German banker Fritz Thyssen through his bank in Holland, which, in turn, owned the Union Banking Corporation (UBC) in New York City. Loftus is president of the Florida Holocaust Museum and the author of several books on CIA-Nazi connections, including *The Belarus Secret* and *The Secret War Against the Jews*.

Two prominent U.S. business leaders who supported Hitler and served on the board of directors of the Union Banking Corporation were George Herbert Walker and his son-in-law Prescott Bush, father of George H. W. Bush and grandfather of President George W. Bush. The attorneys for these dealings were John Foster Dulles and his brother Allen. John later became secretary of state under President Dwight D. Eisenhower while Allen became one of the longest-serving CIA directors before being fired by President John F. Kennedy in 1961. Both were original members of the Council on Foreign Relations.

On October 20, 1942, the office of U.S. Alien Property Custodian, operating under the "Trading With the Enemy Act" (U.S. Government Vesting Order No. 248), seized the shares of UBC on the grounds that the bank was financing Hitler. Also seized were Bush's holdings in the Hamburg-America ship line that had been used to ferry Nazi propagandists and arms. Another company essential to the passing of Nazi money was the Holland American Trading Company, a subsidiary of UBC. It was through Fritz Thyssen's Dutch Bank, originally founded by Thyssen's father in 1916, that Nazi money was passed. This Dutch connection tied the Bush and Nazi money directly to former SS officer and founder of the Bilderberg Group, Prince Bernhard of the Netherlands, who was once secretary to the board of directors of I. G. Farben, with close connections to the Dutch Bank. Loftus noted, "Thyssen did not need any foreign bank accounts because his family secretly owned an entire chain of banks. He did not have to transfer his Nazi assets at the end of World War II, all he had to do was transfer the ownership documents—stocks, bonds, deeds, and trusts—from his bank in Berlin through his bank in Holland to his American friends in New York City: Prescott Bush and Herbert Walker. Thyssen's partners in crime were the father and father-in-law of a future president of the United States."

The leading shareholder in UBC was E. Roland Harriman, son of Edward H. Harriman, who had been an early and important mentor to Prescott Bush. Another son, Averell Harriman, also held ownership in UBC. He was named ambassador to the Soviet Union by President Roosevelt in 1943 and participated in all major wartime conferences. Averell later became ambassador to Great Britain, U.S. secretary of commerce, and governor of New York State. Both Harrimans had been members of the Yale secret society Skull and Bones and were closely connected to the globalists at the Council on Foreign Relations. Averell also was a close advisor to President Lyndon Johnson.

On November 17, 1942, U.S. authorities also seized the Silesian-American Corporation, managed by Prescott Bush and his father-in-law, George Herbert Walker, and charged the firm with being a Nazi front company that was supplying vital coal to Germany. But according to government documents that have recently come to light and were published by the *New Hampshire Gazette* in 2003, "the grandfather of President George W. Bush failed to divest himself of more than a dozen 'enemy national' relationships that continued as late as 1951." The newly released documents also showed that Bush and his associates routinely tried to conceal their business activities from government investigators and that such dealings were conducted through the New York private banking firm of Brown Brothers Harriman. Brown Brothers Harriman, the oldest privately owned bank in America, was formed in 1931 when the Brown brothers, originally importers of Irish linen, merged with railroad tycoon Edward H. Harriman.

"After the war," according to the *Gazette* report, "a total of 18 additional Brown Brothers Harriman and UBC-related client assets were seized under the Trading with the Enemy Act, including several that showed the continuation of a relationship with the Thyssen family after the initial 1942 seizures. The records also show that Bush and the Harrimans conducted business after the war with related concerns doing business in or moving assets into Switzerland, Panama, Argentina and Brazil—all critical outposts for the flight of Nazi capital after Germany's surrender in 1945."

Why was Prescott Bush not more openly and aggressively prosecuted for his Nazi dealings? This may be due to the fact that the patriarch Bush was "instrumental in the creation of the USO in late 1941," according to a news release from the United Service Organization in 2002. After all, it would have looked very bad during wartime to publicly prosecute as a Nazi asset the man who helped create the USO, so beloved by U.S. servicemen in all subsequent wars.

"The story of Prescott Bush and Brown Brothers Harriman is an introduction to the real history of our country," said publisher and historian Edward Boswell. "It exposes the money-making motives behind our foreign policies, dating back a full century. The ability of Prescott Bush and the Harrimans to bury their checkered pasts also reveals a collusion between Wall Street and the media that exists to this day."

It was rumored that the trial transcripts of the 1942 prosecution of Prescott Bush were destroyed in the September 11, 2001, collapse of World Trade Center 7, which housed offices of the Securities and Exchange Commission. The SEC admitted that more than seven thousand prosecution files were lost with the building, including files on Enron and World.com.

Prescott Bush's banking connection to Nazis was not the only object of U.S. investigations during the war. Rockefeller-owned Standard Oil also came under scrutiny for a series of complex business deals that resulted in desperately needed gasoline reaching Nazi Germany. "None of these transactions was ever made public," reported journalist Charles Higham. "The details of them remained buried in classified files for over forty years." However, it was established that Standard Oil shipped oil to fascist Spain throughout World War II, paid for by Spanish dictator Francisco Franco from funds that had been unblocked by the Federal Reserve Bank and passed to Nazi Germany from the vaults of the Bank of England, the Bank of France, and the Bank for International Settlements. Such shipments through Spain to Hamburg indirectly but materially assisted the Axis. "While American civilians and the armed services suffered alike from restrictions, more gasoline went to Spain than it did to domestic customers," noted Higham.

Questioned about this by the *New York Times,* a spokesman for U.S. Secretary of State Cordell Hull explained that the oil was coming from the Caribbean, not the United States. What was not explained was that Standard Oil, under the leadership of William Stamps Farish, had early on changed the country of registration for Standard's tanker fleet to Panama. Higham claimed that both Standard chiefs, Farish and Teagle, were "mesmerized by Germany" and were close associates of I. G. Farben's president Hermann Schmitz. The person who authorized the masking of Standard's shipping through Panamanian registry was then-undersecretary of the navy James V. Forrestal, also a vice president of General Aniline and Film (GAF).

Another aspect of Bormann's flight capital program concerned Hermann Schmitz, who, as a director of Thyssen's steel empire, owned companies in neutral Sweden, along with other German firms. Schmitz's Swedish firms built ships and transported coal and coke. "A further example of masked investment was the money paid into the Swedish shipping firm of Rederi A/B Skeppsbron, which received a German-guaranteed loan of $3 million . . . in which the vessels were mortgaged to the lender," explained Paul Manning. "Although the Swedish company remained officially the owner of the vessels, the Hamburg-Amerika line [part of Prescott Bush's holdings] was the real owner."

"It is bad enough that the Bush family helped raise the money for Thyssen to give Hitler his start in the 1920s, but giving aid and comfort to the enemy in time of war is treason," declared Nazi prosecutor Loftus. "The Bush's bank helped the Thyssens make the Nazi steel that killed Allied soldiers. As bad as financing the Nazi war machine may seem, aiding and abetting the Holocaust was worse. Thyssen's coal mines used Jewish slaves as if they were disposable chemicals. There are six million skeletons in the Thyssen family closet, and a myriad of criminal and historical questions to be answered about the Bush family complicity."

ALONG WITH THE desire to create a Nazi-directed European economy, Martin Bormann and his henchmen also drew up plans to create a new generation of National Socialists, beginning in Germany but with an eye toward other nations.

Heimschulen, or home schools, were created within Germany to train youngsters in the techniques of explosives and sabotage as well as how to live and act in foreign countries. "In the spring of 1943, the curriculum of these schools was changed slightly," stated Curt Reiss. "This was logical, for since the leaders of the Third Reich no longer expected to win this war, they now began to put the accent on the work that would have to be done after the war. Instead of developing spies and saboteurs, these schools were put to the task of developing workers for the coming underground."

In a move that has been duplicated within the modern U.S. intelligence community, many SS members seemingly resigned from the Black Shirts but secretly retained their loyalty and affiliation. In today's intelligence parlance, this is called "sheep dipping." "These men will leave the SS for good. Some of them will even leave the party, so as to be completely neutralized. These latter may officially disavow the party before public witnesses, who can be used later to testify how anti-Nazi they have been for a long time," wrote Reiss.

"Several intelligence services have commented on the sudden disappearance of important personalities from [Germany's] political and party life," wrote Reiss in 1944. "And it has become quite the accepted thing to everybody in Germany. But what has not yet become known is that all this also applies to a much greater number of anonymous persons all over Germany, those on the second and third levels of the Nazi strata.

"These unknown personalities may be used later by the underground. Party functionaries who may be known locally, but certainly not nationally, can easily be transferred to another city or town, where they will suddenly appear as anti-Nazis. The party helps in their masquerades. These men get new documents which 'prove' that they have always been anti-Nazi. Notes are inserted in their personal files saying they must be watched on account of their anti-Hitler attitudes and 'unworthy' behavior. Some of them will undoubtedly be sent to concentration camps for crimes which they have never committed, but which will make them look dependable in the eyes of the Allies; some have perhaps already succeeded in joining anti-Nazi circles and are pretending to conspire against Hitler. Later on they will be able to use such activities as alibis."

Nazi sympathizers across the world were brought into the plan for resurrecting National Socialism through the Auslandsorganisation (AO), or the League of Germans Abroad. Various forms of this organization had been in existence since the 1800s and have manipulated thousands of persons in many different countries. In Czechoslovakia, Holland, Belgium, and Norway, members aided the Nazi invasions, becoming known as "fifth columnists."

"[T]he Nazis, long before they came to power, put their men or men they trusted into these leagues. At a party meeting in Hamburg it was decided to set up Nazi cells within all these organizations. That was in 1930. One year later Rudolf Hess formed a special Foreign Department of the [Reich Leadership] of the National Socialist Party, which established card files on every member who lived abroad or traveled abroad. This was the basis of the gigantic files which the AO was to organize later," stated Reiss.

In 1944, Reiss pondered when the flight capital program might bear fruit. "How long it will take for the Nazis to come back, to emerge on the surface—if they succeed in their aims—is a question which cannot be answered at all. Even under the most favorable circumstances—that is, most favorable for them—it will take ten or fifteen years. Even then it would be a blitzkrieg, an underground blitzkrieg with somewhat different conceptions of time," he wrote. "The Italian underground needed a half century to achieve its goal, the Irish a whole century, the Bonapartists thirty-five years, and the Russian Socialists twenty-five. The Russians needed two lost wars to bring about their revolution. The Nazis cannot wait for another lost war. They want to come to power so that they can start World War III."

Reiss, with his accumulated knowledge of the Nazis and their methods, issued this warning in 1944:

> It is not the relative strengths of the different powers that must change, but the relations of the human beings within all the countries of this world. Some call it revolution. Some call it a new order. Whatever we call it, it must come about. If it does not, the Nazi underground will live and flourish. In due time, it will make itself felt far beyond the borders of Germany. It will certainly make itself felt in this country—and no ocean will be broad enough to stop it.
>
> For Nazism or Fascism is by no means an Italian or German specialty. It is as international as murder, as greed for power, as injustice, as madness. In our time these horrors were translated into political and cultural actuality in Italy and in Germany first.
>
> ... If we don't stamp out the Nazi underground, it will make itself felt all over the world; in this country too. We may not have to wait ten years, perhaps not even five.
>
> For many years in the past we closed our eyes to the Nazi threat. We must never allow ourselves to close them again. The danger to the world, to this country will not diminish. But it is possible to fight this danger if we know it, if we remain aware of its existence.

Armed with super-science and technology, plus the loot of Europe—to include perhaps Solomon's treasure—the Nazis and their ideology were well placed to begin their Fourth Reich.

PART TWO

THE
REICH
CONSOLIDATES

CHAPTER 6

THE RATLINES

CONSIDERING THE VAST ECONOMIC RESOURCES AT THEIR DIS-
posal—especially if they held dark secrets concerning advanced technology such as tactical nukes, flying saucers, or a device for manipulating
energy—it is now certain that the surviving Nazi leadership wielded
enough power to misdirect investigations and silence foreign governments and news organizations.

There remain long-standing controversies over the proclaimed deaths of
prominent Nazis, including Martin Bormann; the notorious SS Dr. Josef
Mengele, the "Angel of Extermination" at Auschwitz; and Gestapo chief,
SS Gruppenfuehrer Heinrich "Gestapo" Mueller. According to various
unsubstantiated reports, Mengele suffered a stroke while swimming, and
drowned in Brazil in 1979 after hiding there, as well as in Argentina and
Paraguay, for decades. Another ranking SS official, Obersturmbannfuehrer Adolf Eichmann escaped to Argentina but was abducted to Israel
in 1960, where he was convicted of war crimes and executed. Toward the
end of the war, Mueller distanced himself from his boss Himmler and
moved closer to Bormann. After he slipped away from Hitler's bunker in
the last days of the Reich, Mueller's family declared him dead and erected a
tombstone in a Berlin cemetery, with the inscription OUR DEAR DADDY.
However, in 1963, a court-ordered exhumation revealed that the grave

contained three skeletons, none of which matched Mueller's height or bone structure.

But the biggest fish to get away was Reichsleiter Bormann, the ultimate power behind the Nazi super-science projects and the architect of Operation Eagle Flight. In 1972, Munich bishop Johannes Neuhausler made public a postwar church document stating that Bormann had escaped Berlin during the final days and gone to Spain by airplane. The next year, after journalist Paul Manning published an article in the *New York Times* detailing Bormann's escape from justice, West German officials held a news conference proclaiming that Berlin workmen had unearthed two skeletons near the ruins of the Lehrter railroad station and that one of the skeletons had been identified as Bormann. He died in 1945 trying to escape Berlin, they stated.

However, the entire case for the Berlin death of Bormann rested on dental records prepared from memory by a dentist who had been a loyal Nazi for many years, and the sole statement of a dental technician who had been imprisoned in Russia due to his proclaimed knowledge of Bormann's dental work. Adding to suspicions that Bormann's death announcement was most convenient for anyone wishing to cover Bormann's tracks was the fact that Willy Brandt's government canceled all rewards and warrants for Bormann and instructed West German embassies and consulates to ignore any future sightings of the Reichsleiter.

These suspicions were compounded by statements from several persons who told Paul Manning that the body found near the railroad station was placed there in 1945 by SS troops commanded by "Gestapo" Mueller, who was known to have used decoy bodies on other occasions.

Bormann's death notice did not convince the late Simon Wiesenthal of the Documentation Center in Vienna, who said, "Some doubts must remain whether the bones found in Berlin are really those of Bormann." One of Bormann's relatives had no doubts. In 1947, Walter Buch, the father of Bormann's wife, Gerda, declared on his deathbed, "That damn Martin made it safely out of Germany."

According to Manning, Bormann was escorted from dying Berlin by selected SS men who passed him along a series of "safe houses" to Munich, where he hid out with his brother, Albert. In early 1946, Bormann was escorted on foot over the Alps to the northern Italian seaport of Genoa. There Bormann was housed in a Franciscan monastery. All this was arranged by "Gestapo" Mueller.

In mid-1946, a steamer ship carried Bormann, provided with false identification papers, to Spain, where he entered the Dominican monastery of San Domingos in the province of Galicia, once the home of Spanish dictator Francisco Franco, a supposed neutral who covertly supported Hitler. Manning noted that in 1969, when Bormann became aware that Israeli agents were sniffing along his escape route, there was a fire in San Domingos. Curiously, the fire started on the very shelves where the monastery kept its book of visitors, which contained Bormann's name. This incriminating record suspiciously was destroyed.

In the winter of 1947, a large freighter carrying Bormann and several SS officers anchored in the harbor of Buenos Aires, where an organized network of supporters awaited them.

Even before the shooting war ended, lesser-known SS members and hardcore Nazis were fanning out across the world through covert distribution systems. The means was a loosely knit collection of escape routes from Europe, called "ratlines." Chief among these ratlines were the *Kameradenwerk* and the ODESSA, the *Organization der ehemaligen SS-Angehorigen,* or the organization of former SS members. ODESSA was created by Bormann and Mueller, but later administered by Otto Skorzeny, who had escaped war crimes convictions.

Documentation of these ratlines is so incomplete and fragmentary that some historians, taking their cue from the corporate world, have denied that ODESSA existed outside the fevered dreams of fanatical SS men. Ladislas Farago, author of popular histories as well as an acclaimed biography of General George Patton, also wrote that he had proof of Bormann's postwar survival. He acknowledged ODESSA's existence but wrote it was "actually little more than a shadowy consortium of a handful of freelancers and never amounted to much in the Nazi underground."

But then, in 1976, Louis L. Snyder, professor of history at the City College and the City University of New York, produced the mammoth

Encyclopedia of the Third Reich. Snyder described ODESSA as a "vast clandestine Nazi travel organization" to aid the escape of SS members and top Nazis. He noted that the main terminal point for ODESSA was Buenos Aires.

According to Farago, the *Kameradenwerk* was the real focal point for escaping Nazis. It was founded by Luftwaffe colonel Hans Ulrich Rudel, an air ace who lost a leg flying 2,530 combat missions for Germany. After the war, Rudel almost alone put together one of the most far-reaching and best financed of the rescue groups—the *Kameradenwerk*. Rudel's group had help but, according to Farago, it did not come from the Bormann underground but from "the vast organization and the enormous resources of the one agency that, in the end, took care of more Nazis than all the others combined—the refugee bureau of the Vatican."

To understand the seemingly puzzling relationship between Hitler's Nazis and the Holy Roman Church, one must look back to a 1929 agreement signed between the Vatican and the government of fascist Italy. Under this concordat, known as the Lateran Treaty, the Italian government bought favor from the Church by paying almost a billion lire in gold as compensation for church property taken during the nineteenth-century *Risorgimento*, or reorganization, that helped create modern Italy. The Lateran Treaty also established Vatican City in Rome as a sovereign state, as well as making Roman Catholicism the only state religion in Italy.

On July 20, 1933, a similar concordat was reached between Pope Pius XI and Nazi Germany. Treaty negotiations were handled by Cardinal Eugenio Pacelli, who signed on behalf of the pope and later became Pope Pius XII. This concordat, still in effect today, was signed by Franz von Papen on behalf of German president Paul von Hinderburg. Von Papen was tried at Nuremberg but released despite being denounced as a primary mover in Hitler's aggression in Europe.

According to the 1933 concordat, there was to be no interference by the Church in political affairs. It also required all bishops to take a loyalty oath to the state and required all priests to be German citizens and subordinate to government officials. Prior to the concordat's ratification, the Nazi government also reached similar agreements with the major Protestant churches. Hitler, who at a young age trained at a Catholic monastery school and strived to reach accommodations with the German churches, once proclaimed, "I believe today that my conduct is in accordance with the will of the Almighty Creator."

Rumors have circulated for years that a secret codicil of the concordat involved papal leniency toward National Socialism in exchange for Catholicism being proclaimed the state religion of Europe after an appropriate period of time of total Nazi control. Regardless, it mattered little, as Hitler quickly took steps against all churches, including the Catholics. His sterilization laws, attempts to dissolve the Catholic Youth League, and arrests of priest, nuns, and lay leaders all angered the Catholic community. In March 1937, Pope Pius XI issued an encyclical letter titled *"Mit Brennender Sorge,"* or "With Burning Sorrow." In the letter, the pope accused the Nazis of both violating and evading the concordat and even foresaw "threatening storm clouds" of war and extermination. A year later, Pius XI addressed the Nazi persecution of the Jews by proclaiming worldwide, "Mark well that in the Catholic Mass, Abraham is our patriarch and forefather. Anti-Semitism is incompatible with the lofty thought which that fact expresses. . . . I say to you it is impossible for a Christian to take part in anti-Semitism. It is inadmissible. Through Christ and in Christ we are the spiritual progeny of Abraham. Spiritually, we are all Semites."

But if Pius XI's turn against National Socialism was legitimate, it unfortunately was short-lived. On February 10, 1939, the day before Pius XI was scheduled to deliver yet another scathing public attack on fascism and anti-Semitism, he died, reportedly of a massive heart attack. Copies of his planned antifascist speech have never been found. Vatican officials have stated they may have been misfiled. Rumors implicated Dr. Francesco Saverio Petacci in the pope's sudden death. Petacci, one of the Vatican physicians at the time, was the father of Clara Petacci, the longtime mistress of Fascist leader Benito Mussolini. The whispers were that Petacci gave the pope an injection that caused his fatal attack. Strong support for this rumor came some years later, when the same allegation was found in the personal diary of French cardinal and former French Army intelligence agent Eugene Tisserant.

Pope Pius XII, born Eugenio Maria Giuseppe Giovanni Pacelli, was certainly less antagonistic toward fascism and, in fact, had been an honored guest at the society wedding of Clara Petacci and Italian Air Force Lieutenant Riccardo Federici in 1934. The marriage did not last long, and Clara was soon visiting Mussolini at night via a secret staircase in the Palazzo Venezia.

Catholic historian and journalist John Cornwell, in 1999, stunned the Catholic world with his book *Hitler's Pope*. A former seminary student, Cornwell explained that he originally intended to defend the actions of Pope Pius XII but as his research in Vatican archives progressed, his attitude changed. "By the middle of 1997, nearing the end of my research, I found myself in a state I can only describe as moral shock. The material I had gathered, taking the more extensive view of Pacelli's life, amounted not to an exoneration but to a wider indictment," Cornwell wrote. The author eventually saw that this pope's actions—or inaction—actually aided in Hitler's rise to power and the ensuing Holocaust.

Needless to say, Cornwell's perception was immediately and savagely attacked as inaccurate reporting and misinterpretation. In the December 9, 2004, edition of *The Economist,* Cornwell waffled, writing, "I would now argue, in light of the debates and evidence following *Hitler's Pope,* that Pius XII had so little scope of action that it is impossible to judge the motives for his silence during the war, while Rome was under the heel of Mussolini and later occupied by Germany."

Regardless of motives, it is historical fact that many top Nazis and SS men escaped Europe with passports issued by Catholic officials. Luftwaffe ace Rudel admitted as much in 1970, stating, "In Rome itself, the transit point of the escape routes, a vast amount was done. With its own immense resources, the Church helped many of us to go overseas."

One of those helpful clerics was Bishop Alois Hudal, who voiced opinions comparable to those of Hitler's Viennese friend Jorg Lanz von Liebenfels, publisher of *Ostara,* a magazine with occult and erotic themes. A Cistercian monk who founded the anti-Semitic secret Order of the New Templars, von Liebenfels and his mentor Guido von List sought to revive the medieval brotherhood of Teutonic Knights, those heroes of Hitler's youth, who had used the swastika as an emblem. While von Liebenfels headed his Order of the New Templars, Bishop Hudal was named procurator general of the Catholic Order of German Knights. On May 1, 1933, in a Nazi-sanctioned celebration of the pagan Walpurgis holiday, Hudal made a particularly impassioned speech in Rome before assembled Church and Nazi leaders as well as the expatriate German community. "German unity is my strength, my strength is German might," he told the crowd.

It was, in fact, a Franciscan friar serving under Bishop Hudal who helped arrange a Red Cross passport and visa to Argentina in 1950 for Obersturmbannfuehrer Adolf Eichmann, the exterminator of Jews who had managed to slip away from American captors at the end of the war. Bishop Hudal, in his later memoirs, thanked God he was able to help so many escape with false identity papers.

Many of these "false identity papers" were documents issued by the Commissione Pontificia d'Assistenza, or the Vatican Refugee Organization. While not full passports themselves, these Vatican identity papers were used to obtain a Displaced Person passport from the International Red Cross, which, in turn, was used to gain a visa. Supposedly, the Red Cross checked the backgrounds of applicants, but usually it was sufficient to have the word of a priest or a bishop. This method of aiding escaping Nazis—the one favored by Bishop Hudal—came to be known as the "Vatican ratlines."

For example, Ante Pavelic, the wartime pro-Nazi fascist dictator of Croatia, who was given a private audience with Pope Pius XII shortly after taking power in 1941, escaped to South America after the war with a Red Cross passport gained through a Vatican document.

ONE OF THE countries in which the Auslandsorganisation worked with particular success was Argentina. "There it has been able to operate without any disguise or front. All of the more than 200,000 Argentine Nazis are members, not of an Argentine suborganization of the Nazi Party, but of the German Party itself, and hold membership cards signed by Robert Ley, leader of the German Workers' Front—which means, quite obviously, that Berlin considered, and still considers, Argentina not so much an independent foreign country as a German *Gau* [district]," noted Curt Reiss.

Although many Nazis found safe havens in Brazil, Paraguay, Chile, and Uruguay, no South American nation was more accommodating than the Argentina of Dictator Juan Domingo Peron and his lovely second wife, Eva Maria Duarte de Peron, popularly known as Evita.

After participating in a successful military coup in 1943, Peron was voted in as president in 1946 by a majority of voters, who lauded his efforts to eliminate poverty and dignify workers. He was elected against the intense and overt opposition of the United States. Such opposition appeared justified, for soon after his election Peron began to nationalize and expropriate British and American businesses. As their influence in Argentina dwindled, that of the Germans grew.

Luftwaffe pilot Rudel, who created the *Kameradenwerk* ratline, became a trainer for Peron's air force and in the process brought with him about one hundred members of the wartime Luftwaffe staff. Likewise, many Nazi SS and Gestapo fugitives from justice served in the Argentine Army and police forces. Among them was Kurt Tank, who headed a large group of Nazi scientists. Tank, a fighter plane designer and former director of the Focke-Wulf aircraft factory, had slipped away from Germany in disguise and, armed with false identity papers provided by Peron himself, arrived in Buenos Aires with microfilm of aircraft designs hidden in his pants. Soon, about sixty of his old Nazi comrades had joined him, using the same system.

THE MAN MOST responsible for fostering pro-Nazi feelings in South America was General Wilhelm von Faupel. In 1900, Faupel went to China as a member of the German military legation. He later went to Moscow in the same capacity. In 1911, he joined the staff of the Argentine War College in Buenos Aires. Faupel returned to Argentina after serving Germany in World War I and obtained the job of military counselor to the inspector general of the Argentine Army. Von Faupel not only imparted military theories to the armies of Argentina, Brazil, and Peru, he also instilled in them the political theories of National Socialism. "Hating the [Weimar] republic passionately, he did not return to Germany until the Nazis were about to take power," wrote Curt Reiss. "But while he was away he had kept up excellent relations with industrialists such as Fritz Thyssen, Georg von Schnitzler, and Herr von Schroeder. After all, these gentlemen had elaborate interests in Latin America. And so had von Faupel. In fact, he boasted openly among German military and industrial men that he could conquer the whole of Latin America." Faupel most likely was speaking for many globalists who had significant holdings in South America and did not wish to relinquish them to leftists, communists, nationalists, or reformers.

Although he is widely seen as a dictator, many "Peronists" still view Juan Peron as a champion of the working man. Few realized at the time that he was stashing away an estimated $500 million in Swiss bank accounts. According to Manning, at least $100 million came from the Bormann organization. Peron reciprocated for this generosity by allowing many war criminals to immigrate, legally and illegally, to Argentina. He reportedly provided more than a thousand blank passports for escaping Nazis.

Peron was an admirer of Hitler. He had learned German at a young age so he could read *Mein Kampf*. His private secretary, Rudolfo Freude, also was chief of internal security. The Argentine dictator was greatly honored to shelter Deputy Fuehrer Bormann. After several low-key meetings with Bormann, Peron saw Bormann's flight capital program as a means of boosting the Argentine economy.

"Both realized that the capture of Bormann was a clear and ever-present danger," noted Paul Manning, "and so Peron instructed the chief of his secret police to give all possible cooperation to Heinrich Mueller in his task of protecting the party minister, a collaboration that continued for years."

Evita took on the role of liaison between her husband and Nazis seeking asylum. "Born in 1919 as an illegitimate child, she became a prostitute to survive and to get acting roles," wrote investigative reporter Georg Hodel. "As she climbed the social ladder lover by lover, she built up deep resentments toward the traditional elites. As a mistress to other army officers, she caught the eye of handsome military strongman Juan Peron. After a public love affair, they married in 1945."

In June 1947, Eva Peron embarked on a much-publicized "Rainbow

Tour" of Europe, greeted royally by Spain's Franco and a private audience with Pope Pius XII. While in Spain, she reportedly met with Otto Skorzeny, who headed a ratline known as *die Spinne* or the Spider, and arranged the transfer of millions in Nazi loot to Argentina. She also traveled to Genoa, where she met with Argentine shipping fleet owner Alberto Dodero, who within a month was ferrying Nazis to South America. But the primary purpose of the trip appeared to be the meetings Evita held with bankers in Switzerland.

"According to records now emerging from Swiss archives and the investigations of Nazi hunters, an unpublicized side of Evita's world tour was coordinating the network for helping Nazis relocate in Argentina," wrote Hodel. "Though Evita's precise role on organizing the Nazi 'ratlines' remains a bit fuzzy, her European tour connected the dots of the key figures in the escape network. She also helped clear the way for more formal arrangements in the Swiss-Argentine-Nazi collaboration."

In 1955, Peron was ousted in a military coup and forced to flee to neighboring Paraguay and later to exile in Madrid, Spain. He left without the body of Eva, who had died from cancer in 1952, at age thirty-three. Her popularity was such that eight persons were trampled to death in the tumultuous crowds who flocked to see her embalmed body lying in state.

According to Manning, the relationship between Bormann and Peron "became somewhat frayed around the edges after Peron left for Panama and then exile in Spain in 1955, but [Gestapo] Mueller today [1981] still wields power with the Argentinian secret police in all matters concerning Germans and the [Nazis] in South America."

THE IMPACT OF transplanted Nazis continues to be felt in South America. "Those aging fascists accomplished much of what the ODESSA strategists had hoped," noted Georg Hodel, adding, "The Nazis in Argentina kept Hitler's torch burning, won new converts in the region's militaries and passed on the advanced science of torture and 'death squad' operations. Hundreds of left-wing Peronist students and unionists were among the victims of the neo-fascist Argentine junta that launched the Dirty War in 1976."

SS Hauptsturmfuehrer Klaus Barbie, the "butcher of Lyon," after working for the U.S. Army's Counterintelligence Corps following Germany's defeat, ensconced himself in Bolivia under the name Klaus Altmann. Using his contacts in the *Kameradenwerk,* he began running guns between Bolivia, Peru, and Chile. "The gun trade eventually led them into the drug trade," wrote Levenda.

On July 17, 1980, Barbie abetted right-wing officers in the army during a coup over the left-of-center Bolivian government. "Barbie's team hunted down and slaughtered government officials and labor leaders, while Argentine specialists flew in to demonstrate the latest torture techniques," wrote Hodel. With Barbie's aid, Bolivia was soon a primary and secure source of cocaine for the emerging Medellín drug cartel. Two years later, Barbie was captured and extradited to France, where he died in 1991 from cancer while serving a life sentence for crimes against humanity.

ODESSA also turned to gun-running as a means of financing its operations. In fact, it was never intended only as an escape route for Nazis, but, at Bormann's instructions, it was set up as a profitable business enterprise as well. The plentiful supply of surplus arms in Europe turned out to be an immediately profitable commodity.

In late 1945, U.S. military authorities became aware of a huge black market enterprise being operated out of Passau, a picturesque city located ninety miles northeast of Munich at the confluence of the Danube, Inn, and Ilz Rivers. It was a connecting point between Germany, Austria, and Czechoslovakia, and a collection center for Axis arms. It was here that the weapons from three complete armies—the Hungarian Fifth Army and the German Fifth and Twelfth Armies—were stockpiled in an American-run depot. Rifles, machine guns, and ammunition were gathered at Passau while millions of dollars worth of vehicles were cached at Mattinghoffen, Austria.

About $10 million worth of these war materials went missing, sold by black marketers, mainly ODESSA agents, German officials, and criminals, aided by a few Allied soldiers. In one of the greatest ironies of history, the bulk of this material was being shipped to Palestine for use both by Jews trying to create the state of Israel and by Arabs who violently opposed such an effort.

On January 5, 1946, U.S. military intelligence officers under the command of Colonel William Weaver of General Patton's G-2 staff were sent to Passau to make arrests. Instead, the agents were murdered and the house in which they were staying was burned.

One agent, Lieutenant William H. Spector, survived. Hospitalized with a kidney stone, Spector narrowly missed the massacre at Passau. Vowing to avenge his slain fellow officers, Spector was nonplussed to find that the entire affair was hushed up on orders of superior officers who declared it a "security issue." Spector did learn that one of the men involved with the stolen arms as well as the agents' deaths was a Romanian national named Robert Abramovici. Abramovici later changed his name to Robert Adam and started an arms company called Intermecco Socomex, which became closely associated with the CIA's arms company, Interarmco.

The arms deals in Palestine were handled by Joseph Beidas and Eduardo Baroudi and his brothers, who sold weapons to both Arabs and Jews. Baroudi later became a vice president of Intrabank, based in Beirut, Lebanon, a major conduit for black market funds.

Millionaires were made immediately after the war by both war surplus and black market arms deals. "But none were to achieve the profitability of ODESSA, whose agents ranged throughout Europe and even behind the Iron Curtain," explained Manning, adding:

They bought and sold surplus American arms to Arab buyers seeking to strengthen the military capabilities of Egypt and other Middle Eastern Arab nations. Palestine was to be partitioned into a Jewish homeland, and they intended to destroy it at birth. But now, Jewish buyers, funded from America and elsewhere, entered the marketplace. They were barred from purchasing guns and American surplus P-51 Mustang fighter planes by President Truman, and their only recourse for survival was to trade on the European black market, which, unknown to them, was rapidly coming under the control of ODESSA agents. However, the Jewish agency's buyers might have purchased from the devil himself if it meant survival of the small, defenseless nation, just come into being on May 14, 1947.

Again, the behind-the-scenes maneuvering of wealthy globalists can be seen in the creation of modern Israel. This began in 1917, when 2nd Baron Rothschild, Lionel Walter Rothschild, received a letter from British foreign secretary Arthur Balfour replying to his query regarding Balfour's position on Palestine. Balfour expressed approval for the establishment of a home for Jews there. This letter later became known as the Balfour Declaration. In 1922, the League of Nations approved the Balfour mandate in Palestine, thus paving the way for the later creation of Israel. Lord Rothschild was an ardent Zionist, who had served as a member of the British Parliament. The Zionist movement, composed of both Jews and non-Jews, had been working toward the creation of Israel since the late 1800s. Lord Rothschild was the eldest son of Nathan Rothschild, who had controlled the Bank of England and funded the Cecil Rhodes diamond (and secret society) empire. Another Rothschild, Baron Edmond de Rothschild, who built the first pipeline from the Red Sea to the Mediterranean and founded the Israel General Bank, was called "the father of modern Israel."

It also appears that the Zionists employed blackmail to aid in the formation of Israel. Their most famous target was Nelson Rockefeller, who in 1940 was named to the intelligence position of coordinator of inter-American affairs by Secretary of Defense Forrestal. In 1944, Rockefeller was selected to serve as assistant secretary of state for Latin American affairs. It was a post most suitable to Rockefeller, whose primary purpose, according to authors Loftus and Aarons, was "to monopolize Latin America's raw materials and exclude the Europeans." Due to the extensive business dealings between the German Nazis and American globalists, as detailed previously, Loftus and Aarons noted that during the war, the Germans in South America got anything they wanted, from refueling stations to espionage bases, while the British had to pay cash. "Behind Rockefeller's rhetoric of taking measures in Latin America for the national defense stood a naked grab for profits," they wrote. "Under the cloak of his official position, Rockefeller and his cronies would take over Britain's most valuable Latin American properties. If the British resisted, he would effectively block raw materials and food supplies desperately needed for Britain's fight against Hitler."

Soon, Rockefeller controlled much of South America and was able to bring that influence to the newly created United Nations. But when Rockefeller pushed through UN membership for pro-fascist Argentina over the objections of President Truman, he lost his government position. He returned to the world of business. According to Loftus and Aarons, his "partner in moneymaking just happened to be John Foster Dulles, a trustee of the Rockefeller Foundation and a fellow conspirator in smuggling Axis money to safety."

In 1947, when Zionist leader David Ben-Gurion was desperate for votes to ensure the passage of a UN resolution partitioning Palestine and thus creating the state of Israel, he turned to Nelson Rockefeller. According to several former U.S. intelligence agents, Ben-Gurion "blackmailed the hell out of him."

Rockefeller had been able to deflect several investigations into his family's prewar and wartime dealings with the Nazis, but according to Loftus and Aarons, "Then the Jews arrived with their dossier. They had his Swiss bank records with the Nazis, his signature on correspondence setting up the German Cartel in South America, transcripts of his conversations with Nazi agents during the war, and, finally, evidence of his complicity in helping Allen Dulles smuggle Nazi war criminals and money from the Vatican to Argentina."

Loftus, as a U.S. attorney with unprecedented access to classified CIA and NATO files as well as former intelligence operatives, in 1994 joined with Australian broadcast journalist Mark Aarons to produce a national bestseller titled *The Secret War Against the Jews,* which probed the role of Western intelligence agencies in the affairs of Israel. These authors interviewed one of the Jews present at the meeting with Rockefeller. He gave this account:

Rockefeller skimmed through the dossier and coolly began to bargain. In return for the votes of the Latin American bloc, he wanted guarantees that the Jews would keep their mouths shut about the flow of Nazi money and fugitives to South America. There would be no Zionist Nazi-hunting unit, no testimony at Nuremberg about the bankers or anyone else, not a single leak to the press about where the Nazis were living in South America or which Nazis were working for Dulles. The subject of Nazis was closed. Period. Forever. The choice was simple, Rockefeller explained, "You can have vengeance, or you can have a country, but you cannot have both."

The deal was made and Rockefeller delivered. On November 29, 1947, the UN General Assembly approved a resolution recommending the partition of Palestine. The vote shocked the Arab world, which had not foreseen several Latin American countries switching their vote at the last minute. The Jews had traded silence for their new country's security, but they didn't take it lying down. To this day, Israeli leaders have in turn blackmailed the Western employers of Nazi refugees and war criminals, guaranteeing nearly unconditional support for Israel and its policies.

The creation of Israel also explained the inability of U.S. officials to interdict the flight capital out of Germany. John Pehle worked with Orvis Schmidt in the U.S. Treasury's Foreign Funds Control office. "In 1944 emphasis in Washington shifted from overseas fiscal controls to assistance to Jewish war refugees," Pehle explained. "On presidential order I was made executive director of the War Refugee Board in January 1944. Orvis Schmidt became director of Foreign Funds Control. Some of the manpower he had was transferred, and while the Germans evidently were doing their best to avoid Allied seizures of assets, we were doing our best to extricate as many Jews as possible from Europe."

It is apparent that the globalists in both Europe and America were more concerned with gaining a foothold in the oil-rich Middle East than in pursuing escaped Nazis and their treasure.

ALONG WITH THIS outpouring of Nazi assets, capital, scientists, SS men, and former officials within ODESSA were other, more secretive, assistance groups such as the Die Spinne, Sechsgestirn (or the Constellation of Six), and the Deutsche Hilfsverein (or German Relief Organization).

Through such organizations, SS men and Nazi officials escaped through southern France and across the Pyrenees into Spain. These were not-for-profit enterprises like ODESSA, but still they received funding and orders from the Bormann group.

Die Spinne was a creation of commando Otto Skorzeny and was largely composed of troops from his old wartime commando unit. It was funded through the Bormann program. After the meeting with Eva Peron in 1947, many Spinne members made their way to Argentina.

"[T]he number of Germans who went to South America, both along these . . . routes and by less organized means after Martin Bormann had declared his flight capital program in August 1944, totaled 60,000, including scientists and administrators at all levels, as well as the former SS soldiers commanded by General Mueller," noted Paul Manning.

Even before the end of the war, the Nazis used concentration camp prisoners and hired specialists to manufacture respectable-looking but phony identification papers. With these and aided by the general chaos in Europe at war's end, they developed their own effective witness protection program. Many of these false identities have withstood the test of time and are still in use.

While on the run at the end of the war, Bormann controlled his vast commercial empire through an elaborate but well-planned communications system. "Wherever positioned, he turned his hiding place into a party headquarters, and was in command of everything save security," explained Manning. "Telephones were too dangerous, but he had couriers to bear documents to Sweden, where a Bormann commercial headquarters was maintained in Malmö [Sweden] to handle the affairs of a complex and growing postwar business empire. From Malmö, high-frequency radio could transmit coded information to listening posts in Switzerland, Spain, or Argentina to form a continuous line of instructions."

The deputy fuehrer's escape had not gone unnoticed. It was substantiated by a file on Bormann sent to the FBI and obtained by Paul Manning. "When the file . . . was received at FBI headquarters it revealed that the Reichsleiter had indeed been tracked for years," he wrote. "One report covered [Bormann's] whereabouts from 1948 to 1961, in Argentina, Paraguay, Brazil, and Chile. The file revealed that he had been banking under his own name from his office in Germany in Deutsche Bank of Buenos

Aires since 1941; that he held one joint account with the Argentinian dictator Juan Peron, and on August 4, 5, and 14, 1967, had written checks on demand accounts in First National City Bank [now Citibank] (Overseas Division) of New York, the Chase Manhattan Bank, and Manufacturers Hanover Trust Co., all cleared through Deutsche Bank of Buenos Aires."

Then there was a police report from Cordoba Province dated April 22, 1955, in which a police agent with special knowledge of Bormann spotted the Nazi in the company of two other men in a hotel and trailed them. He overheard one of the men acknowledge the short, balding man who obviously was the superior of the three by saying, *"Jawohl, Herr Bormann."*

By 1972, it was apparent to anyone who desired to know that Martin Bormann had been operating in South America for some time. Researcher and author Ladislas Farago caused a minor sensation in that year with his articles published in England's *Daily Express,* detailing Bormann's activities. In 1974, Farago used his findings on Bormann in his book *Aftermath: Martin Bormann and the Fourth Reich.*

His investigative work led to a *New York Times* story published on November 27, 1972, and datelined "Buenos Aires." It stated, "Argentine secret service sources said today that Martin Bormann was sheltered in the country after World War II, but could not confirm reports that he still lived there. Sources in Salta confirmed that the ranch where Bormann was said to have lived was owned by German industrialists. The intelligence sources said other Nazis arrived in Argentina with Bormann and were sheltered there, particularly by Vittorio Mussolini, son of the Italian dictator."

Of course, only Bormann, Mengele, Eichmann, and a few other leading Nazis garnered occasional news headlines. Thousands managed to slip through the hands of authorities unnoticed, thanks to business connections or passports provided by the Vatican.

It seems apparent that it was not only business interests protecting the Nazis but individuals within the American government. For example, someone with access to U.S. archives later took steps to obliterate any record of Kammler or his fate. Nick Cook tried to trace Kammler and hit a brick wall. "Protracted searches by archivists at the U.S. National Archives for any data on Kammler had failed to locate a single entry for him," wrote Cook. "Given Kammler's range of responsibilities in the final months of the war, this absence of evidence was remarkable; so much so, that one archivist at Modern Military Records, College Park, Maryland, said . . . Somebody . . . had been in and cleaned up [the records]."

THE NAZI ORGANIZATION that may have made the greatest impact on the United States was not a ratline but a spy network created by General-major Reinhard Gehlen. This Nazi network was to become America's eyes and ears in the early days of the Cold War.

The son of a Catholic bookshop owner, Gehlen was born in 1902 and joined the German Army in 1920. His middle-class family nevertheless boasted military officers on both sides. In the 1930s, Gehlen moved from the German Staff College to the Army General Staff with the rank of captain. In 1940, he was promoted to major and served on the staffs of two German generals. By 1942, Gehlen, now a Lieutenant colonel, be-came the head of Fremde Heere Ost, or Foreign Armies East (FHO), a curious title for the section of the German General Staff analyzing all in-telligence on the Russian Front.

In an attempt to avoid conflicts with the Abwehr, Germany's counter-intelligence service, Gehlen created his own network of spies and inform-ers. This system soon began making major contributions to the Nazi war effort by upgrading the level of intelligence on the Soviets. Gehlen made use of whatever anticommunists could be found and in particular the anti-Soviet spy network of Russian general Andrei Vlasov, a Russian of-ficer who began working with Gehlen and the Nazis against the Stalin regime. (With Germany's defeat, the Allies turned Vlasov and his "Rus-sian Liberation Army" over to Stalin, who had them all executed.) Gehlen soon put together a remarkable network of agents and spies, all sworn to utmost secrecy, even from their own families. This combined Vlasov/ Gehlen operation became known as the Gehlen Organization, a spy net-work that was continued by U.S. authorities long after the war.

But Gehlen's accurate and realistic intelligence soon rankled Hitler, who toward the end of the war cried, "Gehlen is a fool!" Such vitriol may have led to Gehlen lending a sympathetic ear to plotters against Hitler. But whatever his role, if any, in the failed July 1944 bomb plot against the fuehrer, Gehlen managed to survive.

By then, he had a new plan—one that was to have lasting effect on the Allied nations and particularly the United States and Russia.

In April 1945, realizing that the war was lost for the Germans, Gehlen offered his spy network in Russia to the British but received no answer. "Taking everything into consideration, it seemed more expedient to make our approach to the American military forces," Gehlen recalled. "I suspected that once the shooting stopped the Americans would probably recover a sense of objectivity toward us more rapidly then their European allies, and subsequent history bore me out on this point."

Gehlen also showed no signs of being anything other than an unrepentant National Socialist. In his 1971 memoirs, he stated, "I still believe that we could have achieved our 1941 campaign objectives, had it not been for the pernicious interventions of Adolf Hitler." In other words, Gehlen's only objection to Nazi aggression was that Hitler lost.

Gehlen and his organization stashed their voluminous intelligence files in more than fifty sealed steel containers and buried them as they retreated westward—one cache was stored near the Wendelstein Mountains, another in the Algau province of southwest Bavaria, and the third in the Hunsruck mountain range in the Rhineland. After hiding out in a mountain lodge for some time, Gehlen made his move. "We were determined not to be taken prisoner," he later recounted. "We wanted to surrender on our own initiative to the Americans. It was all part of the plan."

Initially spurned by American officers who failed to recognize his importance, including a member of the Counterintelligence Corps (CIC), Gehlen finally arrived in front of Brigadier General Edwin L. Sibert, senior intelligence officer of the American occupation zone in Germany. "While fighting was still in progress in France, [Sibert] had been prepared to make use of Adolf Hitler's officers in the cause of American strategy," wrote Gehlen chroniclers Heinz Hoehne and Hermann Zolling, adding, "The idea came from . . . the adviser to Allen W. Dulles, the U.S. secret-service officer in Berne."

Sibert listened attentively as Gehlen detailed "the actual aims of the Soviet Union and its display of military might," despite U.S. Army regulations that prohibited personnel from listening to any remarks made by a German against their erstwhile ally in the East. "My later discussions with General Sibert in Oberursel ended with a 'gentlemen's agreement' which for a variety of reasons we never set down in black and white," Gehlen stated.

The terms of this "gentlemen's agreement" were as follows:

- A clandestine German intelligence organization was to be created.
- This organization would work "jointly" with the Americans, but would not be subordinate to them.
- The organization would operate exclusively under German leadership with only assignments coming from the Americans.
- The organization would be funded by the Americans but not from occupation costs.
- The organization would remain in American hands until a sovereign German government was created and agreed to take responsibility for the group.
- Should the organization at any time find German and American interests in conflict, it would consider the interests of Germany first.

"The political risk [of this agreement] to which Sibert was exposed was very great," conceded Gehlen, who was most pleased with the arrangement. "Anti-German feeling was running high, and he had created our organization without any authority from Washington and without the knowledge of the War Department. I understand that he informed his opposite number in the British zone, Major General Sir Kenneth Strong, of our existence, but he asked him not to inquire too closely into the matter for fear that the press might discover our activities."

Gehlen and some of his staff members were soon flown to Washington in a military plane belonging to Walter Bedell Smith, General Eisenhower's chief of staff, who went on to direct the CIA from 1950 to 1953 and also succeeded Averell Harriman as U.S. ambassador to the Soviet Union. Smith was to become an early member of the secretive Bilderberg Group initially headed by former SS officer Prince Bernhard of the Netherlands.

Gehlen's self-serving proposal was accepted by Sibert's military superiors, who did not know of the globalists' control over the Soviet Union and, therefore, were easily frightened by Gehlen's description of this militarily ambitious "evil empire." Under the proposal, Gehlen would operate independently and as an equal, offering the Americans only the information they requested or he decided to share, but never in any way conflicting with the interests of his Fatherland. In other words, virtually everything the United States learned about Soviet aims and capabilities at the end of World War II came from an anticommunist underground filtered through a Nazi organization with connections to the international financial elite.

Carl Oglesby, author of *The Yankee and Cowboy War,* wrote that by 1948, following the formation of the CIA, "Gehlen had grown tight with Dulles and his organization and become in effect the CIA's department of Russian and East European affairs. Soon after the formation of NATO [in 1949], [the Gehlen organization] became the official NATO intelligence organization." It has been made public in recent years that the Gehlen organization received an aggregate of $200 million in CIA funds from Allen Dulles.

Much of Gehlen's intelligence proved questionable, although this was not known at the time, since the Russians had tight control over information behind the Iron Curtain. "Gehlen flooded the Americans with 'authentic' documents provided by the Byeorussians," noted Loftus. "Because the information pertained to Soviet activity in areas where verification was impossible, the Americans had no choice but to view Gehlen's information as genuine. In reality, most of the secret intelligence that Gehlen furnished came from recently arrived émigrés, Soviet newspapers, and mail from Belarus and the Ukraine."

Gehlen went on to an illustrious career in spy craft. In 1946, he returned to Germany and began forming an intelligence organization that evolved into the Bundesnachrichtendienst (BND), or federal intelligence service, in West Germany. His cadre of 350 former comrades grew to more than 4,000. True to the initial agreement, Gehlen became president of the BND from its inception in 1956 until 1968, when he was forced to resign in the wake of a political scandal. Following his death in 1979, Gehlen, a member of the Knights of Malta, was hailed as the consummate Cold War spymaster.

In 2000, the CIA finally admitted its relationship with Gehlen. As the result of a Freedom of Information Act request from Oglesby, the agency filed an affidavit in a U.S. District Court "acknowledging an intelligence relationship with German General Reinhard Gehlen that it has kept secret for 50 years."

ANOTHER FORGOTTEN CONNECTION between U.S. authorities and the Nazis was the International Criminal Police Organization, known as Interpol, which was created as the International Criminal Police Commission in 1924, the same year J. Edgar Hoover became director of the FBI. It was headquartered in Vienna, Austria. It was established to assist in international police cooperation. In 1938, following the *Anschluss,* or unification, of Germany and Austria, the organization came under Nazi control and until the end of World War II functioned as an intelligence and enforcement arm of the Gestapo.

During those years, Interpol was headed by some of the most notorious Nazi war criminals, such as SS Obergruppenfuehrer Reinhard Heydrich, who chaired the infamous Wannsee Conference where the Holocaust was planned; Arthur Nebe, the criminal police chief who also commanded *Einsatzkommandos,* or killer squads, that liquidated "undesirables"; and Ernst Kaltenbrunner, who succeeded the slain Heydrich and was hanged at Nuremberg for war crimes. Working with Heydrich at Interpol was a young SS officer named Paul Dickopf. After the war, Dickopf served as president of Interpol from 1968 to 1972.

At the recommendation of FBI director Hoover, who always seemed more concerned about communists than Nazis, the United States formally joined Interpol just two weeks after the 1938 Nazi *Anschluss* of Austria. Hoover kept up a friendly correspondence with Nazi Interpol leaders until a few days after Pearl Harbor, when apparently he felt such connections might tarnish his image.

After the war, Interpol officials insisted that all its files were destroyed in Allied bombings. However, according to researcher Vaughn Young, a Swedish policeman named Harry Soderman, who had worked with Interpol since its inception, argued that an aborted attempt in 1945 to take the files out of Germany left them in French hands. The next year, Interpol was reestablished with strong support from the French police and headquartered in Paris. Also in 1946, Hoover sidestepped the U.S. State Department by attending a meeting in Brussels to formally reconstitute Interpol, where he was elected vice president. Former U.S. Army intelligence officer William Spector stated Hoover gained blackmail leverage over many prominent American business and political leaders by acquiring the Nazi/Interpol intelligence files at the end of World War II.

To this day, Interpol officials have declined to seek out Nazi war criminals, claiming such action is beyond its jurisdiction.

BY 1980, MARTIN Bormann, then in his eighties, had traveled extensively in South America, often just ahead of Nazi hunters. He lived in a luxurious estate near Buenos Aires, writing his memoirs while still under the protection of "Gestapo" Mueller.

Paul Manning said this aging recluse remained the guardian and silent manipulator of a gigantic industrial pyramid centered in Germany. Bormann also had become mentor to a new generation of lawyers, bankers, and industrialists. In an undated interview following the 1981 publication of his book *Martin Bormann: Nazi in Exile,* Manning stated, "The Bormann organization is not merely a group of ex-Nazis. It is a great economic power whose interests today supercede their ideology."

It is estimated that as many as 100,000 ranking Nazis remained at large after the war. "As such, it constitutes one of the largest—and best-funded, best-trained, best-equipped, and best-connected—cults in the world today," stated Peter Levenda. "And the second generation is being trained and indoctrinated in the streets of London, Berlin, New York, Buenos

Aires...and in secret, heavily armed estates like Colonia Dignidad [in Chile]."

Colonia Dignidad, or Dignity colony, today is called Villa Baviera, or villa Bavaria. It was founded in 1961 by Paul Schaefer, formerly of the Nazi Luftwaffe, and was made up of German immigrants who had been living there since the early 1950s. The large compound boasted its own power plant, two runways, and a restaurant, all surrounded by barbed wire, searchlights, and guard towers. In 1986, an inspection by Amnesty International discovered underground cells where prisoners suffered remote-control torture by means of electronic sound systems and electric shock. "It was a torture and execution center during the regime of Augusto Pinochet who was placed into power in Chile by Henry Kissinger in 1973 to protect Rockefeller interests there," stated Peter Levenda.

The compound was run by approximately three hundred Nazi exiles, some of whom still live there today. An estimated three thousand persons died and thirty thousand were tortured during the violent overthrow of Chile's democracy by Pinochet, which included the still-disputed circumstances of President Salvador Allende's death. In 1997, Schaefer fled Chile after being accused of sexually molesting two young boys at the colony. In 2005, large caches of arms and ammunition were found there.

While there can be no doubt that Bormann's surviving Nazi empire still exerts tremendous control over world economies and politics even today, the full extent may never be known.

What is known is that many of Nazi Germany's most brilliant minds continued their work outside Europe after the war, most notably in the United States.

CHAPTER 7

PROJECT PAPERCLIP AND THE SPACE RACE

ON MAY 19, 1945, JUST TWELVE DAYS AFTER GERMANY'S UNCONDI-
tional surrender, Herbert Wagner, creator of the first Nazi guided missile
used in combat, landed in Washington, D.C., in a U.S. military aircraft
with blacked-out windows.

Wagner was the first of a stream of Nazi scientists, technicians, and
others to arrive in the United States in a program that came to be known
as Project Paperclip. It began as Operation Overcast, a program authorized by the Joint Chiefs of Staff to exploit the knowledge of Nazi scientists. (Overcast was mentioned but not clearly explained in the 2006 film
The Good German starring George Clooney.) This operation was renamed
Paperclip and formally authorized in August 1945 by President Harry
Truman, who was assured that no one with "Nazi or militaristic records"
would be involved.

By mid-November, more Nazi scientists, engineers, and technicians
were arriving in America, including Wernher von Braun and more than
seven hundred other Nazi rocket scientists.

By 1955, nearly a thousand German scientists had been granted citizenship in the United States and given prominent positions in the American scientific community. Many had been longtime members of the Nazi

Party and the Gestapo, had conducted experiments on humans at concentration camps, used slave labor, and committed other war crimes.

Von Braun, who in later years became the head of the National Aeronautics and Space Administration (NASA), is one of the more recognizable names of the Paperclip scientists. Others included Major General Walter Dornberger, a close associate of von Braun's; Werner Heisenberg, physicist and Nobel laureate who founded quantum mechanics; gaseous uranium centrifuge expert Dr. Paul Harteck; Nazi atomic bomb physicist and military project leader Kurt Diebner; uranium enrichment expert Erich Bagge; 1944 Nobel Prize winner Otto Hahn, called the "father of nuclear chemistry"; scientists Carl Friedrich von Weizsaecker, Karl Wirtz, and Horst Korsching; and physicist Walter Gerlach.

CNN reporter Linda Hunt's 1991 book *Secret Agenda: The United States Government, Nazi Scientists, and Project Paperclip, 1945–1990* first revealed the extent to which Nazi infiltration was aided by persons within the U.S. federal government and military. Like other researchers, Hunt found many government files pertaining to recruited Nazis "missing" or otherwise unavailable. Despite government claims that Paperclip was ended in 1947, according to Hunt, this project was "the biggest, longest-running operation involving Nazis in our country's history."

"The project continued nonstop until 1973—*decades* longer than was previously thought. And remnants of it are still in operation today," she wrote. By the 1990s, when details of Paperclip finally reached the public's ears, the infusion of Nazis into America's military-industrial complex was complete.

IN 1952, NEWLY elected President Dwight D. Eisenhower was persuaded to name John Foster Dulles as his secretary of state, and his brother, Allen Dulles, as the director of the CIA. "The reigning Dulles brothers were the 'Republican' replacements for their client and business partner, 'Democrat' Averell Harriman. Occasional public posturing aside, their strategic commitments [to the globalists] were identical to his," stated authors Webster Griffin Tarpley and Anton Chaitkin in their well-documented 1992 unauthorized biography of George H. W. Bush. It should be noted that the Dulles brothers were both attorneys for and business partners with Averell Harriman. It should also be noted that Allen Dulles, as OSS station chief in Bern, Switzerland, sat at the nexus of U.S. intelligence as well as Soviet intelligence, such as the infamous Rote Kapelle, or Red Orchestra, spy network. It was during his stint as assistant to the U.S. ambassador that Dulles used SS Brigadefuehrer, or brigadier general, Walter Schellenberg to communicate with his immediate superior, Reichsfuehrer Heinrich Himmler. Dulles constantly sent intelligence reports to Washington, although, as stated by Adam LeBor, "there are questions as to whether his motive was supplying genuine economic intelligence or merely building a complicated empire of information and disinformation that reached from Bern to Berlin and back again."

Dulles moved from Bern to become OSS station chief in Berlin at the end of the war. In 1947, when the OSS was rolled into the newly created CIA, Dulles's translator was an army intelligence officer named Henry Kissinger, who would go on to become secretary of state under President Richard Nixon, a lifelong friend to Dulles.

Project Paperclip quickly came under the control of an "old boy" network encompassing members of the globalists centered in the Council on Foreign Relations.

After its inception, Paperclip was run by the intelligence division of the U.S. Army's European Command, directed by Robert Walsh operating out of Berlin. "The Paperclip office operated out of the intelligence division's headquarters in Heidelberg, under Deputy Director Colonel Robert Schow, who would become assistant director of the CIA in 1949 and assistant chief of staff for intelligence in 1956," Hunt wrote in *Secret Agenda*. She added that officers of the Joint Intelligence Objectives Agency (JIOA) who managed Project Paperclip soon began receiving security reports from Schow's office, regarding the Germans recruited into the program. All reports on these men had been altered from a determination of "ardent Nazi" to read "not an ardent Nazi."

After assuming the directorship of the newly created CIA, Allen Dulles, who, as attorney for the Shroeder bank, had brokered the deals allowing Hitler's rise to power, assumed control over Project Paperclip and increased the flow of National Socialists into the United States.

After former Nazi spymaster General Reinhard Gehlen met with the CIA director Dulles and offered to turn over his extensive spy network to the CIA in exchange for non-prosecution of their Nazi pasts, the scientists' dossiers were rewritten to eliminate incriminating evidence of their work for the Nazis. "For over forty years . . . Paperclip's dark secrets lay safely hidden in cover-ups, lies, and deceit," stated Hunt.

Hunt uncovered documents showing that even Wernher von Braun, who in 1947 had been described as "a potential security threat" by the military governor, was reassessed only months later in a report stating, "he may not constitute a security threat to the U.S." Likewise, von Braun's brother, Magnus, who had been declared a "dangerous German Nazi" by counterintelligence officers, was brought to America and his pro-Nazi record expunged.

"The effect of the cover-up involved far more than merely whitewashing the information in the dossiers," noted Hunt. "Serious allegations of crimes not only were expunged from the records, but were never even investigated."

In a 1985 exposé in the *Bulletin of the Atomic Scientists,* Hunt wrote that she had examined more than 130 reports on Project Paperclip subjects—and every one "had been changed to eliminate the security threat classification." President Truman, who had explicitly ordered no committed Nazis to be admitted under Project Paperclip, was evidently never aware that his directive had been violated. Again, this is evidence of control at a level higher than the president.

By the late 1940s, the now-ascendant Cold War added new impetus. Potential intelligence assets were recruited from all across Europe, many of them zealous Nazis who could be relied upon to be anticommunists. In an effort to avoid the negative publicity that had burdened some of the early Paperclip activities—some Americans just did not think it proper to bring former Nazis to the United States and place them in responsible positions—JIOA officers began bringing Nazis from Argentina, that haven for the Bormann organization.

According to Hunt, all of this activity was almost totally unknown to the public but rested in the hands of certain top-ranking government officials like the Dulles brothers and John J. McCloy. The agenda of the globalists was moving ahead.

Paperclip had several spinoff projects. Expanding on Paperclip, the National Interest program was tightly connected to the new CIA and provided a means of bypassing close scrutiny by anti-Nazi elements within military intelligence. No longer were Nazi scientists the sole objective; recruitment of Nazis now included Eastern Europeans thought to be helpful against the communists, and even convicted Nazi war criminals. Anyone, regardless of their past, was eligible as long as someone within the U.S. government deemed their presence in the national interest. Linda Hunt wrote:

> Prevailing myth has it that the first group in National Interest, the German scientists, were employed solely because of their scientific expertise, but there were other reasons as well. First, defense contractors and universities could hire German scientists for substantially less money than they could American employees. Salary statistics show that the Germans signed contracts for approximately $2,000 a year less than their American counterparts received in comparable positions. Of course, the Germans were unaware of the salary discrepancy, since they had earned even less money in West Germany. The JIOA, however, took advantage of the situation by promoting cheap salaries to convince corporations to participate in the project. Second, because of the Joint Chiefs of Staff connection with the National Interest project, German scientists could obtain necessary security clearances more easily than could American scientists. Defense contractors looking for new employees to work on classified projects found this aspect of National Interest to be particularly advantageous. By 1957, more than sixty companies were listed on JIOA's rosters, including Lockheed, W. R. Grace and Company, CBS Laboratories, and Martin Marietta. . . .

National Interest placed German scientists at major universities in research or teaching positions, regardless of their Nazi pasts. Even the U.S. Office of Education helped JIOA send fliers to universities all over the country touting the advantages of hiring the Germans on federally financed research projects, since they could obtain security clearances more easily than Americans. The University of Texas, Washington University School of Medicine in St. Louis, Missouri, and Boston University were among the participants.

It should be noted that Yale University, alma mater of the Bush and Harriman families and home of Skull and Bones, also received Paperclip Nazis as employees.

Another program, code-named simply Project 63, was designed specifically to get German scientists out of Europe and away from the Soviets. "Most went to work for universities or defense contractors, not the U.S. government," noted Hunt. "Thus the American taxpayer footed the bill for a project to help former Nazis obtain jobs with Lockheed, Martin Marietta, North American Aviation, and other defense contractors during a time when many American engineers in the aircraft industry were being laid off."

The Project 63 effort to import Nazis grew so public—in 1952, JIOA deputy director Air Force colonel Gerold Crabbe and a gaggle of military officers, Paperclip members, and civilians journeyed to West Germany on a recruitment drive—that even McCloy expressed concern over a "violent reaction" by West German officials. West Germans complained to U.S. ambassador James Conant, demanding that Paperclip be ended. Conant appealed to then secretary of state John Foster Dulles to shut down Paperclip "before we are faced with a formal complaint by the West German government against a continuing U.S. recruitment program which has no parallel in any other Allied country." But the project was not stopped. As usual, there was simply a name change. Paperclip became the Defense Scientists Immigration Program (DEFSIP). Conant may not have realized his appeal was aimed at one of the architects of the very program he was trying to end.

As the Paperclip project began to lose momentum, yet another stimulus arose. On October 4, 1957, the Russians launched *Sputnik I* into orbit around the Earth and the space race was on. The Nazi scientists were in demand more than ever.

Paperclip again began to grow. Specialists were imported from Germany, Austria, and other countries under Project 63 and National Interest and gained positions at many universities and defense contractors, including Duke University, RCA, Bell Laboratories, Douglas Aircraft, and Martin Marietta. "Information about the number of defectors and other individuals brought in by the CIA and military intelligence agencies is unknown, since JIOA records concerning them were either shredded or pulled during the FBI's investigation in 1964," noted Hunt, adding, "It had taken the greatest war in history to put a stop to an unspeakable evil, and now the cutting edge of that nightmare was being transplanted to America."

Interestingly, even as we were bringing foreign defectors into the USA, we discovered a traitor within our own government. Lieutenant Colonel William Henry Whalen, who from 1957 served as deputy director of JIOA, the agency that commanded Paperclip, became the highest-ranking American ever recruited as a mole by the Soviet intelligence service. Only four months after he began spying for the Soviets, Whalen was promoted to the directorship of JIOA. When arrested in 1962, Whalen was an intelligence adviser, which permitted him access to any information pertaining to the Joint Chiefs of Staff planning and allocation of military forces, including communications and electronic intelligence-gathering.

Whalen, who suffered from alcoholism and debt, was recruited by the Soviets in the mid-1950s by Colonel Sergei Edemski, a loquacious Soviet military attaché in Washington, D.C. Although it became publicly known that Whalen admitted providing the Communists with our utmost secrets concerning U.S. nuclear weaponry and strategies, his connection with Paperclip was not revealed. Yet, author Hunt raised a relevant question about Whalen by asking, "Did he use blackmail to recruit a spy or saboteur from among the approximately 1,600 Paperclip specialists and hundreds of other JIOA recruits brought to this country since 1945? It certainly is clear from the evidence that many of them had a lot to hide." Though charged with espionage, in 1966 Whalen made a deal with the Justice Department, in which he pled guilty to a lesser charge in exchange for his cooperation. Federal judge Oren Lewis, while accusing Whalen of "selling me and all your fellow Americans down the river," nevertheless sentenced the spy to a mere fifteen years in prison. He was paroled after serving only six years.

But it was not just homegrown spies like Whalen who were slipping information out of Paperclip. Imported Nazis had every opportunity to pass national security information out of the country. According to Hunt, there was no further army surveillance over the Nazi Paperclip specialists after just four months of their signing a contract with the U.S. government. Furthermore, anyone receiving any more than 50 percent support from a Paperclip specialist constituted a "dependent," according to their contracts. "The large number of so-called dependents—including mistresses and maids—brought to Fort Bliss [Texas] as a result of [this situation] were subject to no off-the-post surveillance, even though it was assumed that they had access to at least some classified information because of their close contact with Paperclip personnel," wrote Hunt.

Incidents of information being passed out of Paperclip were presented to authorities, yet nothing was done. A Fort Bliss businessman reported Paperclip engineer Hans Lindenmayr to the FBI, claiming the German had been using his business address as an illegal letter drop. According to Hunt, at least three other Nazis maintained illegal mail drops in El Paso, "where they received money from foreign or unknown sources and coded messages from South America." It was also learned that many Paperclip Nazis received cash from foreign sources. "Neither Army CIC not FBI agents knew where that money came from, and by all appearances, no one cared to know how more than a third of the Paperclip group suddenly were able to buy expensive cars," noted Hunt.

When word was passed that Nazi scientists working for the French were suspected of receiving orders from Germany to work toward a reemergence of the Reich, army intelligence officers finally began to take a closer look at Paperclip. Amazingly, the biggest catch was Wernher von Braun. It was revealed that at the end of the war, the rocket scientist had been caught sending a map overseas to General Dornberger and concealing information from U.S. officials. Further investigation revealed that Paperclip specialists were allowed to make unsupervised trips off base and even out of town, the only requirement being that they report when they arrived at their destination. Several had their own telephones that were never monitored.

President Truman was once notified by the CIA that the Nazi scientists working for the Soviets were using a postal address in the U.S. sector of West Germany as a cover for communications with the Paperclip scientists in America. One General Electric manager working with Paperclip specialists told the FBI that the Army's lax security at White Sands Proving Grounds bordered on "criminal neglect," especially since about 350 of the Germans' former coworkers were serving the Russians. He believed that it was reasonable to assume that friendly contacts between the two groups still existed.

Apparently, overseas communication between the Nazis in America and the Nazis in Russia continued unabated, which has raised the possibility of a parallel space race controlled or manipulated by the very globalists who had created and financed both communism and the Third Reich.

ALMOST EVERYONE WHO was of age in 1969 recalls vividly the pride and excitement of the U.S. *Apollo* mission's moon landing on July 20. It is difficult then for them to seriously consider the many contradictions and anomalies of the six moon landings. What may be even more difficult is to consider that the space race was never a true competition between the United States and the former Soviet Union; rather, it was a combined space program run by Nazi scientists and controlled by high-level globalists.

As the Allies closed in on Nazi Germany in the spring of 1945, top American commanders were given orders to leave all the rockets and their plans at the Nazi facility at Nordhausen for the Russians. However, some commanders unofficially absconded with about a hundred V-2s, along with a large collection of plans, manuals, and other documents. According to one American officer, "We gave the Russians the key to *Sputnik*. . . . [F]or ten weeks, the American army had in its hands the rocket plant that gave the Russians their head start in the missile race." Here was more evidence of the collusion taking place at the level of the globalists who were already directing activities that would lead to the Cold War.

After the war, at NASA's George C. Marshall Space Flight Center in Huntsville, Alabama, the Nazi rocket scientists established nearly a carbon copy of their organization at the wartime secret Nazi rocket facility at Peenemunde. According to Linda Hunt, James Webb, NASA administrator during the Kennedy years, complained that the Nazi scientists were circumventing the system to the extent of attempting to build their own *Saturn V* rocket in-house at the Marshall Center.

"... the Germans dominated the rocket program to such an extent that they held the chief and deputy slots of every major division and laboratory. And their positions at Marshall and the Kennedy Space Center at Cape Canaveral, Florida, were similar to those they had held during the war," wrote Hunt. "The Peenemunde team's leader, Wernher von Braun, became the first director of the Marshall Space Center; Mittlewerk's head of production, Arthur Rudolph, was named project director of the *Saturn V* rocket program; Peenemunde's V-2 flight test director, Kurt Debus, was the first director of the Kennedy Space Center."

Rudolph, who gained American citizenship after entering the USA with his boss von Braun under the program that was to become Paperclip, was credited with helping to place Americans on the moon. He retired with a NASA pension in 1979 but was stripped of his American citizenship in 1983, after he conceded to the Justice Department that he had "participated under the direction of and on behalf of the Nazi government of Germany, in the persecution of unarmed civilians because of their race, religion, national origin, or political opinion." However, a West German investigation of Rudolph stated there was no factual basis for charging him with war crimes and granted him German citizenship. Several Americans, including Lieutenant Colonel William E. Winterstein Sr., who was commander of the Technical Service Unit at Fort Bliss, Texas, which supported the German scientists, claimed Rudolph was railroaded by the Justice Department's Office of Special Investigations, some members of which "had the full cooperation of the Soviet Union; therefore, close coordination with the KGB."

AS RECENTLY RELEASED files from behind the Iron Curtain have revealed, many of the scientists in Paperclip as well as some on the Manhattan Project indeed were spying for the Soviet Union. Their motivations were many. Some spied for pay, some for ideology, but all were manipulated by intelligence chiefs far above them.

The flow of information between the scientists in the Soviet Union and the United States has led some researchers to suspect that a covert space program—a third program—was in effect. Joseph P. Farrell, who holds a doctorate degree in patristics (the study of early Christian writers and their work) from the University of Oxford, also has researched ancient history and physics, to include the space program. In his 2004 book *Reich of the Black Sun,* Farrell wrote, "[I]t is perhaps significant that some contemporary observers of the American space program and its odd thirty-year-long 'holding pattern' and tapestry of inconsistencies, lies and obfuscations have long suspected that there are indeed two space programs inside the U.S. government, the public NASA one, and a quasi-independent one based deep within covert and black projects."

This idea is somewhat supported by the fact that the space programs of Russia and America moved along different paths. At the start, the Russians proved more capable of attaining space flight than the Americans. Under the leadership of their brilliant engineer Sergei Korolev, the Russians produced giant heavy-lift rockets while their American counterparts were busy developing the internal technology for guidance and control.

The Soviet Russians were first to launch a satellite, *Sputnik,* into Earth's orbit (1957); to orbit a man, Colonel Yuri Gagarin, and return him safely (1961); to place a live animal, the dog Laika (1957), and Valentina Tereshkova, the first woman into orbit (1963); to land unmanned vehicles on the moon (1970); to conduct an extravehicular "space walk" by cosmonauts; and to place nuclear warheads on Intercontinental Ballistic Missiles (ICBM).

Both nations used captured Nazi V-2 rockets to begin their space programs. A common joke in the 1950s involved an argument between a Russian and an American. "Our German scientists are better than your German scientists," they shouted at each other.

Yet, the evidence indicates that the American rocket scientists were indeed placed into a holding pattern while their Soviet counterparts caught up with their technology. William E. Winterstein Sr., a retired U.S. Army lieutenant colonel and one of the rocket engineers on NASA's

Apollo team, noted in his 2002 book *Gestapo USA,* "The space history of this country reveals that during the 1950s, the von Braun team had developed a multistage rocket by adding solid propellant rocket stages to a Redstone rocket as booster. In 1956, such a rocket with two solid stages made successful high-speed rocket reentry tests with model warheads covered with ablative heat protection. With three solid stages, such a rocket could have placed a satellite into orbit more than a year before the U.S. was defeated by *Sputnik.* However, and almost unbelievably, the von Braun team was given direct orders from Washington to stop further development. The team was restricted to the development of rockets whose range was less than 200 miles. It was only after President Kennedy announced the lunar mission in 1961 that the German rocket team was finally released from agonizing bureaucratic blunders from Washington, and was given a free hand, and even orders, to accomplish von Braun's lifelong goal to travel into space."

It has been argued that a primary incentive of the German scientists was the sheer desire to continue their work. "Some of these would stop at nothing, even resorting to duping their colleagues and superiors in order to ensure the continuance of their research," commented British authors Mary Bennett and David S. Percy. However, in some cases, such as that of von Braun, the connection between the work and the Nazis was close and continuous. Von Braun, the son of a well-connected Prussian minister who founded the German Savings Bank, was brought into Germany's rocket program by Luftwaffe general Walter Dornberger, who, although charged as a war criminal for the rocket attacks on London and Antwerp, was never brought to trial. Instead, he came into the United States as part of Project Paperclip. Likewise, von Braun, revered as the father of the U.S. space program, was found to have been a Nazi Party member, a member of the SS with the rank of major, a friend to SS Reichsfuehrer Heinrich Himmler, and, according to Linda Hunt, was accused by survivors of the rocket factories at Mittlewerk and Peenemunde of at least once ordering the execution of slave laborers. Kurt Debus, who became the first director of the Kennedy Space Center at Cape Canaveral, was both a member of the Nazi SS and the SA. According to documents obtained by Hunt, in 1942 Debus turned a colleague over to the Gestapo for making anti-Hitler remarks.

The Soviet manner of dealing with their Nazi scientists greatly differed from the laxity of Project Paperclip. "With hindsight, it would seem that the Soviets demonstrated a more humanitarian approach toward their technical prisoners than did the Americans," noted Bennett and Percy. "Moreover, the way in which the technical information was passed from teacher to pupil was very different. The Soviet experts and the Germans worked side by side in the same factory, but in separate areas. Information was passed between these teams without the Germans ever meeting their Soviet counterparts. They only spoke directly to Korolev, who was far curter with them than he had been in Bleicherode [the V-2 test area in Germany's Harz Mountains, where Korolev had first debriefed the scientists at war's end]."

Some serious researchers have opined that the space programs of both the USSR and the USA, despite the political posturing, were actually the same program, one far ahead of the current joint Russian-American space efforts such as the International Space Station. "This [overall] project was conceived and designed as a collaboration between two superpowers," wrote Bennett and Percy. "The Cold War was a convenient cover under which aspects of this program could be implemented and hidden. All these machinations were orchestrated at the very highest level, with only a select and *hidden* few ever knowing *the overall objectives* of the project [emphasis in original]."

As noted by Farrell, "This, of course, implies some entity or agency of coordination existing both within the Soviet Union and the United States." If this were so, who were these hidden manipulators?

To begin with, there were the German rocket scientists themselves. In 1945, Lieutenant Walter Jessel was assigned to investigate how much trust to give the scientists before bringing them to America. According to author Hunt, the lieutenant "uncovered evidence of a conspiracy among von Braun, Dornberger, and Dornberger's former chief of staff, Herbert Axster, to withhold information from U.S. officers."

Secret codicils within the 1945 Yalta Agreement between Roosevelt, Churchill, and Stalin allowed for the partitioning of Europe between the Allied powers.

Dr. Wilhelm Voss, the former head of the Skoda Munitions Works in

Pilsen, had handled much of the material for Kammler's *Kammlerstab* Special Projects Group. In May 1945, when elements of the American Army arrived in the Czech city of Pilsen, Voss attempted to hand over a truckload of *Kammlerstab* documents but was told by the commanding U.S. officer that he was under orders to give everything to the Russians.

While entire German divisions were trying desperately to surrender to the Western Allies, it is well known that Patton's Third Army had reached the outskirts of Berlin before being ordered back a hundred miles to await the arrival of the Russians, who were required to fight desperately for every block of the city. Such a withdrawal is clear evidence of the deals being made at the highest levels.

As has been noted, there was communication between the two groups of Nazi scientists even though they were half a world away from each other. This could have been facilitated by the interconnected business and banking interests already described.

As detailed in the section "Communism versus National Socialism" the same Western bankers and financiers who funded Hitler's National Socialism also supported Communism in Russia. The U.S. federal government's leniency toward communism has been well documented, beginning with President Franklin D. Roosevelt, who began his career as a Wall Street attorney specializing in corporate law. Roosevelt echoed his Wall Street cronies' warm regard for both Stalin and communism. According to historian Thomas Fleming, the U.S. government was rife with globalist agents conveying secrets back to Russia. "There was scarcely a branch of the American government, including the War, Navy, and Justice Departments, that did not have Soviet moles in high places, feeding Moscow information. [William] Wild Bill Donovan's Office of Strategic Services, the forerunner of the CIA, had so many informers in its ranks, it was almost an arm of the NKVD. Donovan's personal assistant, Duncan Chaplin Lee, was a spy," Fleming wrote.

Another possible crossing point for aerospace information may have been the British Interplanetary Society (BIS), according to Bennett and Percy. While the BIS was reportedly created in September 1945 by combining several existing organizations interested in the future of space exploration, it was not officially inaugurated until December of that year.

At that time, Wernher von Braun, the man behind the V-2 rockets, was named as an honorary fellow. Arthur C. Clarke, an early member of the BIS, claimed the society had been in existence long before the war and was merely in "suspended animation" from 1939 to 1945.

Noting that the Soviet embassy in London subscribed to no less than twenty copies of the bimonthly BIS journal, Bennett and Percy asked, "Why was it necessary to reform a society already in existence? Why did the British hasten to grant such an award to the man who only nine months before [as technical director of the Nazi V-1 and V-2 rocket programs], was responsible for the annihilation of so many people in London and the Home Counties? Why did both the British (and von Braun) wish to play down the real timing, if everybody felt comfortable with the reasons for honoring [von Braun]?" Their insinuation is that valuable rocket technology information was passed along via the BIS, possibly with the approval of von Braun.

President Eisenhower, aware of the American public's concern that the Soviets might be winning the space race, ordered a Manhattan Project approach to the problem. This mandate resulted in a structure that became compartmentalized and shrouded with secrecy. All relevant information was on a strictly "need to know" basis, controlled by members of the self-styled globalist elite, the plutocrats who owned the emerging multinational corporations.

To fully understand how this control over parallel space programs worked, one must look past the Eisenhower administration and study the National Security Act of 1947.

On September 15 of that year—only three months after pilot Kenneth Arnold saw flying discs over Mount Rainier, and just two months after something crashed near Roswell, New Mexico—President Harry S. Truman signed into law the National Security Act of 1947, which, among other things, created the National Security Council (NSC) and the Air Force as a separate branch of service, united the military branches under a Department of Defense, and created America's first peacetime civilian intelligence organization, the Central Intelligence Agency.

An important example of the tight inner government control by secret society members may be found in the NSC, which has come to dominate

U.S. policy decisions, including the use of armed force. Most Americans have no idea who exactly comprises the powerful NSC. The council principals are the president, vice president, and secretaries of state and defense, positions predominantly held throughout the later twentieth century by members of the globalist societies, the Council on Foreign Relations, or the Trilateral Commission. The NSC staff is directed by the president's national security adviser. To coordinate covert operations, the NSC created the 5412 Committee, also called the Special Group, which has changed names several times to avoid public exposure. In 1964, it was known as the 303 Committee and in 1970 it was renamed the 40 Committee. Within this organization—which included such familiar names as Nelson Rockefeller, Robert McNamara, McGeorge Bundy, Gordon Gray, and Allen Dulles—was a subcommittee dealing with science and technology. It is here that the connection between the corporate and financial world and government-held technological secrets may be found. Here is centered control over rocketry, space, alternative energy sources, and even UFOs. And it is here that researchers have tracked the mysterious group known as Majic Twelve, later known as Majestic Twelve or simply MJ-12.

The MJ-12 issue was first publicly raised in 1984 when a TV producer and UFO researcher received an undeveloped roll of 35-mm black-and-white film in his mail. The film contained eight pages of what appeared to be official U.S. government documents stamped TOP SECRET/ MAJIC EYES ONLY and dated November 18, 1952. The pages were a "briefing document" prepared for president-elect Dwight D. Eisenhower, concerning "Operation Majestic 12." There has been ongoing controversy over the legitimacy of these and the subsequent release of other MJ-12 documents, including a Standard Operations Manual (SOM 1-01) marked "Top Secret/MAJIC," dated April 1954, and titled "Extraterrestrial Entities and Technology, Recovery and Disposal."

The documents listed twelve prominent men as members of Operation Majestic 12, "a TOP SECRET Research and Development/Intelligence operation responsible directly and only to the President of the United States," who were to deal with the UFO issue at the highest level. The papers went on to detail how a "secret operation" was begun on July 7, 1947, to recover the wreckage of a disc-shaped craft from a crash site "approximately 75 miles northwest of Roswell Army Air Base." Also, "four small human-like beings [who] had apparently ejected from the craft" were found dead about two miles east of the wreckage site. The document added, "Civilian and military witnesses in the area were debriefed, and news reporters were given the effective cover story that the object had been a misguided weather research balloon." Later, when the weather balloon story became discredited, the story was changed to a Mogul balloon, used to monitor the upper atmosphere for Soviet A-bomb testing, though why such a monitoring device would be launched from New Mexico was never explained.

The "briefing" papers ended by stating, "Implications for the National Security are of continuing importance in that the motives and ultimate intentions of these visitors remain completely unknown. In addition, a significant upsurge in the surveillance activity of these craft beginning in May and continuing through the autumn of [1952] has caused considerable concern that new developments may be imminent. It is for these reasons, as well as the obvious international and technological considerations and the ultimate need to avoid a public panic at all costs, that the Majestic-12 Group remains of the unanimous opinion that imposition of the strictest security precautions should continue without interruption into the new administration."

These MJ-12 documents created a storm of controversy within the UFO research community. Debunkers claimed to have found all sorts of discrepancies—from misspellings to identical signatures. However, no one has been able to definitively disprove all the MJ-12 documents as fakes and, in fact, there is much evidence to indicate their authenticity. For example, Dr. Robert M. Wood, who managed research and development at McDonnell Douglas for forty-three years, found that the typeface and style of the SOM 1-01 manual matched that of U.S. government printing presses in use during the 1950s.

If the information in the MJ-12 documents is proven correct, it is strong evidence that certain persons within the United States had access to remarkable technology, both taken at Roswell and similar to that described as being in Nazi hands toward the end of the war.

A cursory look at the men identified as the original MJ-12 group, as well as their corporate and intelligence connections, makes clear the potential for high-level control over exotic technology—groundbreaking technology that could upset the monopolies over energy, transportation, and communications held by the wealthy globalists who financed Hitler. As listed in the documents, MJ-12 members included:

◆ ***Adminstrator Roscoe H. Hillenkoetter,*** a 1919 graduate of the Naval Academy, who was familiar with both intelligence work and the Nazis, having worked undercover for a year in Vichy, France. After serving as the third director of Central Intelligence Group, he became the first director of the CIA upon its formation in September 1947, obviously a good choice for a top-secret group like MJ-12. After his retirement from government, Hillenkoetter joined the National Investigations Committee on Aerial Phenomena (NICAP), a private UFO group, and stated publicly that UFOs were real and "through official secrecy and ridicule, many citizens are led to believe the unknown flying objects are nonsense."

◆ ***Dr. Vannevar Bush,*** an eminent American scientist, who in 1941 organized the National Defense Research Council, and in 1943 the Office of Scientific Research and Development that led to the production of the first atomic bomb. Dr. Bush was another prime candidate for a high-level group dealing with space. He also was a close friend to Averell Harriman, the U.S. ambassador to the Soviet Union, who had ownership in Union Banking Corporation along with Prescott Bush. (It is reported that Vannevar Bush was unrelated to the political Bush family.) In 1949, the U.S. Intelligence Board asked Bush to study ways of combining intelligence from all agencies. Bush's plan was initiated by America's first secretary of defense, James V. Forrestal, who also is listed as an MJ-12 member. Bush's connections to the corporate world were deep and many. In 1922, Bush, along with his former roommate Laurence K. Marshall and scientist Charles G. Smith, formed American Appliance Company, today known as the powerful Raytheon Corporation heralded in its company literature as an "industry leader in defense and government electronics, space, information technology, technical services and business aviation and special mission aircraft." Bush joined the Massachusetts Institute of Technology (MIT) as a professor in 1919, and in 1936 was awarded a major grant by the Rockefeller Foundation. His work during World War II resulted in the development of the Rockefeller Differential Analyzer, an analog mainframe computer composed of 2,000 vacuum tubes and 150 motors. Bush also served on the board of directors of the Metals and Controls Corporation, which in 1959 merged with Texas Instruments to become the first U.S. government–approved fabricator of uranium rods. Bush also was a presence in the corporate world of pharmaceuticals, eventually becoming chairman of the board of Merck and Company, one of the world's most powerful drug companies. Merck has been among the leaders in researching the human genome, the DNA structure that forms cells into humans. Bush also was connected to the Carnegie wealth, serving as president of the Carnegie Institute from 1935 to 1955.

- *James V. Forrestal,* who, prior to World War I, was a bond salesman for William A. Read and Company, later to become Dillon, Read and Company. After the war, he returned to Read and Company and by 1937 was named president. This was at a time when Dillon and Read were the most profitable of all Wall Street syndicate managers handling German industrial issues in the U.S. capital market. In 1957, *Fortune* magazine named Clarence Dillon as one of the wealthiest men in America, with a fortune estimated to have been between $150 and $200 million. Russell A. Nixon, the young attorney for the U.S. Military Government Cartel Unit who tried to break up the Nazi corporate syndicates at the end of the war, was blocked in his efforts by Brigadier General William H. Draper, who along with Forrestal was an officer of Dillon and Read. According to Sutton, "Banker William

Draper, as Brigadier General William Draper, put his control team together from businessmen who had represented American business in prewar Germany." Forrestal also sat on the board of General Aniline and Film (GAF), a subsidiary of I. G. Farben with 91.5 percent ownership by the brother-in-law of Farben chairman Hermann Schmitz. Heading GAF was Rudolf Ilgner, who near the outbreak of war offered the U.S. Army Agfa film at a low price for photographing the Panama Canal and other defense installations. "Ilgner has a sense of humor," noted Charles Higham, the *New York Times* writer who traced the Nazi-American money plot in his 1983 book *Trading with the Enemy.* "He gave the American government copies of the movies and still photographs and kept the originals, which were shipped via the Hamburg-Amerika steamship line [partly owned by Prescott Bush]. The president of this company was Julius P. Meyer, head of the Board of Trade for German-American Commerce, whose chairman was—Rudolf Ilgner." Forrestal became secretary of defense in July 1947—the time of the Roswell incident—but resigned in March 1949, a month before he reportedly committed suicide at Bethesda Naval Hospital. He claimed he was being followed by Zionist agents. His MJ-12 position was permanently filled by General Walter B. Smith.

- ◆ *General Walter Bedell Smith,* who had been Eisenhower's chief of staff and former U.S. ambassador to Moscow, replacing Averell Harriman. In 1950, Smith replaced Admiral Hillenkoetter as Director of Central Intelligence. Most intriguing was Smith's close relationship as friend and business partner with Prince Bernhard of the Netherlands, the former SS officer who, with Smith's help, founded the secretive Bilderberg Group. Before leaving for England prior to hostilities, the German-born Bernhard was employed in I. G. Farben's Intelligence Department, NM7.

- ◆ *General Nathan F. Twining,* commander of the Air Material Command based at Wright-Patterson, who was already heavily involved in the UFO issue by the time of MJ-12. He had canceled a scheduled trip on July 8, 1947, "due to a very important and sudden matter." This was the day the Roswell Air Base press release regarding the recovery of a flying saucer was issued. UFO researcher William Moore claimed that Twining actually made a two-day trip to New Mexico. On September 23, 1945, just as the air force became a separate service, Twining sent a letter to the chief of staff of the Army Air Force, Brigadier General George Schulgen, who had requested information on "flying discs." In a letter stamped SECRET, Twining began by stating without equivocation, "The phenomenon reported is something real and not visionary or fictitious." He recommended that a permanent group be established to study UFOs.

◆ *General Hoyt S. Vandenberg,* a West Point graduate and military man, who served as U.S. Air Force chief of staff and Director of Central Intelligence. As a named MJ-12 member, Vanderberg did not appear to have solid Wall Street connections. However, he was a close relative to the powerful U.S. senator Arthur Vandenberg, who served as president pro tempore of the Senate, third in line of succession to the presidency, and chaired the Senate Committee on Foreign Relations. Senator Vandenberg also participated in the creation of the United Nations. In January 1945 the senator made headlines by announcing his conversion from isolationism to internationalism. As such, he orchestrated bipartisan support for the Truman Doctrine and the Marshall Plan. The "Vandenberg Resolution," passed by the Senate in 1948, paved the way for mutual Allied security through the creation of NATO. In the early 1950s, it was General Vandenberg who ordered the destruction of the original Project Sign Air Force report stating that UFOs were real. Many UFO researchers believe Vandenberg's role was to maintain security for MJ-12.

◆ *Dr. Detlev Bronk,* a physiologist and biophysicist with international credentials, who chaired the National Research Council and was a member of the medical advisory board of the Atomic Energy Commission. From 1953 until 1968, he was president of the Rockefeller Institute for Medical Research, during which time he was given a $600,000 mansion. Bronk maintained a long correspondence with Vannevar Bush and also was on the Scientific Advisory Committee of the Brookhaven National Laboratory along with Dr. Edward Condon, who later debunked UFOs in a major UFO study for the Air Force.

◆ ***Dr. Jerome Hunsaker,*** an aircraft designer, who chaired the departments of mechanical and aeronautical engineering at the Massachusetts Institute of Technology, and the National Advisory Committee for Aeronautics. In 1933, the year Hitler came to power, Hunsaker became vice president of the Goodyear-Zeppelin Corporation, which manufactured airships that offered passenger flights to various countries, including Germany, Brazil, and the USA. It should be noted that less than three months after the 1984 death of Dr. Hunsacker, the last survivor of those named in the MJ-12 documents, the disputed documents suddenly arrived at the home of a UFO researcher. Many feel Hunsacker's death may have signaled to someone in the official world that it was now permissible to leak the MJ-12 Eisenhower briefing document.

◆ ***Sidney W. Souers,*** a retired rear admiral who in 1946 became the first Director of Central Intelligence, appointed by President Truman. He was executive secretary to the National Security Council in 1947 and remained a special consultant on security matters for a time after leaving that post. Souers also had a lifelong connection to American corporate business. Between 1920 and his death in 1973, Souers held executive positions in the Mortgage and Securities Company of New Orleans, First Joint Stock Land Bank, the Canal Bank and Trust Company of New Orleans, the Aviation Company, and the General American Life Insurance Company.

♦ ***Gordon Gray,*** an heir to the R. J. Reynolds Tobacco Company fortune, who was assistant secretary of the army in 1947, became secretary of the army in 1949, and a year later was named a special assistant on national security affairs to President Truman. After that, Gray was named director of the government's Psychological Strategy Board (PSB), established in 1951 to undertake disinformation and psychological warfare against enemies. During his stint on the PSB, Gray's chief consultant was Henry Kissinger, who was also a paid consultant to the Rockefellers. According to one source, Gray directed a psychological strategy study of UFOs, consulted by CIA director Walter B. Smith. Gray was also a member of the Council on Foreign Relations from the Truman through the Ford administrations. He also was chairman of the board for the communications companies Piedmont Publishing Company, Triangle Broadcasting Company, and Summit Communications.

♦ ***Dr. Donald Menzel,*** a director of the Harvard College Observatory, a respected astronomer who led a double life. He became a widely known debunker of UFOs after writing three books in which he explained away most reports and dismissed others, saying, "All non-explained sightings are from poor observers." However, physicist Stanton Friedman, after studying Menzel's unpublished biography and interviewing his widow, discovered that Menzel had been a covert consultant for both the CIA and the NSA with a top-secret ultra security clearance. This was verified in a letter Menzel wrote to President John F. Kennedy, in which he mentioned his intelligence work stating, "I have been associated with this activity for almost thirty years and probably have the longest continuous record of association with them." Menzel also worked closely with the State Department, especially on Latin American affairs. Just before the outbreak of World War II, Menzel unsuccessfully tried to interest the Rockefeller Foundation and Howard Hughes in funding a high-altitude observatory at Boulder, Colorado. In Menzel we find a man who, while publicly known simply as a notable astronomer, had intriguing and high-level intelligence connections.

◆ *General Robert M. Montegue,* a military man with no known corporate links, who nevertheless was the base commander of Fort Bliss near El Paso, Texas, in 1947, during the time the Paperclip scientists worked there. He also served as director of the Anti-aircraft and Guided Missile Branch of the army's Artillery School as well as commanding general of the Sandia Atomic Energy Commission facility in Albuquerque, New Mexico, from July 1947 to February 1951. His responsibilities included security at the White Sands Proving Ground. Montegue was at the center of the controversies concerning both the Roswell crash and the Paperclip scientists.

◆ *Dr. Lloyd V. Berkner,* who worked under Vannevar Bush as executive secretary of the Joint Research and Development Board in 1946 and headed a study that resulted in the creation of the Weapons Systems Evaluation Group. Berkner was also a member of the 1952 CIA-sponsored panel headed by Dr. H. P. Robertson, which deflected public attention away from UFOs by concluding that they did not constitute any direct threat to national security. Berkner was also president of Associated Universities, Incorporated (AUI), established in 1946 to "acquire, plan, construct and operate laboratories and other facilities that would unite the resources of universities, other research organizations, and the federal government." Funding for the AUI came from such luminary institutions as Cornell, Harvard, Johns Hopkins, MIT, and Yale. One of the institutions closely connected to the AUI is Brookhaven National Laboratory on Long Island, long-rumored to be involved with both defense weaponry and UFOs.

These distinguished men appeared to have two things in common— they were all connected to the highest levels of the national security as well as American corporate business. They were also all dead at the time the MJ-12 papers surfaced, thus unable to answer any questions about their role, if any, in such a group.

The agenda of this control group may have been best expressed by Senator Lyndon B. Johnson who, speaking to the Senate Democratic Caucus on January 7, 1958, stated, "Control of space means control of the world. . . . From space, the masters of infinity would have the power to control the Earth's weather, to cause drought and flood, to change the tides and raise the levels of the sea, to divert the Gulf Stream and change the climate to frigid. . . . There is something more powerful than the ultimate weapon. That is the ultimate position—the position of total control over the Earth that lies somewhere in outer space. . . . And if there is an ultimate position, then our national goal and the goal of all free men must be to win and hold that position." Johnson, who in 1954 became the youngest Senate majority leader in U.S. history, was in a position to serve those in both the military and the corporations. In 2007, President George W. Bush echoed Johnson's remarks by calling for new space missions and the weaponization of space.

ADDING TO JOHNSON'S puzzling statement about "the masters of infinity" are facts indicating an astounding connection between the well-documented occultism of the Nazis, the NASA space program, and the Soviet space program.

Richard C. Hoagland, a former science adviser to Walter Cronkite and CBS News during the Apollo program, astounded conspiracy researchers in the 1990s with his assertion that the time and date of many NASA space launches, including the *Apollo* moon missions, were set to coincide with astrological alignments of the stars and planets. In 1992, Hoagland briefed UN officials on the mathematical and geometric linkage connecting the siting of "Cydonia" on Mars, the location of the Martian "Pyramids" and "Face," with the Egyptian location of the pyramids and sphinx on Earth.

"This remarkable new evidence that 'all is not as we have thought regarding NASA' is distinctly different from the official NASA imagery that [I have] been analyzing for almost fifteen years," reported Hoagland. "This new evidence is of a 'pattern' [shown by] an official, undeniable log of NASA mission planning, mission priorities, and space agency decisions extending back to when the agency was officially formed by Act of Congress on July 29, 1958. This log has been carefully compiled from recorded network mission broadcasts from, among others, 'my' old network—CBS; officially published NASA mission time lines; and documented testimonials of former NASA scientists."

According to Hoagland, these cross-correlated public records now provide firm evidence of an astonishing, official link between NASA's supposedly strictly "scientific" missions and millennia-old occult beliefs. In fact, the original official NASA Apollo Lunar Program logo of the 1960s clearly depicted the "belt" in the constellation of Orion, long thought to represent Osiris, a central figure in Egyptian celestial mythology. "Curiously, immediately after the *Apollo 13* 'accident,' NASA quietly changed this official Apollo program logo—adding random stars to the existing constellation, thereby cleverly obscuring its direct derivation from Orion," noted Hoagland.

He concluded that the extraordinarily complex and expensive mission-planning for the entire Apollo Lunar Program, far from being merely "represented" by this "interesting" Egyptian mythological connection, was in fact completely controlled by, and designed around, this crucial Orion symbolism. In other words, someone with enough authority to set the launch date and time for an Apollo space mission, as well as many others, was guided by the astrological alignment of the stars and planets rather than an objective scientific basis. This occult aspect has been kept carefully hidden not only from the American taxpayers, who paid for these missions, but from the vast majority of NASA personnel as well.

"[I]magine the astonishment that you would feel if you learned that *Apollo 11*'s historic lunar touchdown . . . took place at the one location on the entire lunar surface—Tranquillity—and within minutes out of an entire solar year (8:17 P.M. GMT, July 20, 1969), where and when Sirius, the brightest star in the sky, and the central stellar figure, Isis, in the Egyptian triumvirate of Isis, Osiris, and Horus, could have been seen hovering above the airless Eastern horizon—precisely at 19.5 degrees elevation!" said Hoagland. Yet, this is exactly what happened.

Mary Ann Weaver, a former Boeing engineer and computer professional, was intrigued by the data produced by Hoagland and his associate Michael Bara. As a former researcher with the antenna division of Boeing, she was experienced in 3-D computer modeling, computational analysis, and developing equations and analytic methods for problem-solving. Weaver set out to confirm or deny the findings of Hoagland and Bara.

After a careful study of their data, Weaver concluded that the star alignments for the mission activities and launches did not happen by accident. "[T]hey must happen by design," she stated. "To try and explain them via random processes results in odds of *billions* to one. I would not bet on the 'random' side of these kind of odds."

"The significance of these findings is [that] I have shown there to be a pattern throughout eighty-two launches that were part of the Apollo preparation phase—and Apollo itself. Additionally, I have shown that the Apollo missions follow this same pattern on a day-to-day, mission-activities level, which is even more improbable because of its consistency with the launch data. Furthermore, it is improbable that the frequency of these stellar alignments are [*sic*] tied to weather or lighting conditions, because of the fact that they occur for a variety of mission events, even those that do not require specific lighting or weather conditions," she concluded.

Michael Bara pointed to a disturbing similarity with Russian space flights. He noticed that the launch of the first module for the new International Space Station was launched from the Baikonur Cosmodrome located in Kazakhstan, now an independent nation that borders Russia and the Caspian Sea, and was apparently designed to coincide with a number of the significant celestial alignments already found in NASA's long-established ritual pattern. This would indicate some connection between the former Soviet space program and NASA, perhaps through the Nazi scientists working in both.

After studying the August 1998 mission, Bara noted, "[The module] *Zarya,* which translates into 'sunrise' or 'rising sun' in English, was launched from pad 333 at its precisely scheduled time despite Russian requests to have the launch delayed. NASA, citing a number of minor technical considerations, refused the Russian request and the launch went off as originally scheduled and was witnessed by NASA administrator Dan Goldin. Considering that the [International Space Station] program was already a year behind schedule, another minor delay would not, despite NASA protestations to the contrary, have led to a significant problem. Only when you consider the symbolic significance of the moment does this steadfast insistence make sense." It is not necessary for one to believe in astrology. The point is that someone high enough in power in the U.S. government—and, apparently, in Russia—to be able to order the dates and times of space missions does believe in such things. Is this evidence of Nazi occultism in our space program?

President John F. Kennedy may have been aware of a parallel space program and decided to make it public policy. On November 12, 1963, ten days before his assassination, he instructed NASA administrator James Webb to develop a program of "joint space and lunar exploration" with the Soviet Union. This proposal, which may startle Americans today, was verified by Sergei Khruschev, the eldest son of the former Russian premier, in 1997. The importance of Kennedy's step toward reconciliation with the Soviet Union and his control over NASA will become apparent in the upcoming chapter, "Kennedy and the Nazis."

Who would have wanted to stop joint U.S-Soviet space missions that might have ended the Cold War in the early 1960s? And who has been orchestrating the launches of space missions in both the United States and Russia with an eye toward occult astrological alignments? Does this mean that someone with the power to set space mission launches in both nations truly believes in the power of the position of the stars? And whom did Johnson mean by the "masters of infinity"?

Strong evidence suggests they may well be the subject of this book—those global National Socialists and their minions, who have a goal of controlling the entire world. Nicholas "Nick" Rockefeller, a participant in the World Economic Forum, member of the Council on Foreign Relations and the International Institute for Strategic Studies, may have revealed the overall globalist agenda when he said, "The end goal is to get everybody chipped, to control the whole society, to have the bankers and the elite people control the world."

But this control goes far beyond military and space hardware. In modern warfare, there is also the struggle for control over the hearts and minds of whole populations, whether by psychological or chemical means.

CHAPTER 8

NAZI MIND CONTROL

NAZI INVESTIGATIONS INTO EXOTIC SCIENCES DID NOT END WITH flight and weapons technology. Since prior to the twentieth century, Germans had delved into psychology and psychiatry with an eye toward their application to warfare, even to the extent of exploring occult practices. As well documented in a number of books, articles, and videos, there was a very definite and underlying occult aspect to National Socialism. As elucidated in *Rule by Secrecy, Unholy Alliance, The Occult and the Third Reich, The Spear of Destiny,* and other works, World War II was largely the result of infighting between secret occult societies composed of wealthy businessmen on both sides of the Atlantic. Eventually the tensions between these groups provoked open warfare that consumed the entire world.

Sir Winston Churchill "was insistent that the occultism of the Nazi Party should not under any circumstances be revealed to the general public," stated author Trevor Ravenscroft, who wrote that he worked closely with Dr. Walter Johannes Stein, a confidential adviser to Churchill. "The failure of the Nuremberg trials to identify the nature of the evil at work behind the outer facade of National Socialism convinced him that another three decades must pass before a large enough readership would be present to comprehend the initiation rites and black-magic practices of the inner core of Nazi leadership."

This remarkable statement was corroborated by Airey Neave, one of the Nuremberg prosecutors, who said the occult aspect of the Nazis was ruled inadmissible because the tribunal thought that such beliefs, so contrary to Western public rationalism, might allow Nazi leaders to go free by pleading insanity.

Even Hitler acknowledged that Nazi ideology ventured into a spiritual realm, when he stated, "Anyone who interprets National Socialism merely as a political movement knows almost nothing about it. It is more than religion; it is the determination to create a new man."

To attempt this creation, the Nazis turned to occultists such as Baron Rudolf Freiherr von Sebottendorff, Jorg Lanz von Liebenfels, Guido von List, Dietrich Eckart, and Karl Haushofer, all of whom had immersed themselves in the philosophies of the Theosophical Society. Theosophy, derived from the Greek *theos* (god) and *sophia* (wisdom), was an attempt to blend Christianity with Cabalistic and Eastern mysticism. One tenet of theosophy was that "Great Masters," sometimes called the "Great White Brotherhood," are secretly directing humankind's evolution.

"The rationale behind many later Nazi projects can be traced back . . . to ideas first popularized by [Theosophical Society founder Helen] Blavatsky," wrote Peter Levenda, who detailed connections with other European secret organizations, such as the Ordo Templi Orientis, or Oriental Templars (OTO), Dr. Rudolf Steiner's Anthroposophical Society, and the Order of the Golden Dawn.

Such groups were "concerned with raising their consciousness by means of rituals to an awareness of evil and non-human Intelligences in the Universe and with achieving a means of communication with these Intelligences. And the Master-Adept of this circle was Dietrich Eckart [the man Hitler called "spiritual founder of National Socialism"]," noted Ravenscroft. Hitler wrote of his own occult experiences as a soldier in World War I: "I often go on bitter nights, to Wotan's oak in the quiet glade, with dark powers to weave a union."

As previously mentioned, the deeply occulted *Germanenorden* contrived the Thule Society as a cover organization. "The original conception of the modern Thulists was extremely crude and naive," Ravenscroft explained. "The more sophisticated versions of the legend of Thule only gradually developed in the hands of Dietrich Eckart and General Karl Haushofer, and were later refined and extended under the direction of Reichsfuehrer SS Heinrich Himmler, who terrorized a large section of the German academic world into lending a professional hand at perpetuating the myth of German racial superiority."

In light of the occultism apparent in modern space missions mentioned earlier, recall that General Karl Haushofer, who used astrology to provoke the strange flight of Rudolf Hess to England, was a member of the mysterious Vril, an occult society that practiced telepathy and telekinesis.

It is surmised that it was perhaps through such occult practices that psychic contact was made with nonhuman intelligences, thus providing the Nazis with the concepts that led to their futuristic technology. Nazi occult researcher Nicholas Goodrick-Clarke, in his 1992 book *The Occult Roots of Nazism: Secret Aryan Cults and Their Influence on Nazi Ideology,* wrote that the power that motivated the occultists surrounding Hitler and Himmler "is characterized either as a discarnate entity (e.g., 'black forces,' 'invisible hierarchies,' 'unknown superiors'), or as a magical elite in a remote age or distant location, with which the Nazis were in contact."

Although rumors have floated about for years that the Nazis captured a UFO, no credible evidence has ever been produced. Some of those who have studied this issue have come to suspect that any such knowledge of nonhuman technology may instead have come through Nazi occultists using psychic means similar to remote viewing, a psychic ability studied, taught, and used operationally by the U.S. Army, the CIA, and the National Security Agency beginning in the early 1970s. It was Soviet interest in psychic experimentation that led to experiments in the United States and the eventual creation of a unit of psychic spies within the U.S. Army. Remote viewing, known in parapsychological terms as clairvoyance, is the ability to discern persons, places, and things at a distance by means other than the normal five senses.

According to former U.S. military intelligence agent Lyn Buchanan, who at one time trained the U.S. Army's remote viewers, the Nazis formed a unit of psychics and called it Doktor Gruenbaum. This name was for the psychic project, not a person, although apparently a German psychic who assumed the name Gruenbaum may have lived in the United States after the war. The name Gruenbaum, or green tree, apparently was a reference to the green-tree symbol in the Cabala, which relates to the "tree of knowledge" in the Garden of Eden.

Buchanan reported: "When Adolf Hitler lost the war and the victors began to divide the spoils, the U.S. and other countries dividing up the nuclear and rocket scientists had little or no concern for the mystical research. Russia, however, did, and so they took the scientists from the 'Doktor Gruenbaum' project back to Russia." Interestingly enough, it was reported that the Nazi Doktor Gruenbaum unit was connected to a broader program called Majik. This name has prompted comparisons to America's original UFO-secrecy group, Majic Twelve.

Could it be that psychic viewing by the Nazi Doktor Gruenbaum unit tipped off Hitler to the possible pending secret attack on Europe from the Soviet Union, resulting in his preemptive Barbarossa attack? Since this ability is intuitive and not always crystal-clear, German viewers may have perceived the buildup of Soviet forces but been unable to foresee the end result of Barbarossa—the eventual defeat of Germany.

Whether or not the Nazis used psychic mental abilities to acquire exotic technology, it is beyond question that the study of the human mind began in earnest in Germany, with far-reaching consequences.

Behind the horrors of the Nazi regime rested a foundation of European study of the human mind. Justification of euthanasia and extermination programs was provided by some of Germany's most learned men. "Hitler's philosophy and his concept of man in general was shaped to a decisive degree by psychiatry. . . . an influential cluster of psychiatrists and their frightening theories and methods collectively form the missing piece of the puzzle of Hitler, the Third Reich, the atrocities and their dreadful legacy. It is the overlooked yet utterly central piece of the puzzle," wrote Dr. Thomas Roeder and his coauthors Volker Kubillus and Anthony Burwell in their 1995 book *Psychiatrists—the Men Behind Hitler.*

Psychiatry in general can trace its origins to five prominent European scientists in the 1800s—Thomas R. Malthus, the British economist who viewed war, disease, and starvation as beneficial survival mechanisms against unchecked population growth; Charles Darwin, the naturalist whose 1859 book *The Origin of Species* convinced whole generations that survival of the fittest is a law of nature; Friedrich Wilhelm Nietzsche, philosopher and close friend to Hitler's paragon, composer Richard Wagner, who declared "God is dead" and advocated the superiority of the *Übermensch,* or superman, over lesser races, virtues, and values; Joseph Arthur, Comte de Gobineau, a French diplomat who championed the concept of an Aryan aristocracy and its preeminence over others; and Houston Stewart Chamberlain, the British-born philosopher who moved to Germany, married the daughter of Richard Wagner, and also promoted an "Aryan world philosophy."

"Darwin entangled his theory of natural selection with the assumptions of Malthus's population theory. The result was a strange, incongruous marriage of Darwin's observations of the animal world with Malthus's emotional assumptions about the uncontrollable population growth and social solutions to preserve the British aristocracy," noted Roeder, Kubillus, and Burwell. They added that social Darwinism was perhaps the Nazis' most central theoretical foundation.

"Social Darwinism had a profound and long-lasting effect on the mind of Adolf Hitler," agreed Professor Snyder. "He expressed its ideas in simplified form in the pages of *Mein Kampf* . . . and he made it the theme of most of his major speeches."

From the viewpoint that certain people are more evolved and thus more competent to judge others came the profession of psychiatry. The term itself came from the Greek *psyche,* or soul, and *iatros,* or doctor. However, these doctors of the soul quickly became preoccupied with more material matters—the physical brain and how to manipulate or destroy it.

As the field of psychiatry grew, so did its definitions. In 1871, "The Psychical Degeneration of the French People" was published, a paper that left the impression that simply being French constituted a mental illness. "One of psychiatry's leading figures, Richard von Krafft-Ebing, added to his list of varieties of mental disorders 'political and reformatory insanity'—meaning any inclination to form a different opinion from that of the masses," the trio of researchers stated.

At the time of World War I, the attempt to bring respectability to the emerging psychiatric profession resulted in a certain bonding between psychiatry and the aristocratic German government. The German military was particularly impressed with the "therapy" of Fritz Kaufmann, because it referred to "war neurosis" or "shell shock." Based on the idea that antiwar behavior was a chemo-biological dysfunction, the "Kaufmann therapy" consisted of applied electrical shocks, actually more of a disciplinary measure than true medical therapy. The army was delighted that recalcitrant troops, following electroshock, quickly agreed to return to service.

Psychiatry continued to grow in power even as its agenda continued to widen. Psychiatrist P. J. Moebius, who had lectured on the "psychological feeble-mindedness of the woman," pronounced, "The psychiatrist should be the judge about mental health, because only he knows what ill means."

Such arrogance of belief soon led to the creation of various psychiatric organizations, such as the Gesellschaft fur Rassenhygiene, or Society for Racial Hygiene, which only served to further the ambitions of the profession. Since no one has yet found a significant and general "cure" for insanity, psychiatrists turned to the dubious concept of prevention. This came to be known as "mental hygiene," a bland term for the prevention of mental illness, whatever form that might take. In the Germany of the 1930s, the rush to isolate and "cure" mental defectives quickly was interpreted to include malcontents and dissidents opposed to the Nazi regime. This open-ended concept resulted in the Nazi Sterilization Act, which went into effect in July 1933, just six months after Hitler's ascension to power.

One of the leading and articulate authorities behind the rationale for this act was Dr. Ernst Rudin, a psychiatrist who in 1930 had traveled to Washington, D.C., to present a paper called "The Importance of Eugenics and Genetics in Mental Hygiene." It was well received by those present as many Americans, especially among the globalists, had come to embrace the racist and elitist views of the German philosophers.

Nazi interest in science and psychological warfare was paralleled by their concern with eugenics, the scientific study of selective breeding to improve the human population. The term "eugenics" was coined in the late 1800s by Francis Galton, a British psychologist and half-cousin of

Darwin's, who wanted to extend the theory of natural selection into deliberate social engineering.

Race and genetics were always a top concern to ranking Nazis. We find the same concern exhibited by America's ruling families.

BY THE TIME of his death in 1937, John D. Rockefeller and his only son, John D. Rockefeller Jr., had not only built up an amazing oil empire but had established such institutions as the University of Chicago (1889); the Rockefeller Institute for Medical Research (1901) later renamed Rockefeller University, in New York City; the General Education Board (1903); the Rockefeller Foundation (1913); and the Lincoln School (1917), where the Rockefeller siblings began their education. These Rockefeller-funded institutions ensured their early entry into the fields of medicine, pharmaceuticals, and education.

The Rockefellers were also interested in the eugenics movement, a program of scientifically applied genetic selection to maintain and improve their ideal for human characteristics, which included birth and population control. In 1910, the Eugenics Records Office was established and endowed by grants from Mrs. Edward H. Harriman and John D. Rockefeller. It seems the wealthy elite of America were as concerned with bloodlines as the Nazis.

Another American supporter of German psychiatry was James Loeb, son and by 1894 a business partner to Solomon Loeb, founder of the prominent Kuhn, Loeb and Company, the bankers and backers of railroad tycoon Edward H. Harriman.

In 1917, thanks to financial support from James Loeb, Dr. Emil Kraepelin, a professor at the University of Munich, was able to found the Deutsche Forschungsanstalt fur Psychiatrie, or the German Research Institute for Psychiatry. "Kraepelin was certainly a conservative nationalist," stated Roeder, Kubillus, and Burwell. "But he also was a pioneer of psychiatric atrocities such as racial hygiene and sterilization, who, except perhaps for Rudin, had no equal in his advocacy for a legal foundation for the policies of Nazi extermination."

By 1924, Kraepelin's research institute, rescued from bankruptcy by

Loeb's money, had become incorporated into the prestigious Kaiser Wilhelm Institute, and the growing Nazi leadership was paying attention to its science.

Initially, they went for the most defenseless of the German population—the children. On July 14, 1933, only six months after Hitler was named chancellor of the Reich, the Law for the Prevention of Genetically Diseased Children was passed. A leading proponent for this legislation was Ernst Rudin, by then director of the Kaiser Wilhelm Institute.

Leading the movement to eliminate "mental defectives" from the German population were lawyer Karl Binding and the psychiatrist Alfred Hoche, who popularized the chilling phrase *"lebensunwertes Leben"*—or "life unworthy of life"—in a 1920 tract titled *Die Freigabe der Vernichtung lebensunwerten Lebens,* or "Lifting Controls on the Destruction of Life Unworthy of Life." "This text, arguably more than any other, made available to the Nazi regime an 'ethical' rationale for 'euthanasia.' Although in the early days of the regime the public discussion would focus on the prevention of offspring with hereditary disease, hence sterilization, the destruction of life unworthy of life would spread as an unspoken principle," wrote John Cornwell in his 2003 book *Hitler's Scientists: Science, War, and the Devil's Pact.*

The chosen means of prevention, enforced sterilization, was administered by special "hereditary health courts," made up of two doctors—usually psychiatrists—and one civil official, usually a judge close to the Nazi Party, who acted as chairman. The Nazi euthanasia program was not carried out in the open but instead by secret decrees, as Hitler steadfastly refused to seek a legal ruling, knowing that such a program was illegal under existing laws.

It is estimated that more than 400,000 people were sterilized as "life unworthy of life" between 1934 and 1945. "The projected total of 410,000 was considered only preliminary, drawn mostly from people already in institutions; it was assumed that much greater numbers of people would eventually be identified and sterilized," stated Robert Jay Lifton in *The Nazi Doctors: Medical Killing and the Psychology of Genocide.*

Lifton went on: "Not surprisingly, Fritz Lenz [whose eugenics work was parroted by Hitler] carried the concept farthest in suggesting the advisability of sterilizing people with only slight signs of mental disease, though he recognized that a radical application of this principle would lead to the sterilization of 20 percent of the total German population—something on the order of 20 million people!"

One revealing anecdote involving this medical-driven policy was the release in 1933 of a mental patient who had been imprisoned as a violence-prone hardened criminal, a "dangerous lunatic," according to a local official. His psychiatrist, Dr. Werner Heyde, however, pronounced Theodor Eicke fit for discharge, and Eicke was soon named the first commandant of Dachau concentration camp. In 1934, Eicker was promoted to inspector general and chief of all concentration camps. Eicke, whose influence and spirit within the SS was "second only to that of Himmler," died in 1943, when his plane was shot down behind Russian lines.

Dr. Heyde, whose recommendation released Eicke from prison, went on to become the medical director of the infamous Nazi T4 euthanasia program begun in 1940. (The designation T4 referred to the address of the stone building from which they operated: Tiergartenstrasse 4.) The approved method of killing ordered by Hitler, acting on the advice of Dr. Heyde, was the use of carbon monoxide. In a prototype of the Nazi death camps, a fake shower room, complete with benches, was constructed and used to gas the first victims.

Great pains were taken to employ what Robert Jay Lifton called "bureaucratic mystification," a snarl of red tape and bureaucracy so convoluted that the victims, their families, and even those working within the system did not realize the full extent of the euthanasia program.

Interestingly enough, in 1941 Hitler ordered the official T4 euthanasia program halted for no recorded reason. Some have argued that Hitler may have developed pangs of conscience, while others believe that as more and more of the German population became aware of the killing, cries of objection could have caused Hitler a political problem. Authors Roeder, Kubillus, and Burwell argued that the program was stopped simply because it had achieved its original quota of victims. "The original campaign apparently had accomplished its purpose and was shut down. But that did not mean that a new euthanasia program wasn't waiting to begin," they wrote.

Starting in April 1941, the now-experienced doctors of T4 began visiting the Nazi concentration camps and soon were practicing their newest euthanasia program in earnest—the *Endlosung,* or final solution. "The extermination of the Jews was an exact replica of T4's earlier euthanasia program," stated Roeder, Kubillus, and Burwell.

DURING THE WAR, as today, both medical doctors and psychiatrists were quite vulnerable to peer pressure as well as the goodwill of the state, which provided the credentials and certificates necessary for their practice. So it seemed natural that mind manipulation through psychiatry and psychology was soon joined by a companion therapy—drugs.

German psychiatrists were merely following the lead of medical doctors, who in the twentieth century increasingly moved away from the tradition of homeopathy, which involved using minimal doses of drugs as therapy, to allopathy, the straightforward treatment of disease with drugs. What therapy could not accomplish through psychological means might be accomplished through drugs. This trend to increase the use of prescription drugs set the stage for the rise of the giant pharmaceutical corporations during the twentieth century.

During the time of the Opium Wars of the late 1800s, any type of drug was used for profit. For example, in 1898, the German Bayer Company began mass production of heroin (diacetylmorphine) and used that name to market the new remedy. Bayer described heroin as a nonaddictive panacea for adult ailments and infant respiratory diseases. In the late 1800s, Bayer also promoted cocaine, which until the 1920s was an ingredient in the soft drink Coca-Cola.

But as more easily produced petrochemical drugs made their debut, they prompted the attention of the major global corporations.

The Rockefeller family's interest in pharmaceuticals reaches back to the days of John D.'s father, William "Big Bill" Rockefeller, who sold "Rock Oil," a diuretic medicine that guaranteed "All Cases of Cancer Cured Unless They Are Too Far Gone." "William Rockefeller's original miracle oil survived until quite recently as a concoction called Nujol, consisting principally of petroleum and peddled as a laxative," wrote Eustace

Mullins. "Nujol was manufactured by a subsidiary of Standard Oil of New Jersey, called Stanco, whose only other product, manufactured on the same premises, was the famous insecticide Flit."

Along with the previously documented business alliances between Standard Oil and I. G. Farben, Standard Oil vice president Frank Howard also served as chairman of the research committee at Sloan Kettering Institute, today known as the Memorial Sloan-Kettering Cancer Center, the New York City–based cancer center built in 1939 on land donated by John D. Rockefeller Jr. and financed by the Rockefeller family. According to the center's literature, it has "long been a leader in cancer surgery, chemotherapy, and radiotherapy. It was the first to develop services specifically dedicated to the psychiatric aspects of cancer, to the relief of cancer pain, and to genetic counseling."

Howard, in addition to maintaining relations between Nazi I. G. Farben and Standard Oil, represented Rockefeller interests with the firm of Rohm and Haas, still one of the world's largest suppliers of specialty chemicals. Current company literature states, "From maintaining the freshness in fruits and vegetables to purifying antibiotics, we help customers create products that enhance the way of life for people around the world. Today Rohm and Haas extends to the far corners of the earth, with sales in more than 100 countries."

According to Mullins, the American College of Surgeons maintained a monopolistic control of U.S. hospitals through its Hospital Survey Committee, with members Winthrop Aldrich and David McAlpine Pyle representing the Rockefellers.

In 1909, John D. Rockefeller Sr. extended his reach into the southern states by a $1 million donation to establish the Rockefeller Sanitary Commission, dedicated to eradicating hookworm disease. "[D]espite its philanthropic goals, the Rockefeller Sanitary Commission required financial contributions from each of the eleven southern states in which it operated, resulting in the creation of state departments of health in those states and opening up important new spheres of influence for their Drug Trust," wrote Mullins. The physician who served as director of the Rockefeller Sanitary Commission during World War I was Dr. Olin West, a primary figure in the creation of the Tennessee State Department of

Health. West went on to become a top executive for forty years at the American Medical Association.

The Rockefeller Institute for Medical Research is now Rockefeller University. Another deep penetration of America's education system was made in 1903, when John D. Rockefeller established the General Education Board (GEB). According to the Rockefeller Archives, "The [GEB] program included grants for endowment and general budgetary support of colleges and universities, support for special programs, fellowship and scholarship assistance to state school systems at all levels, and development of social and economic resources as a route to improved educational systems. Major colleges and universities across the U.S., as well as many small institutions in every state, received aid from the Board. The emphasis, however, was on the South and the education of blacks."

"Rockefeller's General Education Board has spent more than $100 million to gain control of the nation's medical schools and turn our physicians to physicians of the allopathic school, dedicated to surgery and the heavy use of drugs," commented Mullins, adding, "America became the greatest and most productive nation in the world because we had the healthiest citizens in the world. When the Rockefeller Syndicate began its takeover of our medical profession in 1910, our citizens went into a sharp decline. Today, we suffer from a host of debilitating ailments, both mental and physical, nearly all of which can be traced directly to the operations of the chemical and drug monopoly, and which pose the greatest threat to our continued existence as a nation."

Mullins pointed to Britain's Wellcome Trust as one of the world's largest medical research charities. It finances research into the health of both animals and humans. It also illustrates the intertwining connections of the globalists.

The knighted Sir Oliver S. Franks, described as "one of the founders of the postwar world," directed the Wellcome Trust as well as serving as British ambassador to the United States from 1948 to 1952. He was a director of the Rockefeller Foundation and its principal representative in England. According to Mullins, he was given a life peerage as Baron Franks of Headington, County of Oxford, in 1962 and was "a director of the Schroeder Bank, which handled Hitler's personal bank account, director

of the Rhodes Trust in charge of approving Rhodes scholarships [Bill
Clinton, among others], visiting professor at the [Rockefeller-endowed]
University of Chicago and chairman of Lloyd's Bank, one of England's
Big Five."

ANOTHER ASPECT OF Nazi medical science employed in America involves the toxic chemical sodium fluoride. The controversy over the fluoridation of municipal water supplies has raged since the 1950s and continues
today.

Aluminum oxide is extracted from clay and bauxite. Through a chemical called cryolite the material is converted into aluminum. A by-product
of this process is sodium fluoride, which for many years was used as a rat
poison. One recent dictionary defined fluoride as a "poisonous pale yellow
gaseous element of the halogen group." Sodium fluoride also acts as an
enzyme inhibitor and has been linked by several studies—such as a 1982
report from the University of Iowa—to Alzheimer's disease, a degenera-
tive and fatal neural disease named for the German doctor Alois Alzheimer. According to the Alzheimer Association, this brain-destroying
disease is the seventh leading cause of death in America today.

Although aluminum has been associated with Alzheimer's, such claims
have been disputed. Some have claimed that Alzheimer's disease is more
common in areas where the aluminum content in the water supply is
highest, but the method and results of these studies have been questioned.
But it is true that, in 1986, the Reagan administration's Environmental
Protection Agency raised the "safe" level of sodium fluoride in public water supplies from 2 parts per million gallons to 4 parts, even though one
part per million has been shown to impair neurological efficiency. In October 2007, despite "heated hearings" in 2003, the Los Angeles–based
Metropolitan Water District began fluoridating the drinking water of 18
million Southern Californians in six counties, including San Diego. According to a report by the Environmental Working Group (EWG), a
Washington-based, nonprofit organization whose mission is to "protect
public health and the environment," the plan to fluoridate water in South-

ern California "will put 14.5 percent of children under one year old, and 12.5 percent of children one to two years old, over the recommended fluoride exposure limits published by the National Academy of Sciences' Institute of Medicine and endorsed by the American Dental Association. In Los Angeles County alone, more than 40,000 children age two and under will exceed the safe dose."

Bill Walker, EWG's vice president for the West Coast, pointed to recent studies that call into question claims that fluoridation is safe. A March 2006 National Academy of Sciences/National Research Council report identified fluoride as a potent hormone disruptor that may affect normal thyroid function; the NAS/NRC report also cited concerns about the potential of fluoride to lower IQ, noting that the "consistency of study results appears significant enough to warrant additional research on the effects of fluoride on intelligence." That finding was echoed by a December 2006 study published in the prestigious, peer-reviewed journal *The Lancet,* which identified fluoride as an "emerging" neurotoxin; and a 2006 peer-reviewed study at Harvard strongly supported concerns that fluoridated water is linked to osteosarcoma, an often-fatal form of bone cancer, in boys. The Harvard study found a five-fold increase in bone cancer among teenage boys who drank fluoridated water from ages six through eight, compared to those drinking nonfluoridated water.

Brain studies reported by the Alzheimer's Society in England show that aluminum accumulates in nerve cells that are particularly vulnerable to Alzheimer's disease, although not all persons exposed to aluminum develop Alzheimer's. Although many studies on animals and on isolated cells have shown that aluminum has toxic effects on the nervous system, it is claimed that the doses of aluminum used were much higher than those occurring naturally in tissues. This obviously raises the question of how much fluoride/aluminum the public is ingesting from nonnatural sources, such as the fluoridation of drinking water.

The human brain contains 100 billion nerve cells (neurons) that form networks to manage our thinking, learning, and remembering. By the mid-1990s, studies were indicating a link between Alzheimer's and aluminum. But just as the debate over the link between cigarettes and cancer lasted decades, due to the obstructionist studies funded by the tobacco industry, the controversy over aluminum—and sodium fluoride—continues today.

Numerous Web sites and periodicals have carried the accusation that sodium fluoride was placed in the drinking water of Nazi concentration camps to keep inmates pacified and susceptible to external control. Such use of fluoridation by the Nazis to dull the senses of prisoners was described by Charles Eliot Perkins, a prominent U.S. industrial chemist whom the U.S. government sent to help reconstruct the I. G. Farben chemical plants in Germany at the end of the war. In a 1954 letter to the Lee Foundation for Nutritional Research, Perkins stated: "The German chemists worked out a very ingenious and far-reaching plan of mass control that was submitted to and adopted by the German General Staff. This plan was to control the population of any given area through mass medication of drinking water supplies. . . . In this scheme of mass control, 'sodium fluoride' occupied a prominent place. . . . However, and I want to make this very definite and very positive, the real reason behind water fluoridation is not to benefit children's teeth. . . . The real purpose behind water fluoridation is to reduce the resistance of the masses to domination and control and loss of liberty. . . . Repeated doses of infinitesimal amounts of fluorine [*sic*] will in time gradually reduce the individual's power to resist domination by slowly poisoning and narcotizing this area of brain tissue, and make him submissive to the will of those who wish to govern him. . . . I was told of this entire scheme by a German chemist who was an official of the great Farben chemical industries and was prominent in the Nazi movement at the time. . . . I say this with all the earnestness and sincerity of a scientist who has spent nearly 20 years research [*sic*] into the chemistry, biochemistry, physiology, and pathology of 'fluorine [*sic*].' . . . Any person who drinks artificially fluoridated water for a period of one year or more will never again be the same person, mentally or physically."

A *Christian Science Monitor* survey in 1954 showed that seventy-nine of the eighty-one Nobel Prize winners in chemistry, medicine, and physiology declined to endorse water fluoridation. Yet, today, two-thirds of all municipal water and most bottled water in the United States contain sodium fluoride, which has long been used as a rat poison. Most people do not realize that fluoride is a key ingredient in Prozac and many other psychotropic drugs. Prozac, whose scientific name is fluoxetine, is 94 percent fluoride. More than 21 million prescriptions for fluoxetine were filled in the United States in 2006, making it one of the most prescribed antidepressants.

Every U.S. Public Health Service surgeon general from the 1950s to this day has supported the introduction of this poison into America's water supply, even though fluoride, this poisonous waste product of aluminum manufacture, accumulates in the human body and has been shown to affect tooth decay only in children under twelve years of age.

It is quite ironic that Prozac, which is 94 percent fluoride and given to hyperactive children, requires a prescription from a licensed physician while the same substance can be placed in our drinking water by dealers who have no medical training, no license to dispense medications, and no idea to whom they are administering this corrosive, toxic, and impairing substance.

In 1946, Oscar Ewing, a Wall Street attorney and former counsel to the Aluminum Company of America (now known by the acronym Alcoa), was appointed by President Harry S. Truman to head the Federal Security Agency, which placed Ewing in charge of not only the U.S. Public Health Service but also the Social Security Administration and the Office of Education.

Congressman A. L. Miller, a physician turned Republican politician, said that Ewing had been placed in his position and highly paid by the Rockefeller syndicate to promote fluoridation. Miller stated, "The chief supporter of the fluoridation of water is the U.S. Public Health Service. This is part of Mr. Ewing's Federal Security Agency. Mr. Ewing is one of the highly paid lawyers for the Aluminum Company of America."

Other opponents were less kind. Leaflets handed out in New York City cried, "Rockefeller agents order fluoride-(rat-)poisoning of nation's water. Water fluoridation is the most important aspect of the Cold War that is being waged on us—chemically—from within, by the Rockefeller-Soviet axis. It serves to blunt the intelligence of a people in a manner that no other dope can. Also, it is genocidal in two manners: it causes chemical castration and it causes cancer, thus killing off older folks. . . . This committee

[Ewing's study of fluoride] did no research or investigation on the poisonous effects of water fluoridation. They accepted the falsified data published by the U.S.P.H.S. [U.S. Public Health Service] on the order of boss Oscar Ewing, who had been 'rewarded' with $750,000 by fluoride waste producer, Aluminum Co. He then developed the 'public spirit' that impelled him to take a $17,500 job as Federal Security Administrator. He immediately demanded of Congress an appropriation of $2,500,000 for promotion of fluorides by his U.S.P.H.S."

It is interesting to note that West Germany banned the use of fluorides in 1971, a time when it was still heavily occupied by Allied soldiers. "Apparently they could no longer silence the German scientists who had proved that fluoridation is a deadly threat to the population," wrote Mullins. "Sweden followed West Germany in banning fluoridation, and the Netherlands officially banned it on June 22, 1973, by order of their highest court."

BUT THE SCIENTIFIC minds encouraged by globalist funding were not content with drugs to merely pacify a population. They wanted direct control. It should come as no surprise that the men behind the documented CIA mind-control projects—MKULTRA, ARTICHOKE, BLUEBIRD, MKDELTA, etc.—had received Nazi medical science passed along by Paperclip doctors and their protégés. The infusion of Nazi mind-control specialists within the fledgling CIA resulted in Project MKULTRA (pronounced M-K-ULTRA), a code name for mind-control research that continued until the late 1960s, when it was said to have been discontinued. Project MKULTRA was created in 1953 by CIA officer Richard Helms, a good friend to CIA psychiatrist Dr. Sidney Gottlieb. It was the brainchild of then CIA director Allen Dulles. Dulles reportedly was intrigued by reports of mind-control techniques allegedly conducted by Soviet, Chinese, and North Koreans on U.S. prisoners during the Korean War.

Published accounts show this project not only used drugs to manipulate a person's personality, but also electronic signals to alter brain functioning. According to a 1975 internal CIA document, "MKULTRA was a group of projects, most of which dealt with drug or counter-drug research and development. The Director of Central Intelligence (DCI) and the Deputy Director of Plans (DDP) were kept informed on the program via annual briefings by Chief Technical Services Division (C/TSD) or his deputy. Most of the research and development was externally contracted. . . . The objectives were behavioral control, behavior anomaly production, and countermeasures for opposition application of similar substances. Work was performed at U.S. industrial, academic, and governmental research facilities. Funding was often through cutout arrangements."

After discussing testing on "volunteer inmates" and the diminished role of the MKULTRA project as fears of Soviet drug use eased, the CIA officer that authored the report noted, "Over my stated objections the MKULTRA files were destroyed by order of the DCI (Mr. [Richard] Helms) shortly before his departure from office."

To study psychochemicals and the possibility of using them to achieve mind control, the CIA, along with military intelligence, launched a program code-named BLUEBIRD, later changed to ARTICHOKE.

THE CIA HAS even admitted that its drug testing on college campuses resulted in the "drug revolution" of the 1960s.

This amazing story began in 1943, when Swiss chemist Dr. Albert Hofmann, working for Sandoz Laboratories in Basel, accidentally absorbed through his fingertips a chemical derived from the cereal fungus ergot. He proceeded to experience a semiconscious delirium complete with kaleidoscopic colors and visions. As this was the twenty-fifth compound of lysergic acid diethylamide, synthetically produced by Sandoz, Hofmann named it LSD-25.

The editors of *Consumer Reports,* in their monumental 1972 book *Licit & Illicit Drugs,* noted: "Psychiatrists were naturally interested from the beginning in LSD effects. Many of them took the drug themselves, and gave it to staff members of mental hospitals, in the belief that its effects approximate a psychotic state and might thus lead to better understanding of their patients."

About the same time Dr. Hofmann was discovering LSD, General

William "Wild Bill" Donovan, a former J. P. Morgan Jr. operative and head of the Office of Strategic Services (OSS), which conducted irregular and unorthodox warfare, began searching for a drug that would loosen the tongues of captured spies and enemy soldiers. Donovan called together a group of psychiatrists, who tested numerous drugs, including alcohol, barbiturates, and even caffeine. Plant extracts such as peyote, scopolamine, and even marijuana were also tested.

In 1947, the old OSS was superseded by the newly created CIA, within which drug experimentation continued, though with mixed results. Liaisons were formed between academics in universities, police departments, criminology laboratories, doctors, psychiatrists, and even hypnotists. Secret CIA funding was provided and experiments were not limited to laboratory animals. Like the Nazis before them, they also used sometimes unsuspecting human guinea pigs. These experiments were carried out in collaboration with hundreds of known Nazi scientists, who had experimented with these same drugs on prisoners in concentration camps. These scientists were brought into the United States after the war to continue their work.

By the mid-1950s, the CIA had managed to secure a monopoly on LSD. At first, the agency personnel tested LSD only on themselves, but later decided they would slip LSD into each other's food or drinks without prior notice, to observe the effects. Such childish experimentation soon got out of hand. Nothing was done to stop this practice until rumors circulated that the annual CIA Christmas party punch was to be spiked with LSD.

By the end of the 1950s, CIA experimentation had grown, with funding running through such CIA fronts as the Geschickter Fund for Medical Research, the Society for the Study of Human Ecology, and the Josiah Macy Jr. Foundation. With the CIA funding masked by such foundations, drug experimentation reached down to university campuses and other public institutions across the United States.

Once again, the U.S. Public Health Service played a role. At the Public Health Service's Addiction Research Center in Lexington, Kentucky, drug addicts would be given morphine or heroin in exchange for participating in drug experiments there, including the ingestion of LSD.

One of the universities involved was Harvard, where Dr. Timothy Leary, along with Richard Alpert, later known as Ram Dass, conducted a series of experiments with LSD and psilocybin. Leary had come to Harvard after serving as director of clinical research and psychology at the Kaiser Foundation Hospital in Oakland, California.

The Kaiser Family Foundation, a "leader in health policy and communications," was named for Henry J. Kaiser, a wealthy industrialist and ship builder who in 1946 began Kaiser Aluminum. According to foundation literature, "Kaiser campaigns are based on a new model of public service programming pioneered by the Foundation—direct partnerships with major media companies and a comprehensive 'multiplatform' communications strategy that goes far beyond traditional 'PSAs' [Public Service Announcements]. Current partners in the U.S. include MTV, BET, Univision, Viacom/CBS, and Fox. Together, Kaiser's campaigns reach tens of millions of people annually, and have won multiple Emmy and Peabody awards in recent years." It is no wonder that the aluminum waste product fluoride has received such favorable media attention over the years.

Many stories circulated around Harvard concerning LSD parties and undergraduates selling LSD-laced sugar cubes on and off campus. Leary was fired from the university in 1963, officially because he missed a committee meeting. Alpert, too, was dismissed, reportedly for violating an agreement not to supply LSD to undergraduates. This was the first time in the twentieth century that Harvard faculty members had been fired. Both Leary and Alpert began writing articles chastising Harvard and extolling the virtues of drugs. Leary became immortalized with his slogan "Turn on, tune in, and drop out."

As the drug culture grew rapidly in the late 1960s, "it was widely observed that young people paid little or no attention to dire warnings against the hazards of marijuana-smoking, LSD-using, and other forms of drug use," noted the editors of *Consumer Reports*. "When the evidence of their own experience contradicts adult propaganda, they (like sensible adults) rely on their own experience—and tend to distrust in the future a source of information which they had found unreliable in the past."

After the major media began to report stridently on the campus drug scene, public interest grew, as did demand, and the campus drug revolution of the 1960s was off and running. Some researchers wondered whether the drug revolution was simply happenstance—or part of the fascist globalist plan to weaken the structure of American society.

Following hearings by a Senate committee on the testing of human subjects in 1977, Senator Ted Kennedy referred to the infamous CIA mind-control experiments by stating, "The deputy director of the CIA revealed that over thirty universities and institutions were involved in an 'extensive testing and experimentation' program which included covert drug tests on unwitting citizens at all social levels, high and low, native Americans and foreign. . . . The intelligence community of this nation, which requires a shroud of secrecy in order to operate, has a very sacred trust from the American people. The CIA's program of human experimentation of the fifties and sixties violated that trust. It was violated again on the day the bulk of the agency's records were destroyed in 1973. It is violated each time a responsible official refuses to recollect the details of the program."

One certain violation involved the death of a scientist working on mind control. According to the government, in 1953, Dr. Frank Olson, a biological and mind-control scientist working for the U.S. Army at Fort Detrick, Maryland, was surreptitiously given an LSD-laced drink by Dr. Sidney Gottlieb, while attending a conference at Deep Creek Lodge in Maryland. Some days later, a distraught and hallucinating Olson threw himself out of a high window of a New York hotel. It appeared to be either a drug-induced accident or a suicide. However, Olson's close friends and family members still believe Olson was murdered to prevent him from speaking out against the MKULTRA program, which he had come to both regret and despise.

RESEARCH BASED ON Nazi pharmaceutical science even spread to more exotic attempts at mind control.

In 2005, at the Eighth Annual Ritual Abuse, Secretive Organizations and Mind Control Conference held at the Doubletree Hotel in Windsor Locks, Connecticut, one speaker was Carol Rutz, author of *A Nation*

Betrayed. Rutz claimed to be the victim of government abuse and mind-control experiments. She mentioned an astounding connection. Referring to a woman who described "eye experiments" at a residential school where "they were trying to change our eye color from brown to blue," Rutz stated, "This particularly caught my attention, since the change of eye color was a pet project of Auschwitz 'Doctor' Josef Mengele, aka Dr. Black, as I knew him, who I allege worked alongside Dr. Ewen Cameron, Sid Gottlieb, and others."

Another seedy side to this experimentation on unwitting subjects involved a CIA contract agent named George Hunter White, who worked under the auspices of Dr. Gottlieb. White would bring unsuspecting men from local bars to a CIA-financed bordello in San Francisco, where he would give them LSD-spiked drinks and then watch the men have sex with prostitutes from behind a two-way mirror. The U.S. taxpayers paid for it all, as White would send bills for his "unorthodox expenses" to Dr. Gottlieb. White once said of this work, "I was a very minor missionary, actually a heretic, but I toiled wholeheartedly in the vineyards because it was fun, fun, fun. Where else could a red-blooded American boy lie, kill, cheat, steal, rape, and pillage with the sanction and blessing of the All-Highest?"

Indeed, where else but in an agency penetrated by displaced Nazis and their philosophies?

Much of the drug experimentation was centered at secret facilities at the Edgewood Arsenal, located on Chesapeake Bay northeast of Baltimore. In 1955, thanks to the influx of Nazi chemists, a new drug-testing program was instituted at Edgewood. "Volunteer soldiers were recruited, but not told what drugs they would be given, nor that men had died as the result of similar experiments. They were told they would only suffer temporary discomfort," wrote mind-control researcher and author Walter H. Bowart. "Seven thousand soldiers underwent the Edgewood Arsenal tests. Five hundred eighty-five men were given LSD; the rest were administered other unspecified drugs."

Carol Rutz produced a letter she received from a U.S. soldier who underwent experimentation at Edgewood Arsenal: "I can see where you don't believe anything coming out of the government. I don't. I am one of the 6,720 enlisted soldiers used at Edgewood in 1955–1975. I was there in '74, when they had just got their brand-new lab. I am glad Congress investigated it in '75 and the army shut the program down—at least they won't do it openly anymore. I was young and stupid when I volunteered; I was eighteen. I am now forty-nine and totally disabled, I have the body of a seventy-year-old, and the VA says the army didn't do anything to me. I don't believe them."

Despite a congressional investigation in the 1970s and a lawsuit by one of the soldier victims, Master Sergeant James Stanley, which went to the U.S. Supreme Court in 1986, the work at Edgewood never reached the public. "For all this, the secret Paperclip connections at the base remain unexposed. The fact that Paperclip scientists worked at Edgewood at various times between 1947 and 1966 has been kept a closely guarded secret," wrote Linda Hunt.

For example, Kurt Rahr, who "should be considered an absolute security threat to the United States," according to an early report by the Public Safety Branch of the Office of Military Government U.S. (OMGUS), nevertheless was hired under Paperclip and set to work at Edgewood. Rahr was deported back to Germany in 1948 after another scientist, Hans Trurnit, accused him of being a communist.

Other Edgewood scientists included Theodor Wagner-Jauregg and Friedrich Hoffman. These men initially studied the Nazi poison gasses, tabun and sarin, the most deadly agents the U.S. military had ever encountered. U.S. soldiers were exposed to tabun and mustard gas in an Edgewood gas chamber reminiscent of those in the Nazi death camps.

"In 1949, the direction of Edgewood's work abruptly changed," noted Hunt. "A consultant of the Chemical Division at [European Command] sent information about an amazing drug, LSD, that caused hallucinations and suicidal tendencies in humans. As a result, Edgewood's [scientific director of the Chemical and Radiological Laboratories Dr.] L. Wilson Greene seized the idea of conducting 'psychochemical warfare.' He then suggested that $50,000 be set aside in the 1950 budget to study psychochemicals."

Friedrich Hoffmann, another Paperclip chemist, traveled the world in search of exotic and new psychochemicals. He used the University of

Delaware's chemistry department as a cover to prevent anyone from connecting him to Edgewood Arsenal. This subterfuge was easy enough to maintain, because both the department chairman William Mosher and Professor James Moore were heavily involved in the MKULTRA program. "We were all being paid by the CIA," Moore told Linda Hunt.

SS Brigadefuehrer, or brigadier general, Walter Paul Emil Schreiber, whom one U.S. Army officer described as "the prototype of an ardent and convinced Nazi who used the party to further his own ambitions," worked for more than a decade for the chemical division of the U.S. European Command. His attempt to join Paperclip scientists in America was thwarted when counterintelligence connected Schreiber to hiding SS officers and unexplained business dealings with both the French and the Soviets.

According to Gordon Thomas, author of *Mindfield: The Untold Story Behind CIA Experiments with MKULTRA & Germ Warfare,* "Walter Schreiber had been the paymaster for all the doctors working the Nazi biochemical warfare programs. Under the cover of Paperclip he was brought to the United States. By 1951 he was working at the Air Force School of Medicine in Texas." A year later, Schreiber, fearing that the media might discover his background, obtained a visa and found a job in Argentina, where his daughter was living. There in 1952, he met his old friend, Dr. Josef Mengele.

"U.S. laws governing the American zone of Germany forbade the Germans from doing research on chemical warfare," noted Hunt. "But that did not stop the Army Chemical Corps or the High Commissioner of Germany [John J. McCloy], the U.S. organization that replaced OMGUS, from hiring chemical warfare experts as 'consultants' or funding German industries to produce chemical warfare materials for the United States."

Hunt may not have noticed the interconnectedness of the personalities and business interests in both Germany and America before, during, and just after the war. But one person who did notice that funny business was taking place within the CIA was John K. Vance, a graduate of Columbia University who had served as a military translator at the Nuremberg trials. Vance stumbled upon MKULTRA in the spring of 1963 while working on an inspector general's survey of the CIA's technical services division.

A resulting inspector general's report concluded, "The concepts involved in manipulating human behavior are found by many people both within and outside the agency to be distasteful and unethical." As the result of Vance's discovery and the inspector general's report, the agency began scaling back the project, which was eventually said to have ended in the late 1960s.

The MKULTRA program, which used patients in psychiatric hospitals, and other unwitting subjects, to develop mind-control techniques, became public knowledge in 1977, during hearings conducted by a Senate committee on intelligence chaired by Senator Frank Church. Some of the most distinguished figures in psychiatry participated in MKULTRA, including Dr. Ewen Cameron, the man whom Allen Dulles sent to study Rudolf Hess and who later served as president of both the American and Canadian Psychiatric Associations as well as the World Psychiatric Association.

Any in-depth study of MKULTRA shows that the CIA, in addition to seeking a truth serum, also was highly interested in the ability to program individuals to act in accordance with someone else's will. The 1962 Frank Sinatra film *The Manchurian Candidate* portrayed a programmed assassin ordered to kill a ranking political figure. This movie came out at the same time the CIA was actively working on just such a program, thanks in great part to the groundwork laid down years earlier by Nazi mind-control experts.

In the words of Kathleen Ann Sullivan, who claimed to have been part of the MKULTRA program as reported by Gordon Thomas, author of *Mind Field*: "I am a survivor of the MKULTRA program. It was run by the CIA and designed to control a subject's mind and will to the point where he or she would become an assassin. To achieve this I was forced to undergo extensive drugging, electroshock, sensory deprivation, hypnosis, partake in pornographic films, act as a prostitute, and much else. I finally have realized I cannot keep hiding that it has left me only a shell of life to live. In going public I want to end the fear all survivors of MKULTRA live with."

Hunt noted: "That both the Army and CIA MKULTRA experiments

stemmed from Nazi science was certainly relevant to understanding the early history of those secret projects."

A full accounting of the Nazi-inspired mind-control experiments—whether failures or successes—will never be known, because in 1973, on orders of Helms, Gottlieb destroyed all MKULTRA files before leaving the agency.

CHAPTER 9

BUSINESS AS USUAL

WITH THE U.S. MILITARY-INDUSTRIAL COMPLEX PENETRATED BY unrepentant Nazis and their ideology, the globalists' attention to business and politics remained unabated.

In the war's aftermath, some of the same personalities who helped place Hitler in power came back into play. John J. McCloy, wartime assistant secretary of war, close friend to Deutsche Bank chairman Dr. Hermann Josef Abs, and attorney for I. G. Farben, was appointed America's high commissioner in Germany. As such, he pardoned more than seventy thousand Nazis accused of war crimes. One example of McCloy's magnanimity came after forty-three SS officers, including their leader, Obersturmbannfuehrer Joachim Peiper, were condemned to death in 1946 for the massacre of more than a one hundred American prisoners near Malmedy, Belgium, during the Battle of the Bulge. As historian William L. Shirer wrote in *The Rise and Fall of the Third Reich*, "In March 1948, 39 of the death sentences were commuted; in April, General Lucius D. Clay reduced the death sentences from 121 to six; and in January 1951, under a general amnesty, John J. McCloy, the American high commissioner, commuted all the remaining death sentences to life imprisonment. At the time of writing [1959], all have been released." Once again, hidden from public scrutiny, high U.S. officials had moved to free convicted Nazis.

According to former U.S. attorney John Loftus, about three hundred ranking Nazi sympathizers from the Belarus region of the Soviet Union were brought to the United States after the war in hopes they could provide intelligence on the motives and abilities of the Soviets. In tracking the activities of such collaborators, Loftus uncovered a secret unit within the State Department, called the Office of Policy Coordination (OPC), which he said was hidden away from normal government operations and answerable only to Secretary of Defense James V. Forrestal and "the Dulles faction in the State Department." Loftus found that the OPC had recruited Nazi collaborators to fight against communism. When it was discovered that their work was ineffectual, these Nazis were allowed to immigrate to the United States, their former activities concealed.

Such men included former Belarusan president Radislaw Ostrowsky, who was offered immunity from prosecution for war crimes; Franz Kushel, an SS general who commanded the Belarus Brigade and was responsible for the execution of more than forty thousand Jews; Stanislaw Stankievich, who organized the execution of seven thousand Jews; and Emanuel Jasiuk, who, as the wartime mayor of Kletsk, supervised the execution of more than five thousand Jews in just one day. Both Ostrowsky and Jasiuk are buried in a South River, New Jersey, cemetery near a monument to Belarusan war veterans, which is topped by the image of an Iron Cross.

In 1948, soon-to-be CIA director Allen Dulles, who had been on hand in 1921 for the creation of the Council on Foreign Relations, authorized Frank Wisner to use false paperwork to maintain the Vatican "ratlines" for escaping Nazis. Within a year, Wisner's Nazis were working for CIA propaganda fronts such as Radio Liberty and the Voice of America.

Dulles had encouraged the choice of Wisner as head of the OPC, which undertook active operations against the Soviet Union as the Cold War developed. Although OPC agents in Europe wore U.S. military uniforms, they were paid by the CIA. According to Loftus, "OPC's program emanated almost entirely from the State Department's policy and planning staff, headed by George Kennan."

EVERY U.S. GOVERNMENT administration since the CFR's inception has been packed with council members. Conservative journalist and CFR researcher James Perloff noted that through 1988, fourteen secretaries of state, fourteen secretaries of the treasury, eleven secretaries of defense, and scores of other federal department heads were members of the CFR. This trend continued through both the Bill Clinton and George W. Bush administrations. Current and former members of the CFR in the 2007 Bush cabinet included Dick Cheney, Condoleezza Rice, Elaine Chao, Robert M. Gates, Joshua B. Bolten, and Susan Schwab.

The like-mindedness of CFR members—one cannot ask to join, you must be invited and pass a stringent vetting process to show that you are in agreement with their worldview—along with their close ties with the corporate business world has caused many conspiracy writers to view the CFR as a group with plans to control the world through multinational business mergers, economic treaties, and global government. The activities of the CFR may have been summarized by sociologist G. William Domhoff, who wrote: "If 'conspiracy' means that these men are aware of their interests, know each other personally, meet together privately and off the record, and try to hammer out a consensus on how to anticipate and react to events and issues, then there is some conspiring that goes on in CFR, not to mention the Committee for Economic Development, the Business Council, the National Security Council, and the Central Intelligence Agency."

"Many of the council's members have a personal financial interest in foreign relations because it is their property and investments that are guarded by the State Department and the military," noted researcher Laurie Strand in the 1981 edition of *The People's Almanac #3*.

Nothing had changed since the 1960s, when President John F. Kennedy's special adviser John Kenneth Galbraith bemoaned, "Those of us who had worked for the Kennedy election were tolerated in the government for that reason and had a say, but foreign policy was still with the Council on Foreign Relations."

EVEN TODAY'S TERRORIST groups may be traced back to the Nazis, which prompts speculation on who is truly behind them.

According to Peter Levenda's *Unholy Alliance,* Otto Skorzeny, the Nazi commando who may have found Solomon's treasure, made his way to postwar Egypt, where he created an Egyptian "Gestapo" staffed almost completely with former SS officers. According to Levenda, this was a measure "that received wholehearted support from CIA director Allen Dulles, who was at that time involved with Reinhard Gehlen in developing an anticommunist espionage service within the ranks of CIA."

This operation encompassed the Nazi-associated Muslim Brotherhood, a progenitor of today's al-Qaeda terrorist organization.

The connection between Muslim fanatics and Nazis, according to John Loftus, began with Muslim Brotherhood founder Hassan al-Banna, who formed a group of Egyptian youth dedicated to social reform and Islamic morals. He was a devotee of Muhammad ibn Abd al-Wahhab, the eighteenth-century Muslim who founded the Wahhabi sect that teaches that any additions or interpretations of Islam after the tenth century are false and should be eradicated, even by violence.

"In the 1920s there was a young Egyptian named al-Banna. And al-Banna formed this nationalist group called the Muslim Brotherhood. Al-Banna was a devout admirer of Adolf Hitler and wrote to him frequently. So persistent was he in his admiration of the new Nazi Party that in the 1930s, al-Banna and the Muslim Brotherhood became a secret arm of Nazi intelligence," said Loftus, who had unprecedented access to secret U.S. government and NATO intelligence files. "The Arab Nazis had much in common with the new Nazi doctrines. They hated Jews; they hated democracy; and they hated the Western culture. It became the official policy of the Third Reich to secretly develop the Muslim Brotherhood as the fifth Parliament, an army inside Egypt. When war broke out, the Muslim Brotherhood promised in writing that they would rise up and help General Rommell and make sure that no English or American soldier was left alive in Cairo or Alexandria." While they obviously failed in this, it is a fact that Arab raiders caused continual problems for Allied forces.

After World War II, the Muslim Brotherhood and its German intelligence handlers were sought for war crimes, as they were not considered regular military units. Following arrests in Cairo, captured Brotherhood members were turned over to the British Secret Service, who hired them to fight against the infant state of Israel in 1948. Loftus stated:

Only a few people in the Mossad know this, but many of the members of the Arab Armies and terrorist groups that tried to strangle the infant State of Israel were the Arab Nazis of the Muslim Brotherhood. What the British did then, they sold the Arab Nazis to the predecessor (the OSS) of what became the CIA (soon to be headed by Allen Dulles). It may sound stupid, it may sound evil, but it did happen. The idea was that we were going to use the Arab Nazis in the Middle East as a counterweight to the Arab communists. Just as the Soviet Union was funding Arab communists, we would fund the Arab Nazis to fight against [them]. And lots of secret classes took place. We kept the Muslim Brotherhood on our payroll. But the Egyptians became nervous. [Egyptian President Gamal Abdal] Nasser ordered all of the Muslim Brotherhood [to get] out of Egypt [in 1954] or be imprisoned, and we would execute them all. During the 1950s, the CIA evacuated the Nazis of the Muslim Brotherhood to Saudi Arabia [still a stronghold of the Wahhabi sect]. Now when they arrived in Saudi Arabia... [one] student was named Osama bin Laden. Osama bin Laden was taught by the Nazis of the Muslim Brotherhood who had emigrated to Saudi Arabia.

In 1979 the CIA drew fanatics from these Saudi Brotherhood members and sent them to Afghanistan to fight the Soviet Russians. "We had to rename them," said Loftus. "We couldn't call them the Muslim Brotherhood, because that was too sensitive a name. Its Nazi cast was too known. So we called them the Maktab al Khidimat il Mujahideen, the MAK.... we left this army of Arab fascists in the field of Afghanistan." Once out of Afghanistan, the mujahideen became known as al-Qaeda, or the Base. While many people still think the term "base" refers to some central headquarters, former British foreign secretary Robin Cook told the House of Commons the term actually referred to a computer database containing the names of Muslim activists, mujahideen, and others long used by the CIA.

Following the Soviet withdrawal from Afghanistan, the Saudis didn't want the fanatics to return, so they bribed Osama bin Laden and his al-Qaeda followers to stay out of Saudi Arabia.

"There are many flavors and branches, but they are all Muslim Brotherhoods. . . . So the Muslim Brotherhood became this poison that spread throughout the Middle East, and on 9/11, it began to spread around the world," concluded Loftus, who added that current CIA members don't know this history. "[T]he current generation CIA are good and decent Americans and I like them a lot. They're trying to do a good job, but part of their problem is their files have been shredded. All of these secrets have to come out."

DURING THE 1950S, while Dulles headed the CIA, his brother, John Foster Dulles, was President Eisenhower's secretary of state. Both were in prime positions to shape the Nazification of America.

"In the chill of the Cold War, few Americans remembered that John Foster Dulles had been pro-Nazi before Hitler invaded Poland," observed Walter Bowart, a journalist and former editor who authored the book *Operation Mind Control*. "No one thought, either, to question the fact that while John Foster Dulles was running the State Department, his brother Allen was running the CIA, which he once described as a State Department for dealing with unfriendly governments. No one seemed at all disturbed by the Dulles dynasty, and only a handful of people realized to what extent the Dulles brothers held power in the Eisenhower administration."

The German banking industry particularly profited from such connections as the Dulles brothers and McCloy. Immediately after the war ended, Allied authorities ordered the breakup of Germany's largest bank, Deutsche Bank. Initially, the banking giant was split into ten regional banks, but by 1953 these were consolidated into three major banks—Nordeutsche Bank AG, Suddeutsche Bank AG, and Rheinisch-Westfalische Bank AG.

A mere four years later, without any opposition, these three major banks merged and began takeovers, including the London Morgan Grenfell investment bank in 1989, the East German central bank following reunification in 1990, Banker's Trust of New York in 1999, and two large Russian banks in 2006. By the turn of the current century, Deutsche Bank was reunited and had become a world banking leader.

Astute authorities in both Britain and the United States undoubtedly recognized Deputy Fuehrer Martin Bormann's flight capital plan. But they were dissuaded from action by the general euphoria at the war's end, the chaotic conditions in Europe, the idea that growth would restore the wrecked European economy, and the increasing belief that the United States would soon have to confront a victorious and powerful Soviet Union. There was also the behind-the-scenes power of men in banking and commerce who had been in league with the Nazis and still carried sympathy for their cause.

"Treasury officials in Washington, as in London, knew what was transpiring; the teams they sent into the field uncovered enough evidence to prove a definite pattern," wrote Paul Manning. "As for the news media, it did not seem important, although long-term it was really the biggest of the postwar stories. Yet so quietly was it handled by the Germans, and so diffident was the reaction by the Allies, that few ripples rose to the surface, and investigators of the U.S. Treasury Department were taken off the case." Investigators were reassigned by their superiors, some of whom had been business partners with the Nazis for years.

One such example was attorney Russell A. Nixon, who, as a member of the U.S. Military Government Cartel Unit, found himself working directly under Brigadier General William H. Draper Jr., an advocate of eugenics who sat with James Forrestal on the board of directors of Dillon, Read Company, the firm that early on had helped finance the German cartels. According to Professor Antony C. Sutton, "Three Wall Street houses—Dillon, Read; Harris, Forbes; and National Citys Bank—handled three-quarters of the reparation loans used to create the German cartel system, including the dominant I. G. Farben and Vereinigte Stahlwerke, which together produced 95 percent of the explosives for the Nazi side in World War II."

Nixon was blocked at every turn in his attempt to break up the Farben cartel. He finally went over the head of Brigadier General Draper, meeting with General Lucius D. Clay, military governor of the U.S. zone in Germany. The general was told Draper had canceled orders to dismantle or destroy Farben facilities, lied about bomb damage to Farben plants, and deliberately violated General Eisenhower's orders to break up the Farben cartel. Despite Nixon's complaints, nothing was done. When Nixon took the initiative and had Nazi industrialist Richard Freudenberg arrested, U.S. ambassador to Germany Robert Murphy ordered him released. One occupation official explained Murphy's decision by commenting, "This man Freudenberg is an extremely capable industrialist: a kind of Henry Ford."

Nixon later told a Senate subcommittee in Washington, "Generally speaking, in spite of the efforts that have been made, at the present time there is a continuation of the dissipation and further concealment of these [Nazi] assets throughout all the neutral countries."

Another Allied investigator, Department of Justice attorney James Stewart Martin, was sent to U.S. Military Command in London to investigate collaboration between the Nazis and American businessmen. Martin bristled when he found his commanding officer was a Colonel Graeme K. Howard, an official with General Motors. After Martin pointed out the cozy relationship General Motors had with the Nazis, Howard was quietly reassigned back to the States.

Martin later investigated the fate of ITT's German chairman Gerhardt Westrick, who, after the war ended, had fled Berlin and hidden out in a castle in southern Germany. After presenting a report on the status of ITT firms to U.S. military authorities, Westrick was given a light prison sentence and released. Martin found that queries concerning General Aniline and Film's connections to I. G. Farben had been referred to Allen Dulles with no results.

"We had not been stopped in Germany by German business. We had been stopped in Germany by American business," Martin wrote in the 1950 book *All Honorable Men*. "The forces that stopped us had operated from the United States but had not operated in the open. . . . Whatever it was that had stopped us was not 'the government.' But it clearly had command of channels through which the government normally operates. The relative powerlessness of governments in the growing economic power is of course not new. . . . national governments have stood on the sidelines while bigger operators arranged the world's affairs."

One of those insiders who helped protect Nazi interests in the wake of World War II was aforementioned William H. Draper Jr., a business partner with Prescott Bush, who, in July 1945, was appointed head of the economic division of the U.S. Control Commission, which decided which Nazi corporations would be saved and who would face war crimes prosecution. This placed Draper in almost as powerful a position as Germany's new high commissioner John J. McCloy, who, besides banking for the Nazis, had spent a year in Italy as a financial adviser to Mussolini and shared Hitler's box at the 1936 Berlin Olympics.

Martin also may have stumbled across the answer to why the international business community turned against Hitler. The leading bankers and industrialists were looking forward to a postwar world composed of intertwining corporate business connections among the nations of the world—a New World Order.

Hitler, on the other hand, was planning to attack the United States just as soon as he had effective rocket and long-range bomber delivery systems in place. The globalists did not desire a continuous war, nor did they want Hitler to control a world National Socialist government. They had their own plans.

"[T]he 750 new corporations established under the Bormann [flight capital] program gave themselves absolute control over a postwar economic network of viable, prosperous companies that stretched from the Ruhr to the 'neutrals' of Europe and to the countries of South America; a control that continues today and is easily maintained through the bearer bonds or shares issued by these corporations to cloak real ownership," stated Paul Manning, who worked with Edward R. Murrow covering the war in Europe for CBS Radio. Manning revealed that "there are U.S. Treasury old-timers of World War II still not aware of the magnitude of the Bormann operation and of its success. Those who know, in Washington, in South America, and in the capitals of Europe, are locked together in a conspiracy of silence."

No one in a position of power within the financial centers of Washington, Wall Street, the City of London, or Paris desired a real search for the scattered German assets. Manning explained: "They had understandable reasons if you overlook morality: the financial benefits for cooperation (collaboration had become an old-hat term with the war winding down) were very enticing, depending on one's importance and ability to be of service to the organization and the 750 corporations they were secretly manipulating, to say nothing of the known multinationals such as I. G. Farben, Thyssen AG, and Siemens."

Without the German industrial base that had been its foundation for years, the European economy was suffering. U.S. Joint Chiefs of Staff directives had ordered that nothing be done to rebuild Germany's industries. But business is business, and by the summer of 1947, President Roosevelt's friend and chief of staff General George Catlett Marshall had become secretary of state under President Truman. The son of a prosperous coal producer and a participant in every high-level policy conference from Casablanca to Yalta and Potsdam, Marshall convinced Truman that it was in the best interests of European prosperity that Germany be allowed to rehabilitate its economy. Thus, the Marshall Plan was born, and millions of dollars of aid began pouring into war-devastated Europe.

Many Americans, including Colonel Robert McCormick, editor of the *Chicago Tribune,* and Senator Joseph McCarthy, attacked the Marshall Plan as merely another Rockefeller scheme to bilk Amerian taxpayers. According to Mullins, "The Marshall Plan had been rushed through Congress by a powerful and vocal group, headed by Winthrop Aldrich, president of Chase Manhattan Bank, and Nelson Rockefeller's brother-in-law, ably seconded by Nelson Rockefeller and William Clayton, the head of Anderson, Clayton Company."

Marshall Plan money flowed through the same banking conduits as before the war, such as the 1936 partnership between the J. Henry Schroeder Bank of New York and Rockefeller family members, described by *Time* magazine as "the economic booster of the Rome-Berlin Axis." Partners in Schroeder, Rockefeller and Company included Avery Rockefeller, nephew of John D., Baron Bruno von Schroeder in London, and Kurt Freiherr von Schroeder of the Bank of International Settlements and the Gestapo in Cologne. Attorneys for the firm were John Foster

Dulles and Allen Dulles of Sullivan and Cromwell. Future CIA director and Warren Commission member Allen Dulles sat on the board of Schroeder. "Further connections linked the Paris branch of Chase National Bank to Schroeder as well as the pro-Nazi Worms Bank and Standard Oil of New Jersey in France. Standard Oil's Paris representatives were directors of the Banque de Paris et des Pays-Bas, which had intricate connections to the Nazis and to Chase," noted *New York Times* journalist Charles Higham.

William Bramley, who researched the causes of war in his 1990 book *The Gods of Eden,* noted these international banking connections: Max Warburg, a major German banker, and his brother Paul Warburg, who had been instrumental in establishing the Federal Reserve System in the United States, were directors of I. G. Farben. H. A. Metz of I. G. Farben was a director of the Warburg Bank of Manhattan, which later became part of the Rockefeller Chase Manhattan Bank. Standard Oil of New Jersey had been a cartel partner with I. G. Farben prior to the war. One American I. G. Farben director was C. E. Mitchell, who was also director of the Federal Reserve Bank of New York and of Warburg's National City Bank.

It is interesting to note that throughout the war, Chase maintained its financial connections with the Nazis through its Paris bank. Further, I. G. Farben chief Hermann Schmitz served as Chase president for seven years prior to the war. According to Manning, "Schmitz's wealth—largely I. G. Farben bearer bonds converted to the Big Three successor firms, shares in Standard Oil of New Jersey (equal to those held by the Rockefellers), General Motors, and other U.S. blue chip industrial stocks, and the 700 secret companies controlled in his time by I. G. [Farben], as well as shares in the 750 corporations he helped Bormann establish during the last years of World War II—has increased in all segments of the modern industrial world. The Bormann organization in South America utilizes the voting power of the Schmitz trust along with their own assets to guide the multinationals they control, as they keep steady the economic course of the Fatherland."

After the war, twenty-four lower I. G. Farben executives stood trial at Nuremberg for crimes against humanity, including the building and maintenance of concentration camps and the use of slave labor. Schmitz was convicted of war crimes in 1948 at Nuremberg, but served a mere two years in prison, although it was under Schmitz's leadership that the giant chemical combine produced and distributed the notorious Zyklon-B gas, used for human extermination in Nazi concentration camps.

Regarding the relationship between Schmitz and Martin Bormann, Manning wrote: "Their association was close and trusting over the years, and it is the considered opinion of those in their circle that the wealth possessed by Hermann Schmitz was shifted to Switzerland and South America, and placed in trust with Bormann, the legal heir to Hitler."

These long-standing banking and business connections coupled with the Schmitz business network allowed Reichsleiter Martin Bormann to forge a formidable Nazi-controlled organization for postwar activities. The late Jim Keith, author of several conspiracy books, wrote, "[I]n researching the shape of totalitarian control during this century, I saw that the plans of the Nazis manifestly did not die with the German loss of World War II. The ideology and many of the principal players survived and flourished after the war, and have had a profound impact on postwar history, and on events taking place today."

Interestingly enough, the Nazis' attempt at central control through the economic sector produced the early stages of a united Europe so sought by Hitler.

AFTER THE WAR, a devastated Europe looked to Germany for economic leadership. The economic steps taken that became the Common Market took the shape of prewar Nazi plans. "[S]omehow the Germans had the answer originally in 1942 when they were melding the economic institutions of the Continent into their own design," noted Manning.

It is interesting to note that the present European Union (EU) began as merely economic measures. From this reasonable beginning as a trade organization, the Common Market evolved into the European Union, a supranational and intergovernmental body with tendrils into all aspects of European life. In 1950, French foreign minister Robert Schuman proposed the joint management of the French and West German coal and steel industries. This "Schuman Declaration," ratified by the 1951 Treaty of Paris, was the first step toward what Winston Churchill termed a "United States of Europe."

The 1951 treaty between Belgium, France, Italy, the Netherlands, Luxembourg, and West Germany provided for the shared production of coal and steel. It was thought that this mutual endeavor would ensure against another war between these nations. It also could be seen as a means of consolidating the Nazi business holdings of the Bormann organization.

The European Economic Community, better known as the Common Market, was established in 1957 by the Treaty of Rome, signed by the same nations as in 1951. George McGhee, a member of the secretive Bilderberg Group and former U.S. ambassador to West Germany, acknowledged that "the Treaty of Rome, which brought the Common Market into being, was nurtured at Bilderberg meetings."

Again, the Common Market was styled as merely a step toward equalizing trade balances and tariffs. But with the 1992 signing of the Treaty Establishing the European Community, popularly known as the Maastricht Treaty, the word "Economic" was deleted from both the treaty and the community. By 2000, the European Union was composed of various economic, political, and judicial institutions including the European Central Bank, the European Parliament, the European Court of Justice, and a unified European currency, the Euro.

The 1942 Nazi vision of a unified Europe had become a reality.

BUT IF THE reality of the Nazi vision of a united Europe appeared to signal success for the emerging Fourth Reich, nothing showed more success than the seeds that had been sown in the postwar United States.

In the postwar period, FDR Democrats, who looked favorably on federal social and economic controls, were firmly in the pocket of the global corporate elite, although there was a small problem with President Harry S. Truman, who was never fully under their command. So Truman, like Jimmy Carter, was elected to only one term. The real problem lay with the Republicans, longtime supporters of less government and more economic freedom.

One willing accomplice of the globalist/Nazi nexus was an obscure

Republican politician from California. According to authors John Loftus and Mark Aarons, "When Truman was reelected in 1948, [future President Richard M.] Nixon became Allen Dulles's mouthpiece in Congress. Both he and Senator Joseph McCarthy received volumes of classified information to support the charge that the Truman administration was filled with 'pinkos.' When McCarthy went too far in his Communist investigations, it was Nixon who worked with his next-door neighbor, CIA director Walter Bedell Smith, to steer the investigations away from the intelligence community [and the Nazis]." "The partnership between Allen Dulles and Richard Nixon was truly a Faustian bargain, but it is hard to tell just who is Faust and who is Mephistopheles in the scenario," wrote Peter Levenda.

Nixon, who joined the Council on Foreign Relations in 1961 but resigned in 1965 after his membership became a campaign issue, got an initial boost to his political career by befriending a Nazi named Nicolae Malaxa. A Romanian who was a former business partner of Nazi Luftwaffe chief Hermann Goering, Malaxa had belonged to Baron Otto von Bolschwing's Gestapo network, as had his associate, Valerian Trifa, who was then living in Detroit. Both were members of the Nazi Iron Guard in Romania and had fled prosecution for their role in the deaths of many Jews. Their common interest was in Richard Nixon, who was elected vice president in 1952.

In late 1946, Nixon's influence, along with that of the Dulles brothers, freed Malaxa's money frozen in Chase National Bank during the war. The Treasury Department official whose firm won Malaxa's case to release the money was the same man who froze these assets in the first place. "Such interrelationships were very common between émigré Fascists and the right wing of the American establishment," noted Loftus and Aarons.

According to the late Mae Brussell, a California conspiracy researcher and hostess of the *World Watchers International* radio program, Trifia came to the United States with the help of the opportunistic Baron von Bolschwing, an early member of the Nazi Party and highly placed intelligence operative. Malaxa had escaped from Europe with more than $200 million in U.S. dollars. In America, following the retrieval of his frozen assets, Malaxa gained another $200 million from Chase Manhattan

Bank. The legalities were handled by the Dulles brothers' law firm of Sullivan and Cromwell. "Undersecretary of State Adolph Berle, who had helped Nixon and star witness Whittaker Chambers convict Alger Hiss, personally testified on Malaxa's behalf before a congressional subcommittee on immigration," noted Brussell.

A company called Western Tube was another effort to further help Malaxa. In 1951, Nixon introduced a private bill in Congress, which would have allowed Malaxa to remain in the United States despite provisions of the Displaced Persons Act, which prohibited persons with Fascist backgrounds. When this effort failed, Malaxa pretended to create Western Tube, which manufactured a seamless tube that he claimed was vital to the Korean War effort. Loftus and Aarons reported, "He set up his company in Nixon's hometown of Whittier, California, and registered it in the same building as the law firm of Bewley, Kroop & Nixon. Nixon's former law partner, Thomas Bewley, was secretary of Malaxa's company, and the vice president was another Nixon crony." They added that Western Tube never made any tubes. "The company was used to evade millions of dollars' worth of taxes and then dissolved," they noted.

It was the $2 million fee Nixon received from Malaxa that funded his successful election campaign for Congress. "In 1946 Nixon had gotten a call from Herman L. Perry asking if he wanted to run for Congress against Rep. Jerry Voorhis. Perry later became president of Western Tube," wrote Mae Brussell.

In 1952, in spite of being granted permanent resident status in the United States, Malaxa moved from Whittler to Argentina, where he joined Argentine dictator Juan Peron and Hitler's chief commando Otto Skorzeny in profitable business dealings with the network created by Martin Bormann and Allen Dulles.

Meanwhile, ranking Republicans were abetting the influx of Nazis. Elmer Holmes Bobst of Warner-Lambert Pharmaceutical, described as "a mentor and father figure" to Nixon, was a power behind Nixon's 1960 presidential campaign. Bobst was a close associate of Otto von Bolschwing, head of the Gestapo network that included Malaxa and Trifia. "In preparation for the 1952 Eisenhower-Nixon campaign, the Republicans formed an Ethnic Division, which, to put it bluntly, recruited the 'displaced Fascists' who arrived in the United States after World War II," wrote Loftus and Aarons.

To mask the wave of Nazis flooding into the United States and to gain tighter federal control over Americans, the prewar specter of the Communist Menace was resurrected. A quick look at America's reaction to liberation movements in postwar Europe demonstrates the hypocrisy of the often-stated Eisenhower-era Republican policies to "roll back communism."

In 1953, half a million citizens of the so-called German Democratic Republic (GDR) staged a revolt against the East German regime, only to be violently suppressed by troops and tanks. No further large-scale demonstrations were to take place in East Germany until 1989, the year before the GDR ceased to exist.

Three years later, Polish workers attempted their own rebellion against the Communist regime. Widespread violence was only averted after China's Mao Tse-tung convinced Russian Premier Nikita Khrushchev to allow Polish Communist Wladyslaw Gomulka to assume authority and institute reforms in Poland.

In late October 1956, the people of Hungary spontaneously rose up against their Soviet occupiers, many in response to broadcasts of support by the BBC, Radio Free Europe, and the Voice of America. Put off-guard by initial offers of compromise by the Russians, the Hungarians were unprepared for the full-scale military assault on Budapest on November 4. Thousands were killed and an estimated two hundred thousand fled the country. This Soviet show of force maintained their control over central Europe and also perpetuated the idea that Communism was monolithic and irresistible.

In discussing three books on the failed Hungarian Revolution—*Twelve Days* by Victor Sebestyen, *Failed Illusions* by Charles Gati, and *Journey to a Revolution* by Michael Korda—*New York Times* book reviewer Jacob Heilbrunn noted, "Despite promiscuous pledges to roll back Communism, President Dwight Eisenhower and Secretary of State John Foster Dulles had no intention of rolling back anything. In a National Security Council meeting, Vice President Richard Nixon even stated that a Soviet invasion would not be an unmixed evil for the West, as it would bolster the alliance against Communism."

CHAPTER 10

KENNEDY AND THE NAZIS

MOST AMERICAN LEADERS THROUGHOUT THE COLD WAR COULD only see the danger of international communism. One exception may have been President John F. Kennedy, who warned of the dangers of un-necessary secrecy and secret societies such as Skull and Bones, the Council on Foreign Relations, and the Bilderberg Group. "The very word 'secrecy' is repugnant in a free and open society; and we are as a people inherently and historically opposed to secret societies, to secret oaths, and to secret proceedings," Kennedy said in a 1961 address to the American Newspaper Publishers Association.

Kennedy was the first American president born in the twentieth century and was one of the best-educated, having graduated from Harvard cum laude. The book that first made him a public figure was the best-seller *Why England Slept,* a treatise on prewar British-German diplomacy. This work showed clearly that Kennedy had a keen understanding not only of geopolitics but of the behind-the-scenes machinations of the globalists.

Interestingly enough, his political career may have come about because of his relationship with an alleged Nazi spy. Early in World War II, the FBI suspected Inga Arvad—a former Miss Denmark, who had attended the wedding of Germany's Field Marshal Hermann Goering and met with Adolf Hitler—of being a Nazi spy. After eavesdropping on her, agents determined that one of her visitors was Naval Ensign John F. Kennedy, then working for Naval Intelligence in Washington. After both the navy and his father had been alerted to the danger presented by Kennedy's involvement with a suspected agent, young Kennedy was quickly transferred to the South Pacific. It was there that he led the survivors of PT-109 to safety, thus becoming a war hero and launching his political career toward the presidency—all thanks to the diligent J. Edgar Hoover.

IN 1960, RICHARD Nixon was expected to be the next president of the United States. Corporate America's hopes were crushed when John F. Kennedy managed to win the closest election to that time.

Corporate heads and their Nazi backers must have been mollified to know that Kennedy was being guided by his father, Joseph P. Kennedy, a pro-Nazi sympathizer. But in December 1961, Joseph Kennedy suffered a stroke that left him totally incapacitated. His son now held the nation's highest office with no real control over him.

By mid-1963, Kennedy was beginning to exert his autonomous influence over the most powerful—and violent—groups in U.S. society. He was threatening to disband the CIA, the homebase of many Nazis; withdraw U.S. troops from South Vietnam; close the tax breaks of the oil-depletion allowance; tighten control over the tax-free foreign assets of U.S. multinational corporations, many with connections to the Bormann empire; and decrease the power of both Wall Street and the Federal Reserve System. In June 1963, Kennedy actually ordered the printing and release of $4.2 billion in United States Notes, paper money issued through the Treasury Department without paying interest to the Federal Reserve System, which is composed of twelve regional banks all controlled by private banks whose owners often are non-Americans.

Obviously, persons affected by these moves felt that something had to be done.

Today, most people agree that the assassination of President Kennedy was the result of a conspiracy, the full details of which are still not known due to a cover-up at the highest levels of the federal government.

It is fascinating to note that the connections between Kennedy's death and Nazi-connected persons, groups, and firms are many and well documented. The CIA, which had passed hundreds of millions of dollars to the Nazi Gehlen Organization, has long been fingered as a major player in the assassination. Operatives such as future Watergate burglars E. Howard Hunt and Frank Sturgis; CIA officer Desmond FitzGerald; mobster Johnny Roselli; Cuban minister Dr. Rolando Cubela; defrocked New Orleans priest David Ferrie; and anti-Castro Cubans Carlos Bringuier, Orlando Bosch, and Carlos Prio Soccaras all played roles in the CIA/intelligence mix surrounding the assassination.

But more pointedly, George DeMohrenschildt, a Dallas oil geologist who was the last known close friend to accused assassin Lee Harvey Oswald, began his intelligence career as a Nazi agent. According to one CIA document, DeMohrenschildt had applied to work for U.S. intelligence as far back as 1942 but was turned down because he was a Nazi espionage agent. His cousin, Baron Constantine Maydell, was one of the top Nazi intelligence agents in North America and, after the war, was recruited into the Gehlen Organization to direct the CIA's Russian émigré programs.

At the same time DeMohrenschildt was befriending Oswald in Dallas and introducing him to the White Russian émigré community there, in which the Gehlen Organization was well represented, he was in close contact with his friend J. Walter Moore, an agent of the CIA's Domestic Contacts Division. In 1957, following a trip to Yugoslavia, according to CIA documents, DeMohrenschildt provided the agency with "foreign intelligence which was promptly disseminated to other federal agencies in ten separate reports."

Oswald's other close associates in Dallas just prior to the assassination were Ruth and Michael Paine. Oswald's wife was staying in the Paine home at the time of the assassination and it was Ruth Paine, a woman with CIA connections, who got Oswald his job at the Texas School Book Depository. And Dallas police found incriminating photos—Oswald claimed they were fabricated—of Oswald in Mrs. Payne's garage, holding a rifle authorities identified as the assassination weapon. Her husband worked for Bell Aerospace Corporation in Hurst, Texas, later Bell Helicopter, where Paperclip Nazi Walter Dornberger was a vice president. "Paine's boss at Bell Aircraft, as director of research and development, was none other than the notorious war criminal General Walter Dornberger," stated Brussell.

Oswald's own connections to the CIA are well documented—his training at Japan's Atsugi base, which housed a large CIA facility, his incredible ability to speak fluent Russian despite lack of evidence of language lessons, testimony of his CIA employment by fellow marines and a former CIA paymaster, his ease in obtaining U.S. passports, his use of the word "microdots" in his diary, and his possession of a miniature Minox "spy" camera with a serial number proving it was not commercially available in America.

The most often pointed out, and controversial, evidence of the involvement of both Nazi mentality and actual Nazis in Kennedy's assassination can be found in a treatise passed around for years under the name "Torbitt Document" or "Nomenclature of an Assassination Cabal." This document first appeared under the pen name of William Torbitt but was actually written by a Texas attorney named David Copeland. Copeland told this author he had received the information from friends in both the FBI and the Secret Service. Based on this information, Torbitt/Copeland spent considerable effort searching for evidence to support the paper's thesis.

According to the "Torbitt Document," the Kennedy assassination was orchestrated through a nexus of Nazi-infiltrated anticommunist organizations, elements of the military-industrial complex, the CIA, and the FBI. Torbitt explained, "The director of the Federal Bureau of Investigation was in charge of NASA's Security Division and the Defense Industrial Security Command [DISC] in his position as head of counterespionage activities in the United States. His agents investigated every employee of the space agency as well as the employees of the pertinent contractors doing business with NASA and also prospective employees of every arms and munitions manufacturer. [It is reported that today DISC operations have been incorporated into the National Security Agency.]"

The Defense Intelligence Agency was headed by Lieutenant General Joseph F. Carroll, a former assistant director of the FBI. Carroll worked closely with FBI intelligence chief William C. Sullivan, J. Edgar Hoover, and Canadian L. M. Bloomfield in directing activities of the munition-makers' police agency, the Defense Industrial Security Command. During this time, Walter Sheridan, who later became NBC's special investigator used to undermine the New Orleans assassination investigation of District Attorney Jim Garrison, was a direct liaison between Carroll and Robert F. Kennedy. Hoover worked directly with Wernher von Braun in connection with NASA's security and it was Lyndon Johnson who, as vice president, served as chairman of NASA. Johnson, along with von Braun, Bobby Baker, and Fred Black worked hard to obtain the $9 billion Apollo contract for North American Aviation in 1961 [North American, which, among other famous aircraft, produced the F-86 Sabre jet based on the Nazi Focke-Wulf Ta-183, today is part of the Boeing Company]. North American won the NASA contract despite the fact that its own source evaluation board had recommended another company.

According to Torbitt, each of the NASA security personnel who were assigned duties in connection with the assassination were employees or contractees for Division Five of the FBI. "It must be borne in mind that this was a relatively small group within all of these agencies. It was not official, and it was not an American operation, but was simply the independent action taken by these men, some of whom happened to hold official positions," he wrote.

Torbitt's tale of a NASA conspiracy was supported by New Orleans District Attorney Jim Garrison, who was investigating the Kennedy assassination. In 1968, Garrison telephoned magazine editor Warren Hinkle to say, "Important new evidence has surfaced. Those Texas oilmen do not appear to be involved in President Kennedy's murder in the way we first thought. It was the military-industrial complex that put up the money for the assassination—but as far as we can tell, the conspiracy was limited to the aerospace wing. I've got the names of three companies and their employees who were involved in setting up the president's murder."

When Garrison attempted to subpoena the testimony of their employees, NASA refused to provide them, citing reasons of national security. Amazingly enough, accused assassin Oswald told New Orleans garage owner Adrian Alba that he expected to soon work at a New Orleans plant of NASA. Why Oswald, who had tried to defect to Russia and voiced hostility toward the USA and its policies, thought he could go to work for the nation's leading aerospace agency was never explained.

According to the Torbitt document, this complex matrix of government agents, NASA employees, and Nazis was managed by Louis Mortimer Bloomfield of Montreal, Canada. According to Canadian newspaper accounts, Bloomfield was an ardent Zionist, an attorney, businessman, and philanthropist who had worked for the OSS during World War II and for the fledgling CIA. Torbitt stated Bloomfield was a longtime friend and confidant of J. Edgar Hoover and had been Hoover's contract supervisor of Division Five since his days in the OSS before World War II. Torbitt said Bloomfield "was the coordinator of all activities, responsible only to Hoover and Johnson in carrying out the plans for John Kennedy's assassination."

In 1967, New Orleans DA Jim Garrison charged Clay Shaw, a former OSS officer and founder of the city's International Trade Mart, with conspiracy to assassinate the president. According to several separate sources—including Garrison's files and an investigation by the U.S. Labor Party—the International Trade Mart in New Orleans was a subsidiary of a shadowy entity known as the Centro Mondiale Commerciale (CMC) or World Trade Center, which was founded by Bloomfield in Montreal in the late 1950s, then moved to Rome in 1961. The Trade Mart was connected with CMC through yet another shadowy firm named Permindex (PERManent INDustrial EXpositions), also in the business of international expositions. According to Torbitt, Bloomfield held half of the shares of Permindex and was in total command of its operation in Europe and Africa as well as the North and South American continents. (In the 1962 edition of *Who's Who in the South and Southwest,* Shaw gave biographical information stating that he was on the board of directors of Permindex. However, in the 1963–64 edition, the reference to Permindex was dropped.)

In the late 1960s, both Permindex and its parent company, Centro Mondiale Commerciale, came under intense scrutiny by the Italian news media. It was discovered that Prince Gutierrez di Spadaforo was on the board of CMC. The prince was a wealthy aristocrat who had been undersecretary of agriculture under the dictator Benito Mussolini, and whose daughter-in-law was related to Nazi minister of finance Hjalmar Schacht; Carlo D'Amelio, an attorney for the former Italian royal family; and Ferenc Nagy, former premier of Hungary and a leading anticommunist. The Italian media reported that Nagy was president of Permindex, and the board chairman and major stockholder was Louis Mortimer Bloomfield, the powerful Montreal lawyer who represented the Bronfman family as well as serving U.S. intelligence services. Reportedly, Bloomfield established Permindex in 1958 as part of the creation of worldwide trade centers connected with CMC.

According to a special investigation by reporters David Goldman and Jeffrey Steinberg in their 1981 book *Dope, Inc.: Britain's Opium War Against the U.S.,* Bloomfield was recruited into the British Special Operations Executive (SOE) in 1938, during the war was given rank within the U.S. Army, and eventually became part of the OSS intelligence system, including the FBI's Division Five, where he became quite close with J. Edgar Hoover.

Whatever the truth behind Centro Mondiale Commerciale and its companion company Permindex, and their connection to Kennedy's assassination, the Italian government saw fit to expel both firms in 1962 for subversive activities identical to those in the much-publicized Propaganda-2 (P-2) Masonic Lodge scandal of the Reagan years.

Today it is clear that Clay Shaw was tightly connected to the CIA and intelligence work despite his denials at the time of the Garrison investigation. It is interesting to note that while serving in the U.S. Army during World War II, Shaw worked as aide-de-camp to General Charles O. Thrasher and as a liaison officer to the headquarters of Winston Churchill.

At the time of his arrest in New Orleans by Garrison, Shaw's personal address book was taken. It revealed the names and contact information of important Europeans, many of them pro-Nazi royalty or Bilderberg members.

———

ANOTHER ODD CONNECTION was SS Obergruppenfuehrer Karl Wolff, who had headed the Gestapo in Italy. As part of Allen Dulles's hide-the-Nazi program, Wolff was sentenced to four years' imprisonment after the war but served only a week.

In 1983, Wolff and some former SS associates gathered in Hamburg on Hermann Goering's former yacht, *Carin II*. The boat then belonged to Goering's widow, Emmy, whose estate attorney was the celebrated Melvin Belli. Belli had also represented the Nazi-connected actor Errol Flynn, as well as Jack Ruby, the man who shot Lee Harvey Oswald.

Even odder is a letter smuggled out of the Dallas County Jail and later bought by early JFK assassination researcher Penn Jones. Written by Oswald's killer, Jack Ruby, the letter stated, "my time is running out . . . they plan on doing away with [me]." Ruby pointed to Lyndon Johnson as one of those behind Kennedy's death.

On June 7, 1964, Ruby told visiting Chief Justice Earl Warren he wished his commission and President Johnson had "delved deeper into the situation . . . not to accept just circumstantial facts about my guilt or innocence, and would have questioned to find out the truth about me before he relinquished certain powers to these certain people. . . . Consequently, a whole new form of government is going to take over our country and I know I won't live to see you another time."

To which "certain people" was Ruby referring? After advising a friend to read J. Evetts Haley's blistering attack on Johnson, titled "A Texan Looks at Lyndon," Ruby wrote from jail, "This man [LBJ] is a Nazi in the worst order."

"Now, here is the plan, someone must get to England and France and Israel and tell the right people what has happened to the Jews so they can prepare themselves from [sic] the same thing happening there. They will know that only one kind of people that would do such a thing, that would have to be Nazis, and that is who is in power in this country right now," he added. Ruby concluded, "The only one who gained by the shooting of the president was Johnson."

Considering Johnson's close connections to the Nazi-riddled NASA, perhaps Ruby was not as unhinged as he was portrayed in the media at the time.

ONE INSIGHTFUL VIEW of the Kennedy assassination, possibly based on inside information, came from Nazi SS officer Helmet Streikher, who worked with both Reinhard Gehlen and Otto Skorzeny as well as for the CIA, including time served under former director George H. W. Bush. On a CIA assignment in Africa in late 1963, Streikher was quoted as saying, "One of the worst-kept secrets in the [CIA], is the truth about the president's murder. It wasn't Castro or the Russians. The men who killed Mr. Kennedy were CIA contract agents. John Kennedy's murder was a two-part conspiracy murder. One was the action end with the killers; the other was the deeper part, the acceptance and protection of that murder by the intelligence apparatus that controls the way the world operates. It had to happen. The man was too independent for his own good."

No serious assassination researcher truly believes that Kennedy was killed solely by German Nazis. But, as previously reported, men with Nazi connections—before, during, and after World War II—who were also members of secret societies, were most opposed to Kennedy's policies. They also had the power and influence to affect such an assassination and certainly were capable of blocking any meaningful investigation—whether by government or the media—right up to today.

It may not be sheer coincidence that the men most closely involved in the Warren Commission investigation of Kennedy's death were John J. McCloy and Allen Dulles, both men, as we have seen, with close Nazi connections, along with Gerald R. Ford, who was spying on the commission for J. Edgar Hoover.

As noted by Professor Donald Gibson of the University of Pittsburgh, "Both of these men [McCloy and Dulles] had always been the Establishment's men in government; they were not the government's men in the Establishment. . . . When McCloy served as a high commissioner in Germany or as president of the World Bank or as a member of the so-called Warren Commission, he did so as a servant of that ruling elite. His most extensive ties to that elite were to various Rockefeller family interests. The other leading figure on that commission, McCloy's longtime close friend Allen Dulles, was also a man of the Establishment, even though his name is almost always associated with the CIA."

In his 1994 book *Battling Wall Street: The Kennedy Presidency,* Gibson made a cogent argument that the primary factor behind the JFK assassination was that he was at loggerheads with the Wall Street Establishment, those same globalist financiers who first promoted communism and then National Socialism.

Prior to the Reagan administration, the Senate Committee on Governmental Affairs in 1980 produced a study titled "Structure of Corporate Concentration." Gibson decribed it as "the most thorough investigation of institutional stockholders and of connections among boards of directors that has ever been done" and wrote, "The basic conclusions [of this study] are concise and quite straightforward: financial institutions, part of, or extensively interrelated with, the Morgan-Rockefeller complex, are the dominant force in the economy."

One example Gibson offered was that the "board of directors of Morgan included individuals serving on the boards of 31 of the top 100 firms. Citicorp was directly tied to 49 top companies, and Chase Manhattan, Chemical Bank, and Metropolitan Life each had 24 other top companies represented on their boards. These and a multitude of other overlaps among the top 100 firms provides a dense network of relationships reinforced by frequent ties through private clubs, educational background, marriages, and membership in organizations such as the Council on Foreign Relations and the Business Council."

After detailing how Kennedy sought to deny total control to this nexus of globalists, Gibson stated, "President Kennedy's commitment to science, technology, and economic progress had led him to adopt policies that were vehemently attacked by people in and around the Morgan-Rockefeller complex."

It was less than two hours after Kennedy was shot—at a time when the Dallas police were unsure of Oswald's identity, since he carried identification in the names of two different persons—that J. Edgar Hoover wrote: "I called the attorney general at his home and told him I thought we had the man who killed the President down in Dallas." FBI documents released in 1977 stated that Hoover had concluded that Lee Harvey Oswald was the assassin and was "a mean-minded individual . . . in the category of a nut." The evening of the assassination, Hoover told Lyndon Johnson's aide Walter Jenkins, "The thing I am concerned about . . . is having something issued so we can convince the public that Oswald is the real assassin."

Within days of Kennedy's death, the same forces opposing his policies began to promulgate the official theory for his death—a lone assassin suffering a "strain of madness and violence" fired at the president from the sixth floor of a book warehouse, striking him twice out of three shots fired within six seconds, despite the fact that the target was 265 feet away, moving laterally and downhill away from the shooter, and an evergreen tree obscured the line of sight.

"Within a couple of days, Alan Belmont of the FBI was pushing the 'Oswald did it alone' conclusion and shortly thereafter McCloy and Dulles were settling the dust with the same conclusion," wrote Gibson. Based on White House telephone transcripts released only thirty years after the assassination, Gibson noted a cohesive effort on the part of prominent people to push for a presidential commission that would cement the lone-assassin theory: "Between November 25th and 28th, LBJ was transformed from opponent of to promoter of a commission. It is clear that a number of people acted to bring about this change. [Dean of Yale Law School] Eugene Rostow brought up the idea initially, to both [presidential special assistant] Bill Moyers and [deputy attorney general Nicholas] Katzenbach. Rostow discussed this with at least one unidentified person in the minutes immediately following Oswald's death. [Journalist] Joseph Alsop applied pressure to LBJ less than 24 hours later. If Alsop is to be believed, and there is no reason to doubt this, Dean Acheson was also involved in developing and promoting the idea [Alsop said he had discussed the matter with Acheson apparently in an effort to impress Johnson that high-level people thought well of the idea of a commission]. Other immediate supporters appear to include both [*Washington Post* managing editor] Alfred Friendly and [*Post* owner] Katherine Graham."

"The venue for the McCloy-Dulles work was the [Warren] Commission created at the instigation of Rostow, Alsop, and Acheson. The cover-up was essentially an operation of private power based in the East Coast Establishment," Gibson concluded, adding the Warren Commission was "essentially an Establishment cover-up. Only an Establishment network could . . . reach into the media, the CIA, the FBI, the military (control of [JFK's] autopsy), and other areas of the government."

And, as has been seen, the inner-core U.S. establishment was filled with supporters of National Socialism, just as they had supported communism before that. This global elite was working to lay the groundwork for their New World Order—worldwide socialism broken into three economic blocs to be played against each other for profit and control.

With the assassination of 1963 and its subsequent cover-up, the globalists, who first created communism and then National Socialism, had finally gained a new empire, a Fourth Reich. Only this time, it was in North America.

PART THREE

THE
REICH
ASCENDANT

CHAPTER 11

REBUILDING THE REICH, AMERICAN-STYLE

WITH KENNEDY DEAD AND LYNDON B. JOHNSON IN THE White House, the Nazification of the United States moved ahead largely unhindered.

During the Johnson years, the president was surrounded by a coterie of advisers, collectively known as his "wise men." All were members of the Council on Foreign Relations. These included John J. McCloy, Averell Harriman, Dean Rusk, William Bundy, Dean Acheson, George F. Kennan, and Robert A. Lovett. "By the early 1960s the Council on Foreign Relations, Morgan and Rockefeller interests, and the intelligence community were so extensively interbred as to be virtually a single entity," remarked Professor Donald Gibson.

With young people, the media, and members of Congress fixated on the Vietnam War, few people were aware of the growing power and influence of the military and the immense war machine assembled behind it. Not that this war machine was designed to actually win battles. On the contrary, it was designed to suck tax dollars from the public treasury, centralize power in the government and its corporate sponsors, and spread the new Reich's influence across the globe.

Under the banner of freedom and democracy, yet pursuing the agenda of the globalists who supported the Nazis, the United States slowly turned from one of the most admired nations in the world to one of the most despised. William Blum, a former State Department employee turned author, stated: "From 1945 to 2003, the United States attempted to overthrow more than forty foreign governments and to crush more than thirty populist-nationalist movements fighting against intolerable regimes. In the process, the U.S. bombed some twenty-five countries, caused the end of life for several million people, and condemned many millions more to a life of agony and despair."

The result of America's empire-building national policy has been dismal at best and catastrophic at worst.

Putting aside the historical aggression displayed by American foreign policy in the Mexican War of 1848 and the Spanish-American War of 1898, a series of misguided foreign-policy adventures since the arrival of thousands of Nazis following World War II includes:

- In 1953, a few years after Iran's prime minister Mohammed Mossadegh engaged in a gradual and lawful nationalization of the oil industry in that Mideast nation, he and his democratic government were deposed by a coup instigated by the CIA. This brought the shah to power, with the monarchy assuming complete control in 1963, and turning Iran into a client state of the United States. Thousands of Iranians, perhaps millions, died during the repressive rule of the shah and his brutal SAVAK secret police. The shah was finally forced out in 1979 by the Ayatollah Khomeini, who quickly became the United States' latest foreign enemy, despite the fact that he had been on the CIA payroll while living in Paris. The shah was granted asylum in the United States, and a medieval version of Islam took control over Iran, which by 2007 was again a targeted enemy.
- In 1954, the CIA toppled the popularly elected government of Jacobo Árbenz in Guatemala, which had nationalized United Fruit property. Prominent American government officials such as former CIA director Walter Bedell Smith, then CIA director Allen Dulles, Secretary of State for Inter-American Affairs John Moors Cabot, and Secretary of State John Foster Dulles were all

closely connected to United Fruit. An estimated 120,000 Guatemalan peasants died in the resulting military dictatorships.

◆ Fidel Castro, with covert aid from the CIA, overthrew the military dictatorship of Fulgencio Batista in 1959 and instituted sweeping land, industrial, and educational reforms as well as nationalizing American businesses. He was swiftly labeled a communist, and the CIA organized anti-Castro Cubans, which resulted in numerous attacks on Cuba and the failed Bay of Pigs Invasion in 1961. The island nation has been the object of U.S. economic sanctions since that time.

◆ In 1965, more than 3,000 persons died in the wake of an invasion of the Dominican Republic by U.S. Marines. The troops ostensibly were sent to prevent a communist takeover, although later it was admitted that there had been no proof of such an attempt.

◆ Also in 1965, the United States began the bombing of North Vietnam after President Lyndon B. Johnson proclaimed the civil war there an "aggression" by the North. Two years later, American troop strength in Vietnam had grown to 380,000, and soon after climbed to more than 500,000. U.S. dead by the end of that Asian war totaled some 58,000, with casualties to the Vietnamese, both North and South, running into the millions.

◆ In 1973, the elected government of Salvador Allende in Chile was overthrown by a military coup aided by the CIA. Allende was killed, and some 30,000 persons died in subsequent violence and repression, including some Americans. Chile was brought back into the sphere of influence of the United States and remained a military dictatorship for the next two decades.

◆ In 1968, General Sukarno, the unifier of Indonesia, was overthrown by General Suharto, again with aid from the CIA. Suharto proved more dictatorial and corrupt than his predecessor. Some 800,000 persons reportedly died during his regime. Another 250,000 persons died in 1975, during the brutal invasion of East Timor by the Suharto regime, aided by the U.S. government and Henry Kissinger.

- In 1979, the powerful and corrupt Somoza family, which had ruled Nicaragua since 1937, was finally overthrown and Daniel Ortega was elected president. But CIA-backed Contra insurgents operating from Honduras fought a protracted war to oust the Ortega government, and an estimated 30,000 people died. The ensuing struggle came to include such shady dealing in arms and drugs that it created a scandal in the United States called Iran-Contra, which involved persons connected to the National Security Council selling arms to Iran, then using the profits to buy drugs in support of the Contras. All of those indicted or convicted of crimes in this scandal were pardoned by then-president George H. W. Bush.

- In 1982, U.S. Marines landed in Lebanon in an attempt to prevent further bloodshed between occupying Israeli troops and the Palestine Liberation Organization. Thousands died in the resulting civil war, including hundreds of Palestinians massacred in refugee camps by right-wing Christian forces while Ariel Sharon, then an Israeli general, looked on with apparent approval. Despite the battleship shelling of Beirut, and the destruction of that great Mediterranean city, American forces were withdrawn in 1984 after a series of bloody attacks on them. More than two decades later, the conflict between Israel and the Palestinians remains as intractable and deadly as ever, in large part due to the virtually unconditional support of Israel by the United States, which has been sustained by the Israel lobby.

- In 1983, U.S. troops invaded the tiny Caribbean island nation of Grenada after a leftist government was installed. The official explanation was to rescue a handful of American students who initially said they did not need rescuing. The only real damage inflicted in this tiny war was to a mental-health hospital partly owned by a White House physician and widely reported to be a CIA facility, possibly used for mind-control experiments.

- During the 1970s and 1980s, the U.S. government gave aid and arms to the right-wing government of the Republic of El Salvador, which represented the financial interests of a tiny oligarchy,

for use against its leftist enemies. By 1988, some 70,000 Salvador-
ans had died.

♦ More than a million persons died in the fifteen-year battle in Angola between the Marxist government aided by Cuban troops
and the National Union for the Total Independence of Angola,
supported by South Africa and the U.S. government.

♦ When Muammar al-Qaddafi tried to socialize the oil-rich North
African nation of Libya, beginning with his takeover in 1969, he
drew the wrath of the U.S. government. In 1981, it was claimed
that Qaddafi had sent hit teams to the United States to assassinate President Reagan, and in 1986, following the withdrawal of
U.S. oil companies from Libya, an air attack was launched, which
missed Qaddafi but killed several people, including his infant
daughter.

♦ In 1987, an Iraqi missile attack on the U.S. frigate *Stark* resulted
in 37 deaths. Shortly afterward, the Iraqi president apologized for
the incident. In 1988, a U.S. Navy ship shot down an Iranian
airliner over the Persian Gulf, causing 290 deaths. The Reagan
administration simply called it a mistake.

♦ As many as 8,000 Panamanians died over Christmas 1989, when
President George H. W. Bush sent U.S. troops to invade that
Central American nation to arrest one-time ally, Manuel Noriega.
The excuse was that the Panamanian dictator was involved in the
importation of drugs to the United States. *U.S. News & World
Report* noted that a year later, the amount of drugs moving
through Panama had doubled.

♦ Iraqi casualties, both military and civilian, totaled more than
300,000 during the short Persian Gulf War of 1991. It has been
estimated that more than a million Iraqis, including women and
children, have died as a result of the continued missile and air
attacks—not including those killed since the U.S. invasion in
2003—as well as of economic sanctions against that nation.

♦ Also in 1991, the United States suspended assistance to Haiti after the election of a liberal priest sparked military action and disorder. Eventually, U.S. troops were deployed. Once again in 2004,

the United States fomented and backed the toppling of the demo-
cratically elected president and replaced him with an unelected
gang of militarists, CIA operatives, and corporate predators.

◆ Other nations that have felt the brunt of CIA and/or U.S. military activity as a result of globalist foreign policy include Somalia,
Afghanistan, Serbia, Kosovo, Bosnia, Brazil, Chad, Sudan, and
many others.

IN EARLY 1974, while President Nixon was desperately trying to find a
way out of impending impeachment, G. Gordon Liddy, ringleader of the
break-in at the National Democratic Party headquarters, was preoccupied
with Nazi imagery. Liddy had named the Watergate "plumbers" after that
vast secretive organization that helped Nazis escape both Europe and justice after the war. "Our *O*rganization had been *D*irected to *E*liminate
*S*ubversion of the *S*ecrets of the *A*dministration, so I created an acronym
using the initial letters of those descriptive words [italics added]. ODESSA
appealed to me because when I organize, I am inclined to think in German terms and the acronym was also used by a World War II German
veterans organization belonged to by some acquaintances of mine," Liddy
wrote in his 1980 book *Will: The Autobiography of G. Gordon Liddy.*

According to author Edward Jay Epstein, in 1971, Liddy invited a
number of White House officials to view Nazi propaganda films to "demonstrate how a few determined men could manipulate the emotions of an
entire nation by invoking a few highly visual symbols of fear." These Nixon
officials included John Ehrlichman, Egil Krogh, Donald Santarelli, and
Robert Mardian. "The cycle of films was climaxed on June 13 by the
showing of *Triumph of the Will,* a Nazi propaganda film made under the
auspices of Hitler and Goering, which graphically depicted the way a 'national will' could be inculcated into the masses through the agency of
controlled fear and frenzied outrage," reported Epstein.

Paul Manning noted: "The German–South American group also had
direct access to the Nixon White House through their representatives in
Washington, and were proud of the fact that Bebe Rebozo was President
Nixon's closest friend. For, knowingly or unknowingly, Rebozo processed

millions of their dollars through his Florida bank as part of normal commercial operations."

And it was during Nixon's presidency that Prescott Bush's son, George Herbert Walker Bush, one of the last of Nixon's Republican loyalists, was named chairman of the Republican National Committee (RNC).

With Nixon's resignation in August 1974, the United States entered a period of further turmoil. The Church Committee uncovered conspiracies, including assassination plots within the CIA, and recriminations started, following the loss in Vietnam. Gerald Ford, a Republican insider, had been appointed vice president with the resignation of Spiro Agnew, who was under indictment for tax evasion. When Ford became president, he promptly pardoned Nixon of all crimes and, at the behest of his secretary of defense, Donald Rumsfeld, appointed George H. W. Bush to head the CIA.

At the time, most people could not understand Bush's appointment, having forgotten that his Nazi-connected grandfather, Senator Prescott Bush, had been one of those instrumental in establishing the CIA.

Meanwhile, the globalists, realizing that the Republican Party in the wake of Nixon's resignation was politically vulnerable, were maneuvering to place a Democrat in the White House. They created an outgrowth of the old Council on Foreign Relations called the Trilateral Commission.

THE CONCEPT OF the Trilateral Commission was brought to David Rockefeller in the early 1970s by Zbigniew Brzezinski, then head of the Russian studies department at Columbia University. While at the Brookings Institution, Brzezinski had been researching the need for closer cooperation between the trilateral nations of Europe, North America, and Asia.

In a book titled *Between Two Ages: America's Role in the Technetronic Era,* Brzezinski foresaw a society "that is shaped culturally, psychologically, socially, and economically by the impact of technology and electronics—particularly in the area of computers and communication." He also declared, "National sovereignty is no longer a viable concept" and predicted "movement toward a larger community by the developing nations . . . through a variety of indirect ties and already developing limitations on national sovereignty." He saw this larger community being funded by "a global taxation system," similar to one that is now being proposed in the United Nations.

Brzezinski's plan for a commission of trilateral nations was first presented during a meeting of the ultrasecret Bilderberg Group in April 1972, in the small Belgian town of Knokke. Reception to Brzezinski's proposal reportedly was enthusiastic. The Trilateral Commission was officially founded on July 1, 1973, with David Rockefeller as chairman. Brzezinski was named founding North American director. North American members included Georgia governor Jimmy Carter, U.S. Congressman John B. Anderson (another presidential candidate), and Time Inc. editor in chief Hedley Donovan. Foreign founding members included Reginald Maudling, Lord Eric Roll, *Economist* editor Alistair Burnet, FIAT president Giovanni Agnelli, and French vice president of the Commission of European Communities Raymond Barre.

Even the establishment-oriented media expressed uneasiness over the preponderance of Trilaterals in government in early 1977. Columnist William Greider writing in the *Dallas Morning News* noted: "But here is the unsettling thing about the Trilateral Commission. The president-elect [Carter] is a member. So is vice president–elect Walter F. Mondale. So are the new secretaries of state, defense and treasury, Cyrus R. Vance, Harold Brown and W. Michael Blumenthal. So is Zbigniew Brzezinski, who is a former Trilateral director and Carter's national security advisor, also a bunch of others who will make foreign policy for America in the next four years." Antony C. Sutton and Patrick M. Wood, authors of *Trilaterals Over Washington,* commented, "If you are trying to calculate the odds of three virtually unknown men [Carter, Mondale, and Brzezinski], out of over sixty [Trilateral] commissioners from the United States, capturing the three most powerful positions in the land, don't bother. Your calculations will be meaningless."

Despite being a creation of the Rockefeller-dominated Trilateral Commission and following some of their aims, such as eliminating price controls for domestic petroleum production by establishing a national energy policy and further drawing power to the federal government by creating the departments of energy and education, Carter apparently failed to satisfy the globalists.

The mass media were already focusing on conservative California government Ronald Reagan as the man of the hour. Reagan's nomination as GOP presidential candidate for the 1980 election seemed assured.

Carter asked for and was granted a national television spot during prime time, and many media pundits predicted that he was about to announce sweeping changes in government as well as new initiatives that would move his upcoming presidential reelection campaign off high center. But before his televised appearance, Carter journeyed to California, where he was to address a Hispanic crowd in the Los Angeles Civic Center Mall celebrating Cinco de Mayo, the day the Mexicans defeated the French Army in 1862. A few days later, a handful of newspapers carried a small story stating that a "grubby transient" had been arrested there and was being held on suspicion of the attempted assassination of the president. A Secret Service spokesman downplayed the arrest, stating the incident was about as "nothing as these things get."

However, a few days later, another news item appeared, which reported that the thirty-five-year-old Anglo suspect was being held in lieu of $50,000 on charges of conspiring to kill the president. Finally, a one-time story in the May 21, 1979, edition of *Newsweek* revealed more details of the incident. According to the news magazine, the suspect was arrested after Secret Service agents noticed him "looking nervous." A .22, eight-shot revolver was found on the man along with seventy rounds of blank ammunition. A short time later, the suspect implicated a second man, a twenty-one-year-old Hispanic, who also was taken into custody and subsequently held in lieu of $100,000 bail. The second suspect at first denied knowing the other man, but finally admitted that the pair had test-fired the blank starter pistol from a nearby hotel roof the night before Carter's appearance. Both men said they were simply local street people hired by two hit men who had come up from Mexico. They were to create a diversion with the blank pistol, and the two hit men were to assassinate President Carter with high-powered rifles.

Lending credence to their story, both suspects led authorities to the shabby Alan Hotel located near the civic center. Here investigators found an empty rifle case and three rounds of live ammunition in a room that had been rented under the name Umberto Camacho. Camacho apparently had checked out the day of Carter's visit. No further trace of the hit men could be found.

The Anglo suspect's name was Raymond Lee Harvey and his Hispanic companion's name was Osvaldo Ortiz. This oddity of their names prompted *Newsweek* reporters to state, "References to Lee Harvey Oswald and the assassination of President John F. Kennedy were unavoidable. . . . But," they added, "it was still far from clear whether the authorities had a real conspiracy or a wild goose chase on their hands."

No further news stories appeared, and the disposition of the case against Lee Harvey and Osvaldo apparently has never been made public. A recent search of the federal prisoner database indicated no such persons are currently incarcerated.

But apparently Carter got the message. He canceled his national TV speech and went into seclusion at Camp David. After seeking advice from a lengthy line of consultants, including the Reverend Billy Graham, Carter was reported to have said, "I have lost control of the government."

Backing away from any serious policy changes, Carter remained indecisive in the public eye. By mid-November the following year, the United States took a conservative turn and elected Republican Ronald Reagan. Reagan's victory was due, in large part, to a failed military attempt to rescue U.S. hostages held by Iranian radicals, followed by the collapse of negotiations regarding their release in mid-October 1980.

REAGAN, A FORMER spokesman for General Electric Company, stocked his administrations with current and former members of globalist groups, the very people he had criticized while campaigning.

During the 1980 presidential campaigns, Reagan verbally attacked the nineteen Trilaterals in the Carter administration and vowed to investigate the group if elected. While competing against George H. W. Bush for the nomination, Reagan lambasted Bush's membership in both the Trilateral Commission and the CFR and pledged not to give Bush a position in a Reagan government.

Yet during the Republican National Convention, a strange series of events took place. With Reagan secured as the presidential candidate, there was a contentious fight to see who would be vice president. In midweek, national media commentators suddenly began talking about a "dream ticket" to be composed of President Ronald Reagan and Vice President (the former president) Gerald Ford. It was even suggested that since Ford had been president, he should choose half of the Reagan cabinet.

Faced with the prospect of a split presidency, Reagan rushed to the convention floor late at night and announced, "I know that I am breaking with precedent to come here tonight and I assure you at this late hour I'm not going to give you my acceptance address tonight. . . . But in watching at the hotel the television, and seeing the rumors that were going around and the gossip that was taking place here . . . [l]et me as simply as I can straighten out and bring this to a conclusion. It is true that a number of Republican leaders . . . felt that a proper ticket would have included the former president of the United States, Gerald Ford, as second place on the ticket. . . . I then believed that because of all the talk and how something might be growing through the night that it was time for me to advance the schedule a little bit. . . . I have asked and I am recommending to this convention that tomorrow when the session reconvenes that George Bush be nominated for vice president."

For one brief moment, the power of those who control the corporate mass media was revealed. Reagan never again uttered a word against the globalist groups such as the Trilateral Commission and the Council on Foreign Relations. Following his election, Reagan's fifty-nine-member transition team was composed of twenty-eight Council on Foreign Relations members, ten members of the elite Bilderberg Group, and at least ten members of the Trilateral Commission. He even appointed prominent CFR members to three of the nation's most sensitive offices—Secretary of State Alexander Haig, Secretary of Defense Caspar Weinberger, and Secretary of the Treasury Donald Regan. Additionally, he named Bush's campaign manager James A. Baker III, who then served as a chairman of the Reagan-Bush campaign committee, as his chief of staff. Baker is a fourth-generation member of a family long connected to Rockefeller oil interests.

After Reagan won in November, it was alleged that Bush, along with CIA Director William Casey, had privately cut a deal with Iranian leaders to hold American hostages until after the November election, thus assuring a Reagan victory. Later testimony confirming this "October surprise" came from several people involved, including Richard Brenneke and Heinrich Rupp, who claimed to have flown Casey to a meeting with the Iranians and the Iranian foreign minister. Because of his damaging testimony, Brenneke was tried for perjury but found not guilty. Jury foreman Mark Kristoff stated, "We were convinced that, yes, there was a meeting, and he was there and the other people listed in the indictment were there." Despite this verdict, no action was taken by the Reagan-Bush administration, thanks primarily to debunking by a House Task Force led by Congressman Lee Hamilton, the same man Bush's son would name to cochair his 9/11 Commission in late 2002.

On January 20, 1981, claims of this "October surprise" conspiracy were further supported by the facts that just minutes after Reagan was sworn into office, the American hostages were released, and within weeks, military supplies that Carter had withheld from Iran began moving to that nation. Then, just two months after taking office in 1981, President Reagan was shot by would-be assassin John W. Hinckley, who exhibited the symptoms of brainwashing and whose brother had scheduled dinner with Neil Bush the very day Reagan was shot. For many weeks, while many Americans prayed for Reagan's recovery, the son of Prescott Bush ran the nation.

Bush had exerted his influence to have Alexander Haig appointed secretary of state, and only days before the attempted assassination of Reagan had named Haig to head a special emergency preparedness committee. Haig, a ranking globalist member of the Council on Foreign Relations, was Nixon's chief of staff from 1973 to 1974. It was Haig who finally advised Nixon to resign. Haig was also NATO commander from 1975 to 1979.

Was it sheer coincidence that Hinckley's brother had scheduled dinner with Bush's son Neil the very night Reagan was shot, or that Hinckley's father, a Texas oil man, and George H. W. Bush were longtime friends? It should also be noted that Bush's name—including his then little-publicized nickname "Poppy," which has caused many to wonder if this referred to his parenthood or the narcotic plant—address, and phone number were found in the personal notebook of oil geologist George De-Mohrenschildt, the last known close friend of Lee Harvey Oswald. Many researchers view these seemingly small, unconnected, and little-reported details as being beyond coincidence. Some saw Hinckley's action as an attempt to bring Bush to power eight years before he was elected president.

"This I believe was a coup," stated assassination researcher John Judge, cofounder of the Coalition on Political Assassinations (COPA). In a 2000 interview, Judge stated his belief that "loyalists won the concession that Reagan will be allowed to stay alive but Bush would come into power and at that point Haig emerged from the situation room to the press and said, his famous quote, 'Gentlemen, I am in charge here until the vice president returns.' That meant two things: number one, that they were going extraconstitutional—beyond the twenty-fifth amendment, a military takeover, and [number two] Haig in this office of preparedness, prior to Bush, and basically he's taking charge. The press was questioning, 'What does this mean?' What they don't understand is all that constitution stuff is pushed aside once they declare national emergencies. Then they go into FEMA and they have whole other orders of succession that have to do more with the military and the Pentagon than with any of the civilian sector." Constitutionally, the next in line in the order of succession is the vice president, then the speaker of the House, then the Senate president pro tempore, then the secretary of state. Vice President George H. W. Bush was flying from Texas at the time of Haig's proclamation.

Hinckley was whisked off to Quantico Marine Base, then sent for psychiatric evaluation at Fort Butner, South Carolina, which Judge described as "the first mind-control experimentation prison in the country." All this time, Hinckley was under military control, not civilian. He was eventually brought to court and declared not guilty by reason of insanity for the assassination attempt.

"The patterns are always the same. You have a patsy that takes the blame. You have a second gunman that never comes to light. And you have an ascendance of power. That's what I think happened after that point: that Reagan was basically allowed to function but Bush was president," said Judge.

And Bush was virtually unassailable, due to his hidden but powerful support base. Robert Parry, a former investigative reporter for the Associated Press and *Newsweek,* noted: "Even when—or maybe especially when—Bush found himself in a corner on what appeared to be an obvious lie, he was a master at turning the tables on his critics. Coming to Bush's defense was an impressive network of friends in high places. They rarely failed him. . . . When that happened, it was wise not to ask too many questions."

ANOTHER INDICATION THAT the Reagan administration may have been under the influence of fascists came in May 1985, when the president laid a wreath at a soldiers' cemetery in Bittburg, Germany, where many Nazi SS officers were buried. It was also the former site of Bergen-Belsen concentration camp. Although a hue and cry went up from veterans organizations and Jewish groups prior to his visit, Reagan followed through with his plan to honor war dead. In his remarks, he placed the blame for Nazi atrocities on "the awful evil of one man," an obvious reference to Adolf Hitler. This effort to foist off all the blame for Nazism onto one person was perhaps an indication of the influence of pro-Nazi elements within his party.

Meanwhile, throughout the 1980s, Republican Party leaders continued their policy of bringing former Nazis and Nazi-minded foreigners into the party's camp. According to investigative reporter Christopher Simpson, author of *Blowback,* Nazi émigrés brought into the USA by the CIA were placed in prominent positions within the Republican Party through "ethnic outreach committees."

Online Journal is a reader-supported Web zine that was established in 1998 to "provide uncensored and accurate news, analysis and commentary." According to their reporter Carla Binion, a convicted Nazi war collaborator named Laszlo Pasztor served as an adviser to Republican Paul Weyrich, who founded the powerful conservative Heritage Foundation and is considered by many to be one of the founders of the "New Right." Weyrich garnered large support by appealing to Christian fundamentalists and anticommunists. Pasztor built up the GOP émigré network and was founding chairman of the Republican Heritage Groups Council. Pasztor reportedly belonged to the Hungarian Arrow Cross, a group that helped liquidate Jews there during the war. Interestingly enough, Pasztor's efforts to make the Heritage Groups Council an effective branch of the GOP coincided with George H. W. Bush's term as head of the Republican National Committee.

"After Nixon's landslide victory in 1972, he ordered a general house cleaning on the basis of loyalty," stated John Loftus and Mark Aarons in *The Secret War Against the Jews*. The authors quote Nixon as telling John Ehrlichman, "Eliminate everyone except George Bush. Bush will do anything for our cause." Indeed, it was the elder Bush who fulfilled Nixon's pledge to make émigrés with Nazi backgrounds a permanent part of Republican politics. "It is clear that George Bush, as head of the Republican National Committee in 1972, must have known who these 'ethnics' really were," the authors concluded.

Based on the research of journalist Russ Bellant, author of the 1991 book *Old Nazis, the New Right, and the Republican Party,* other Nazi collaborators involved with the Republican Party included:

◆ *Radi Slavoff,* executive director of the GOP's Heritage Council and leader of "Bulgarians for Bush," who was a member of a Bulgarian fascist group. Slavoff created a Washington public event for writer Austin App, who in 1987 revealed his pro-Nazi sympathies by writing, "The truth is that in WW II, the Third Reich fought for justice, and the Allies fought to prevent justice."

◆ *Florian Galdau,* who directed a Republican outreach program among Romanians and would head "Romanians for Bush" in 1988. Galdau was a supporter of Valerian Trifa, convicted of war crimes when he headed the Romanian Iron Guard in Bucharest.

◆ *Nicholas Nazarenko,* a former SS officer, who was the head of a Cossack Republican ethnic unit during the Nixon years. Although accused of hanging Jews in Odessa, in the 1980s Nazarenko organized an anticommunist demonstration in New York City.

- *Method Balco,* who headed the Slovak-American Republican Federation of the GOP's Heritage Groups Council and during the 1950s organized annual memorials to the pro-Nazi regime of Slovakian Josef Tiso, a creation of Hitler's after the division of Czechoslovakia in 1939.

- *Walter Melianovich,* head of the GOP's Belarusan ethnic unit, who was closely connected to the Belarusan-American Association, an organization rife with transplanted fascists, and in 1988 became national chairman of "Belarusans for Bush."

- *Bohdan Fedorak,* who during the war was a top U.S. representative of the Organization of Ukrainian Nationalists—Bandera, a group that committed atrocities in the Nazi-occupied Ukraine— and in 1988 became national vice chairman of "Ukrainians for Bush." As a ranking member of the Ukrainian Congress Committee of America, Fedorak lobbied Congress trying to stop Justice Department prosecutions of pro-Nazi Ukrainian war criminals.

Allan A. Ryan Jr., former director of the Justice Department's Office of Special Investigations, said he found Bellant's reporting "well-documented and reliable."

Just weeks before the 1988 election, the *Washington Jewish Week* revealed that several Nazis and Jew-haters were involved in the coalition supporting Bush's Republican campaign. When this news broke, at least four of those mentioned by name were forced to resign. The Nazi connections to the Republican Party cited by Bellant and the Jewish publication were confirmed by an investigation by reporters from the *Philadelphia Inquirer* in September 1988.

Online Journal reporter Carla Binion wondered aloud if Reagan, Bush, or Reagan's CIA director William Casey realized they were being aided and supported by Nazis and Nazi collaborators. The available evidence indicates they were. "One thing is certain," Binion concluded, noting that Bush had preceded Casey as a CIA director, "The intelligence agencies know the scope and extent of Nazi involvement with the political right in this country. It is a shame they keep it hidden from the majority of the American people."

This charge is confirmed by a list of nearly two thousand "Former Nazi and Fascist Individuals Entering the U.S. under Official Auspices," recently released by the National Archives after being locked away for years by presidential order.

Peter Levenda also has studied the connection between old Nazis and ranking Republicans. After noting the prosecution of Prescott Bush for being a financial frontman for Hitler, he wrote, "We cannot, of course, hold former President Bush responsible for the sins of his father; nor can we hold his son responsible. Yet, we can expect a higher degree of moral responsibility in their actions as men and as political leaders. Unfortunately . . . in the 1988 presidential campaign, George H. W. Bush was happy to accept support from a range of Nazis and Nazi-sympathizers in his quest for the White House, and was just as happy to keep them on in the administration even after they had been identified as such."

Writing about a streak of anti-Semitism in the globalist individuals and companies that supported Hitler and continued to support Nazism even after the war, Levenda stated, "I believe that the entire racial theory of Nazism was a comfortable environment for these men. They were, after all, from privileged backgrounds: old money, power, prestige, the right companies, the right schools, the right fraternities (such as the infamous Skull and Bones at Yale, to which generations of the Bush family belonged). The Nazis embodied the secret dreams and unspoken loyalties of these men, the public acknowledgment of all that the American elite held dear."

DURING THE REAGAN years, as most Americans were lulled into a false sense of security, the minions of the fascist globalists took steps to change the power structure of America.

James Mann, former Beijing bureau chief for the *Los Angeles Times* and a senior writer-in-residence at the Center for Strategic and International

Studies, took note that during the 1980s, when Bush was in virtual command of the White House during Reagan's hospitalization and recuperation, Dick Cheney and Donald Rumsfeld were conspicuously absent at least once a year. Cheney and Rumsfeld, along with several dozen federal officials and one member of the cabinet, would travel to Andrews Air Force Base, usually in the middle of the night, and from there would proceed to a remote location in the United States, such as a decommissioned military base or an underground bunker.

Mann reported that "Cheney [Gerald Ford's chief of staff and a former director of the CFR] was working diligently on Capitol Hill as a congressman rising through the ranks of the Republican leadership. Rumsfeld, who had served as Gerald Ford's secretary of defense, was a hard-driving business executive in the Chicago area—where, as the head of G. D. Searle and Company, he dedicated time and energy to the success of such commercial products as NutraSweet, Equal, and Metamucil. Yet for periods of three or four days at a time no one in Congress knew where Cheney was, nor could anyone at Searle locate Rumsfeld. Even their wives were in the dark; they were handed only a mysterious Washington phone number to use in case of emergency."

Cheney and Rumsfeld were involved in one of the most highly classified programs of the Reagan administration, a program that called for setting aside the legal rules for presidential succession. This "continuity of government" program was created by a secret executive order from Reagan.

According to Mann, one of the program's participants told him, "One of the awkward questions we faced was whether to reconstitute Congress after a nuclear attack. It was decided that no, it would be easier to operate without them. For one thing, it was felt that reconvening Congress, and replacing members who had been killed, would take too long."

Mann continued: "Within Reagan's National Security Council the 'action officer' for the secret program was Oliver North, later the central figure in the Iran-contra scandal. Vice President George H. W. Bush was given the authority to supervise some of these efforts, which were run by a new government agency with a bland name: the National Program Office. It had its own building in the Washington area, run by a two-star general, and a secret budget adding up to hundreds of millions of dollars a year. When George H. W. Bush was elected president, in 1988, members of the secret Reagan program rejoiced; having been closely involved with the effort from the start, Bush wouldn't need to be initiated into its intricacies and probably wouldn't reevaluate it. In fact, despite dramatically improved relations with Moscow, Bush did continue the exercises, with some minor modifications."

Although the elder Bush gained his own time in the White House in 1988, it was limited to one term due to the controversies and conspiracies swirling about him, not the least of which was his father's pro-Nazi background. It was to escape this heritage that, in 1949, young George had moved from his ancestral home in Connecticut to the more receptive environs of Texas.

During the Clinton administration, those who knew about the "continuity of government" plan considered it an outdated relic. Though it was neglected, it was not abolished. After September 11, 2001, the creators of this plan moved into action. Mann reported that, in the Presidential Emergency Operations Center under the White House, Cheney told President Bush to delay his planned flight back from Florida, while at the Pentagon, Rumsfeld instructed Deputy Secretary of Defense Paul Wolfowitz to leave Washington for the safety of one of the underground bunkers. "Cheney also ordered House Speaker Dennis Hastert, other congressional leaders, and several cabinet members (including agriculture secretary Ann Veneman and interior secretary Gale Norton) evacuated to one of these secure facilities away from the capital," added Mann.

In the days following 9/11, the American news media finally mentioned the existence of this "shadow government." Of course, there was no mention of the Nazi-connected globalists who had inspired it.

"Their [Cheney's and Rumsfeld's] participation in the extraconstitutional continuity-of-government exercises, remarkable in its own right, also demonstrates a broad, underlying truth about these two men," Mann stated. "For three decades, from the Ford administration onward, even when they were out of the executive branch of government, they were never far away. They stayed in touch with defense, military, and intelligence officials, who regularly called upon them. They were, in a sense, a part of the permanent hidden national-security apparatus of the United States."

ALL OF THIS was a far cry from the fringe rants of neo-Nazis like George Lincoln Rockwell, founder of the American Nazi Party, who was assassinated by a former colleague in August 1967.

Rockwell, a former navy pilot during World War II and failed artist like his hero, Adolf Hitler, saw the telltale signs of conspiracy abroad in America, but, like Senator Joseph McCarthy and the more recent Holocaust deniers, attributed it to "international communism" coupled with the anti-Semitic view of a worldwide Jewish conspiracy. He formed various National Socialist political organizations and made headlines trying to conduct public rallies in various places.

Rockwell, as well as the more recent Nazi skinheads, merely served to focus public attention on these fringe elements of society and away from the military-industrial empire being created all around them.

The United States has long been governed by men connected to secret societies such as the Council on Foreign Relations and the Trilateral Commission, both of which can be traced back to much earlier societies, like the previously mentioned Bavarian Illuminati and Freemasonry. These groups can, in turn, be traced back to even earlier societies, such as the Knights Templar and Rosicrucians, which all had a particular interest in alchemy and the occult. As reported earlier, it was the right-wing German Thule Gesellschaft, or Thule Society, an offshoot of the Teutonic Knights, that formed the nucleus of the fledgling Nazi Party. Whether or not the infamous Illuminati still exists, its credo "the end justifies the means" lives on in the hearts of the corporate owners of today—globalists who value the blending of state and corporate power, the very definition of fascism.

One organization that forms the connective tissue between these various secret groups may be the shadowy Bilderberg Group—powerful men and women, many of European royalty, who meet in secret each year to discuss the issues of the day. This reclusive group is considered by researchers to be the center of the world's social and economic manipulation, yet the Bilderberg meetings receive virtually no coverage in the corporate mass media, even though well-known American journalists, such as William F. Buckley and Bill Moyers, attend.

Founded by Prince Bernhard, the Bilderberg Group is composed of the inner-core elite of the Council on Foreign Relations and the Trilateral Commission. The name "Bilderberg" has been identified with the Bilderberg Hotel in Oosterbeek, Holland, where the group was first discovered, but some researchers claim the name was derived from an I. G. Farben subsidiary, Farben Bilder. Prince Bernhard was a director of Farben Bilder in the 1930s.

Dutch prince Bernhard—full name Bernhard Julius Coert Karel Godfried Pieter, prince of the Netherlands and of Lippe-Biesterfeld—was the primary impetus for the Bilderberg meetings. As previously noted, Bernhard was a former member of the Nazi Schutzstaffel, or Elite Guard, and an employee of Germany's I. G. Farben in Paris. In 1937, he married Princess Juliana of the Netherlands and became a major shareholder and officer in Dutch Shell Oil, along with Britain's Lord Victor Rothschild.

In England, after the war, Rothschild and Polish socialist Dr. Joseph Hieronim Retinger encouraged Prince Bernhard to create the Bilderberg Group, which began as unofficial meetings between members of Europe's wealthy elite. The official creation of this highly secret organization came about in the early 1950s, following discussions between Prince Bernhard and Dr. Retinger, a founder of the European Movement after World War II. Retinger became known as the "father of the Bilderbergers."

Retinger was brought to America by Averell Harriman just after the war, when Harriman was U.S. ambassador to England. In America, Retinger visited prominent citizens, such as David and Nelson Rockefeller, John Foster Dulles, and then CIA director Walter Bedell Smith, all men with close connections to the Nazis.

Previously, Retinger had formed the American Committee on a United Europe, working alongside future CIA director and CFR member Allen Dulles, then CFR director George Franklin, CIA official Thomas Braden, and former OSS chief William Donovan. Donovan began his intelligence career as an operative of J. P. Morgan Jr. and was known as an "Anglophile," a supporter of close British-American relations. Retinger continued his participation in Bilderberg meetings until his death in 1960. Another CIA-connected person who helped create the Bilderberg Group was *Life* magazine publisher C. D. Jackson, who served under President Eisenhower as "special consultant for psychological warfare."

In fact, the list of American institutions that initially supported the Bilderberg Group reads like a list of prewar financiers of Hitler—First National City Bank [now Citibank], Morgan Guaranty Trust Company, Ford Motor Company, Standard Oil, and Du Pont.

The common denominator of these societies seems to be the acquisition of money, which translates into power. Spencer Oliver, the ranking Democratic Party leader whose telephone was bugged as part of the Watergate break-in, has stated, "The biggest weapon in American politics is money, because you can use money to influence people, to influence the media, to influence campaigns, to influence individuals, to bribe people." As has been seen, the fascist globalists have all the money. They are where the buck stops . . . and begins.

In 1991, then Arkansas governor Bill Clinton was honored as a Bilderberg guest, and the next year he ran for and won the presidency of the United States. After his election, Clinton made no mention of the Bilderberg meetings. Hillary Clinton attended a meeting in 1997, becoming the first American first lady to do so. Thereafter, talk steadily grew concerning her future role in politics, and by 2008 she was a leading Democratic presidential candidate.

One illustration of globalist control within the Clinton administration can be found in the person of President Clinton's treasury secretary Robert E. Rubin, a former cochairman of Goldman Sachs, who was named to head Clinton's National Economic Council. Despite Clinton's promises to "reform our politics so that power and privilege no longer shout down the voice of the people," according to Professor Donald Gibson, who lectures on wealth and power at the University of Pittsburgh and is author of *Battling Wall Street,* Rubin, in his capacity as council director, fought "to protect China's preferred trading status, to protect employers' interest in health-care reform, and to pursue a tougher policy in negotiations with Japan."

"At Goldman Sachs, Rubin had been involved in the kind of high-level paper-shuffling that Bill Clinton has said was undermining the economy,"

Gibson wrote. "Goldman Sachs, along with Morgan Stanley, First Boston, Dillon Read, and others had arranged corporate mergers and acquisitions costing hundreds of billions of dollars in the 1980s. Goldman Sachs and other investment banks were paid many millions of dollars to arrange these deals. For example, Goldman Sachs earned $10 million arranging U.S. Steel's 1982 buyout of Marathon Oil. Rubin's firm was paid $18.5 million for its role in the 1984 Texaco takeover of Getty Oil, and it was paid $15 million for facilitating General Electric's 1986 acquisition of RCA/NBC. In other words, Rubin would seem to have been part of the problem."

In 2007, Rubin was vice chairman of the board of directors of the Council on Foreign Relations.

NOT JUST FRINGE conspiracy theorists have spoken out about hidden control in the world. President Woodrow Wilson, who was intimately connected with conspiratorial power, once wrote, "Some of the biggest men in the United States, in the field of commerce and manufacture, are afraid of somebody, are afraid of something. They know there is a power somewhere so organized, so subtle, so watchful, so interlocked, so complete, so pervasive that they had better not speak above their breath when they speak in condemnation of it."

President Franklin D. Roosevelt once wrote, "The real truth of the matter is, as you and I know, that a financial element in the large centers has owned the government ever since the days of Andrew Jackson."

Colonel L. Fletcher Prouty, a liaison officer between the Pentagon and the CIA in the 1960s, was able to witness the control mechanisms over both intelligence and the military. Prouty said the United States is run by a "secret team," answerable only to themselves. Their power is derived from their vast covert intragovernmental infrastructure and its direct connections with private industries, mutual funds, investment houses, universities, and the news media, including foreign and domestic publishing houses. Prouty would have been horrified to learn that this "secret team" might include Nazis brought into the military-industrial complex after the war.

Another insider who confirmed that a plot was afoot was President Truman's choice for America's first secretary of defense, James V. Forrestal, a man intimately connected with the globalists. Forrestal noted, "These men are not incompetent or stupid. They are crafty and brilliant. Consistency has never been a mark of stupidity. If they were merely stupid, they would occasionally make a mistake in our favor."

They do not make mistakes that favor the best interests of the American people. Take, for example, the position assumed by the George W. Bush administration toward the Russian Federation.

Following the collapse of Communism, there was a splendid opportunity to create new friendship and working arrangements with the eighty-six political entities that comprise the new Russian Federation. No real negative mention was made of Russia during the Clinton years. Yet, suddenly, following the arrival of the Bush administration and the attacks of 9/11, Russia has been presented as a potential enemy and the United States has provoked hostility there by aggressive diplomatic and military maneuvers.

"When the Cold War ended, we seized upon our 'unipolar moment' as the lone superpower to seek geopolitical advantage at Russia's expense," noted conservative writer Patrick J. Buchanan. "Though the Red Army had packed up and gone home from Eastern Europe voluntarily, and Moscow felt it had an understanding we would not move NATO westward, we exploited our moment. Not only did we bring Poland into NATO, we brought in Latvia, Lithuania, and Estonia, and virtually the whole Warsaw Pact, planting NATO right on Mother Russia's front porch. Now there is a scheme afoot to bring in Ukraine and Georgia in the Caucasus, the birthplace of Stalin."

Others saw America's reaction to Russian peace overtures as nothing less than aggression, perhaps a continuation of the National Socialist agenda of destroying the old Soviet Union. According to Mike Whitney of the Information Clearing House, a reader-supported Web information service, "Since September 11 [2001], the Bush administration has carried out an aggressive strategy to surround Russia with military bases, install missiles on its borders, topple allied regimes in Central Asia, and incite political upheaval in Moscow through U.S.-backed 'pro-democracy' groups." It was also noted that it was Bush's America, not Russia, that withdrew from the antiballistic missile treaty, a move that reminded some of the USSR's Cold War–era public pledge never to be the first to use nuclear weapons, a pledge never reciprocated by the United States.

In mid-2007, such actions prompted federation president Vladimir Putin to chastise the United States in a major press conference at the Forty-third Munich Conference on Security Policy. Putin's reasoned remarks were little reported in the corporate-controlled U.S. mass media. "For the first time in history," he said, "there are elements of the U.S. nuclear capability on the European continent. It simply changes the whole configuration of international security. . . . Of course, we have to respond to that." Criticizing Bush's "war on terror," Putin also produced a copy of a report from Amnesty International and stated, "The organization has concluded that the United States is now the principal violator of human rights and freedoms worldwide."

Equally disturbing was Bush's announced advocacy of an American first-strike nuclear capability as well as his proposed "missile defense" system, which though defended as a deterrent to rogue nations, such as North Korea, nevertheless will be placed in Europe. Nobel Prize–winner Thomas C. Schelling of Harvard, an early advocate of the U.S. Mutual Assured Destruction (MAD) Cold War strategy, which theorized that equal nuclear capability would deter a nuclear exchange, shifted his rhetoric from "deterrence" to "compellence," a newspeak term for blackmailing nations into submission through the threat of nuclear weapons.

Pat Buchanan asked: "How would we react if China today brought Cuba, Nicaragua, and Venezuela into a military alliance, convinced Mexico to sell oil to Beijing and bypass the United States, and began meddling in the affairs of Central America and Carribbean countries to effect the electoral defeat of regimes friendly to the United States? How would we react to a Russian move to put anti-missile missiles on Greenland?"

Some researchers saw this return to the Cold War by America as yet another sign that the global National Socialists have not given up on trying to coerce Russia, a nation only too familiar with socialist tyranny, into their New World Order.

Following the tempestuous Clinton administration, the Republicans took power and swiftly set out on a new "neoconservative" path for the party. John W. Dean, former Nixon counsel, who was jailed for felonies committed at the time of Watergate, referred to this new Republican conservatism in his 2007 book *Broken Government: How Republican Rule Destroyed the Legislative, Executive and Judicial Branches:* "It has been new on Capitol Hill since about 1997, about three years after the GOP gained control of the House; it has been new to the White House since 2001, with the arrival of George W. Bush and Richard B. Cheney, although its roots first emerged during the Nixon presidency and began blossoming in the Reagan and Bush Senior [*sic*] years." Although Dean never quite identifies the origin of this "new Republican way of thinking," it is possible that it stemmed from the National Socialist philosophy brought into this country after World War II.

BUT NOT ONLY the grandson of financier Prescott Bush, Nixon cronies, or the neoconservatives have shown sympathy for National Socialist ideals.

Arnold Schwarzenegger, the Austrian-born former bodybuilder turned actor turned governor of California, has a background of pro-Nazi statements and friends. In a 1977 interview, Schwarzenegger was asked which person he admired. His response: "I admire Hitler, for instance, because he came from being a little man with almost no formal education, up to power. I admire him for being such a good public speaker and for what he did with it."

His admiration for Hitler may have come from sitting at his father's knee. In 1938, Hitler's Nazis took control of Austria in an *Anschluss,* unifying that country with Germany. Arnold's father, Gustav, one year later joined Hitler's infamous Sturmabteilung storm troopers (SA), known as the Brown Shirts. Gustav even sported a Hitler-like mustache.

Schwarzenegger also caught flak because of his friendship with Kurt Waldheim, former secretary general of the United Nations, who lost the presidency of Austria in 1992, after his Nazi past was revealed. Records showed that Waldheim had hidden his role as a member of the Nazi SA. According to the 1991 book *Arnold: An Unauthorized Biography* by

Wendy Leigh, Schwarzenegger toasted Waldheim at his 1986 wedding to Maria Shriver by proclaiming, "My friends don't want me to mention Kurt's name, because of all the recent Nazi stuff and the U.N. controversy. But I love him and Maria does too, and so, thank you, Kurt."

In an effort to rehabilitate this Nazi background, Schwarzenegger has subsequently renounced Hitler and made hefty contributions to the Wiesenthal Center in Los Angeles, named for the Jewish Nazi hunter Simon Wiesenthal.

But if prominent Americans have tried to distance themselves from their Nazi pasts, this same concern did not apply to Nazi-developed ideals and substances.

CHAPTER 12

GUNS, DRUGS, AND EUGENICS

WHILE NAZI SCIENCE WAS BROUGHT TO AMERICA AFTER WORLD War II, so were attendant Nazi restrictions on scientific liberty. "Many of the standards of scientific freedom and exchange of knowledge were suspended by all the belligerents," noted John Cornwell, author of *Hitler's Scientists*. Since 1940, America's scientists have become faceless members of teams working under the auspices of the military-industrial complex or the corporate world.

Addressing the Nazi-connected men in control of America's scientific establishment after the war, Cornwell explained: "The most dramatic alteration was in the West. The Office of Scientific Research and Development under the government science chief Vannevar Bush commissioned more than 2,000 research programs in the course of World War II. The projects involved industrial research and development units employing tens of thousands of scientists and technicians in companies such as Du Pont and General Electric, as well as major university laboratories like MIT and Caltech.... [A] proposal for a barrier between government and military funding and civilian control of the choice and direction of basic research would prove, however, a vain hope." Such tight inner control over scientific advances was reminiscent of the late-war Nazi SS control over technology in the Third Reich.

Hitler's Germany was not only the first nation to use or advance television, rocketry, and computers but also the first to build a national freeway system, to address occupational health issues, restrict the use of firearms, attack the use of alcohol and tobacco, pass laws for the protection of the environment, and wage war against cancer.

Hitler realized that he needed the support of his wealthy conservative followers, so he directed much of his public statements to them, particularly in the areas of rearmament and foreign policy. But his social programs in many cases were a liberal's dream come true.

For example, gun control was already widespread in a pre-Nazi Europe unaccustomed to the freedom to bear arms. Anti–gun control advocates have long pointed out that it was an unarmed population that allowed the Nazis to both gain and maintain power. Ironically, the Nazis used the Weimar Republic's gun-control laws—intended to restrict private armies such as Hitler's SA—to keep the population disarmed.

Hitler and his ilk were against keeping arms in the hands of citizens, especially conquered peoples. Hitler once declared: "The most foolish mistake we could possibly make would be to allow the subject races to possess arms. History shows that all conquerors who have allowed their subject races to carry arms have prepared their own downfall by so doing. Indeed, I would go so far as to say that the supply of arms to the underdogs is a sine qua non for the overthrow of any sovereignty."

Still, the Nazis were not content with the stringent gun laws already on the books. In 1938, they strengthened these laws by asserting that only loyal Nazis could own weapons. This was codified in the Nazi Weapons Law of March 18, 1938.

A group opposed to gun control, called Jews for the Preservation of Firearms Ownership, Incorporate (JPFO), has made the shocking but well-supported argument that U.S. gun-control legislation is based on this Nazi law. "JPFO has hard evidence that shows that the Nazi Weapons Law (March 18, 1938) is the source of the U.S. Gun Control Act of 1968 (GCA '68)," stated the group on its Web site.

"The Nazi Weapons Law of 1938 replaced a Law on Firearms and Ammunition of April 13, 1928. The 1928 law was enacted by a center-right, freely elected German government that wanted to curb 'gang activity,' violent street fights between Nazi party and Communist party thugs. All firearm owners and their firearms had to be registered. Sound familiar? 'Gun control' did not save democracy in Germany. It helped to make sure that the toughest criminals—the Nazis—prevailed."

JPFO literature noted: "The Nazis inherited lists of firearm owners and their firearms when they 'lawfully' took over in March 1933. The Nazis used these inherited registration lists to seize privately held firearms from persons who were not 'reliable.' Knowing exactly who owned which firearms, the Nazis had only to revoke the annual ownership permits or decline to renew them."

The assassination of President John F. Kennedy precipitated a cry for gun control in the United States, and the corporate media went into high gear promoting this agenda. Yet, resistance was strong and the idea languished until after the 1968 murders of Dr. Martin Luther King Jr. on April 4 and Robert F. Kennedy on June 6. Following these shocking deaths, the Gun Control Act of 1968 (GCA) was passed in October of that year, after strenuous debate and compromise. Some conspiracy researchers see this as a classic example of creating a problem, offering a draconian solution, and settling for a compromise that still fulfills the original agenda.

The gun legislation of 1968 stated only licensed dealers could send and receive firearms across state lines, thus ending mail-order sales. It also allowed bureaucrats in Washington to decide what types of firearms Americans could own. The term "sporting" guns was not clearly defined, allowing whole classes of firearms to be banned.

"Given the parallels between the Nazi Weapons Law and the GCA '68, we concluded that the framers of the GCA '68—lacking any basis in American law to sharply cut back the civil rights of law-abiding Americans—drew on the Nazi Weapons Law of 1938," stated JPFO literature.

There seems to be some support for this argument, because the architect of the 1968 Gun Control Act was Connecticut senator Thomas J. Dodd, a Democrat who lost to Republican Prescott Bush in a 1956 Senate election but gained the state's other Senate seat two years later. Dodd had served as a special agent for the FBI in the 1930s and as executive trial counsel for the Office of the United States Chief of Counsel for the Prosecution of Axis Criminality at the Nuremberg war crimes trials at the end of the war. It may have been during his time in Nuremberg that he became familiar with the Nazi gun laws. A letter from the Library of Congress to Dodd in July 1968 showed that four months prior to his gun-control legislation being passed, he received an English translation of the Nazi Weapons Law based on the original German law document he supplied to the library.

Dodd died of a heart attack in 1971. In 1980, his son, Christopher J. Dodd, a member of the Council on Foreign Relations, won his father's seat in the Senate. The younger Dodd, a liberal, nevertheless took money from and lent considerable support to corporate miscreants like Enron and Arthur Andersen, indicating his willingness to support the globalists.

But control of weapons was not the only item on the fascist globalist agenda.

WHILE EVERYONE KNOWS of the Rockefeller control of oil, most do not know the extent of Rockefeller wealth and influence over modern medicine and drugs.

According to Eustace Mullins, the last surviving protégé of the famous twentieth-century intellectual and writer Ezra Pound, and author of the 1988 book *Murder by Injection: The Story of the Medical Conspiracy Against America*, the drug industry is controlled by a Rockefeller "medical monopoly," largely through directors of pharmaceutical boards representing Chase Bank, Standard Oil, and other Rockefeller entities. "The American College of Surgeons maintains a monopolistic control of hospitals through the powerful Hospital Survey Committee, with members [such as] Winthrop Aldrich and David McAlpine Pyle representing the Rockefeller control," he wrote.

Winthrop Aldrich also served on the Committee on the Cost of Medical Care (CCMC), which was originated by Dr. Alexander Lambert, the personal physician to Teddy Roosevelt and a president of the AMA. According to Dr. Charles C. Smith, who researched the activities of the committee: "[Dr. Lambert] obviously was to be the needed 'figurehead.'

Other notable choices were Winthrop Aldrich, president of Chase National Bank; John Frey, secretary-treasurer, AFL; William T. Foster, director of the Pollack Foundation in Economic Research; Olin West, M.D., executive secretary, AMA; and fifteen physicians plus two dentists in private practice. Five physicians from Public Health were chosen, and the director of research for the Milbank Memorial Fund. Representatives from insurance, hospital, nursing, pharmacy sources were appointed and six members from positions. They numbered forty-nine in all. The full-time staff was headed by Harry H. Moore of Washington, who in 1927 published 'American Medicine and the People's Health' while a member of Public Health Service. His main tenets were the need for a system to distribute medical care and an insurance plan to pay for it."

Smith noted that a minority of the committee recommended, among other things, that government competition in the practice of medicine be discontinued and that corporate medicine financed through intermediary agencies, such as Health Maintenance Organizations (HMO), should be opposed, because they fail to provide high-quality health care and exploit the medical profession. These recommendations were not followed. "The tenor of the [CCMC] report was such that one can read into it the seeds of everything that led to the health-care system we have today. . . . So at last we find ourselves, as always, in a health-care crisis," Dr. Smith wrote in 1984. This health-care crisis continues today.

Rockefeller control over the medical establishment also was exercised through the Rockefeller Sanitary Commission and the Rockefeller Institute for Medical Research, at one time headed by Dr. Detlev Bronk, already named as a suspected member of MJ-12. "Rockefeller's General Education Board has spent more than $100 million to gain control of the nation's medical schools and turn our physicians to physicians of the allopathic school, dedicated to surgery and the heavy use of drugs," commented Mullins.

Mullins also pointed to the Nazi connections of GlaxoSmithKline (GSK), the second-largest pharmaceutical company in the world after Pfizer. The history of the Big Pharm giant can also serve as an example of the consolidation of drug companies in recent years.

Burroughs Wellcome & Company was founded in London in 1880 by two American pharmacists, Henry Wellcome and Silas Burroughs. Glaxo, a New Zealand firm that originally manufactured baby food, became Glaxo Laboratories and went multinational in1935. After the postwar acquisition of other companies, including Meyer Laboratories, Glaxo moved its facilities to the United States. Burroughs Wellcome and Glaxo, Incorporate merged in 1995. The new name of the company was GlaxoWellcome.

In 1830, John K. Smith opened his first pharmacy in Philadelphia. Over the years, Smith, Kline and Company merged with the French, Richard and Company, and changed its name to Smith Kline and French Laboratories in 1929. By 1969, the firm had spread its business worldwide and purchased seven additional laboratories in Canada and the United States. In 1982, it merged with Beckman Incorporate, becoming Smith-Kline Beckman. With the 1988 purchase of its biggest competitor, International Clinical Laboratories, SmithKline Beckman grew by 50 percent. The latest merger took place with GlaxoWellcome in 2000, and the firm became GlaxoSmithKline.

According to Eustace Mullins, the original Burroughs Wellcome drug firm was wholly owned by Wellcome Trust, whose director was the British lord Oliver Franks. "Franks was ambassador to the United States from 1948 to 1952," Mullins wrote. "He [also was] a director of the Rockefeller Foundation, as its principal representative in England. He also was a director of the Schroeder Bank, which handled Hitler's personal bank account; director of the Rhodes Trust in charge of approving Rhodes scholarships; visiting professor at the University of Chicago; and chairman of Lloyd's Bank, one of England's Big Five."

Recalling that John D. Rockefeller's father, William "Big Bill" Rockefeller, once tried to sell unrefined petroleum as a cancer cure, Mullins, who spent more than thirty years researching the "Rockefeller medical monopoly," commented, "This carnival medicine-show barker would hardly have envisioned that his descendants would control the greatest and most profitable medical monopoly in recorded history."

Mullins reported that I. G. Farben and the drug companies it controlled in the United States through the Rockefeller interests were responsible for the suppression of effective drugs until a monopoly could be established. For example, from 1908 to 1936, Farben withheld its discovery of sulfanilamide, an early sulfa drug, until the firm had signed working agreements with the important drug firms of Switzerland, Sandoz and Ciba-Geigy. In one of the largest corporate mergers in history, these two firms joined in 1996 to form Novartis.

It has been previously detailed how the support of globalists and transplanted European fascists helped put the Reagan-Bush team into power in 1980. Against this background, it is instructive to look at one of the many controversial drugs now being used by millions of Americans—aspartame, an additive sugar substitute found in most diet soft drinks and more than five thousand foods, drugs, and medicines. Aspartame is found in most sugar substitutes, such as NutraSweet, Equal, Metamucil, and Canderel.

When heated to more than 86 degrees Fahrenheit—keep in mind that the human body temperature is 98.6 degrees—aspartame releases free methanol that breaks down into formic acid and formaldehyde in the body. Formaldehyde is a deadly neurotoxin. One quart of an aspartame-added beverage is estimated to contain about 56 miligrams of methanol. Dr. Louis J. Elsas explained to the U.S. Senate Committee on Labor and Human Resources: "I am a pediatrician, a professor of pediatrics at Emory, and have spent twenty-five years in the biomedical science[s], trying to prevent mental retardation and birth defects caused by excess phenylalanine. . . . [I] have considerable concern for the increased dissemination and consumption of the sweetener aspartame—1-methyl N-L-a-as partyl-L-phenylalanine—in our world food supply. This artificial dipeptide is hydrolyzed by the intestinal tract to produce L-phenylalanine, which in excess is a known neurotoxin." Countering claims that laboratory tests indicated little harm from small amounts of aspartame, Dr. Elsas noted, "Normal humans do not metabolize phenylalanine as efficiently as do lower species, such as rodents, and thus most of the previous studies in aspartame effects on rats are irrelevant to the question."

Before 1980, the Federal Drug Administration had refused to approve the use of aspartame. FDA toxicologist Dr. Adrian Gross testified to Congress that aspartame caused tumors and brain cancer in lab animals and, therefore, violated the Delaney Amendment that forbids putting anything in food that is known to cause cancer. Aspartame also is blamed for the increase in diabetes as it not only can precipitate the disease but also stimulates and aggravates diabetic retinopathy and neuropathy, which, when interacting with insulin, can cause diabetics to go into convulsions.

Dr. Betty Martini worked in the medical field for twenty-two years. She was the founder of Mission Possible International, working with doctors around the world in an effort to remove aspartame from food, drinks, and medicine. She gave this account of how pharmaceutical interests overcame claims of public welfare:

"Donald Rumsfeld was CEO of Searle, that conglomerate that manufactured aspartame. For sixteen years the FDA refused to approve it, not only because it's not safe but because they wanted the company indicted for fraud. Both U.S. prosecutors hired on with the defense team and the statute of limitations expired. They were Sam Skinner and William Conlon. Skinner went on to become secretary of transportation, squelching the cries of the pilots who were now having seizures on this seizure-triggering drug, aspartame, and then chief of staff under President Bush's father. Some of these people reached high places. Even Supreme Justice Clarence Thomas is a former Monsanto attorney. (Monsanto bought Searle in 1985, and sold it a few years ago.) When [John] Ashcroft became attorney general [in 2001, Larry] Thompson from King and Spalding Attorneys (another former Monsanto attorney) became deputy under Ashcroft. However, the FDA still refused to allow NutraSweet on the market. It is a deadly neurotoxic drug masquerading as an additive. It interacts with all antidepressants, L-dopa, Coumadin, hormones, insulin, all cardiac medication, and many others. It also is a chemical hypersensitization drug, so it interacts with vaccines, other toxins, other unsafe sweeteners, like Splenda that has a chlorinated base like DDT and can cause autoimmune disease. It has a synergistic and additive effect with MSG. Both being excitotoxins, the aspartic acid in aspartame, and MSG, the glutamate, people were found using aspartame as the placebo for MSG studies, even before it was approved. The FDA has known this for a quarter of a century and done nothing even though it's against the law. Searle went on to build a NutraSweet factory and had $9 million worth of inventory. Donald Rumsfeld was on President Reagan's transition team and the day after [Reagan] took office he appointed an FDA commissioner who would approve aspartame."

Searle salesperson Patty Wood-Allott claimed that in 1981 Rumsfeld told company employees "he would call in all his markers and that no matter what, he would see to it that aspartame be approved this year."

Dr. Martini noted: "The FDA set up a board of inquiry of the best scientists they had to offer, who said aspartame is not safe and causes brain tumors, and the petition for approval is hereby revoked. The new FDA commissioner, Arthur Hull Hayes, overruled that board of inquiry and then went to work for the PR agency of the manufacturer, Burson-Marstellar, rumored at $1,000 a day, and has refused to talk to the press ever since. There were three congressional hearings because of the outcry of the people being poisoned. Senator Orrin Hatch refused to allow hearings for a long time. The first hearing was in 1985, and Senator Hatch and others were paid by Monsanto. So the bill by Senator [Howard] Metzenbaum never got out of committee. This bill would have put a moratorium on aspartame, and had the NIH do independent studies on the problems being seen in the population, interaction with drugs, seizures, what it does to the fetus, and even behavioral problems in children. This is due to the depletion of serotonin caused by the phenylalanine in aspartame."

Reagan's FDA commissioner Hayes initially approved aspartame only as a powdered additive. But in 1983, just before he left his position, he approved the additive for all carbonated beverages.

Attempting to study or report on aspartame is a thankless task for mainstream academics. Dr. Janet Starr Hull, an OSHA-certified environmental hazardous-waste emergency-response specialist and toxicologist, in 1991 was diagnosed with incurable Graves' disease (a defect in the immunization system that leads to hyperthyroidism) only to learn through her own research that she had been poisoned by aspartame. She stated: "Many scientists at prestigious American universities will tell you they cannot get grants for continued research on aspartame or Splenda, or their department heads have been told to drop all discussions on the topic. Some will say aspartame research isn't worth the effort because they cannot get published in American scientific journals. Others claim the research centers constructed by the large corporations, such as Duke

University's Searle Research Center, were designed with managed research as a construction proviso."

Illustrating the battle between experts in regard to aspartame was the 2005 research by Dr. Morando Soffritti, scientific director of the European Ramazzini Foundation of Oncology and Environmental Sciences in Bologna, Italy. Soffritti conducted a three-year study on 1,800 rats and concluded that aspartame is a multipotential carcinogen. His work was peer-reviewed by seven world experts, and in April 2007, Dr. Soffritti received the third Irving J. Selikoff Award from the Mount Sinai School of Medicine in New York City, where he presented a more recent study that confirmed the cancer-causing potential of aspartame at even small doses. He noted that only a small amount of aspartame can trigger cancer, and babies of mothers who ingested aspartame could grow up to contract cancer. Other research conducted in Spain, such as the "Barcelona Report" by the staff of the biology department of the University of Barcelona, confirmed that aspartame transformed into formaldehyde in the bodies of living laboratory specimens and spread throughout vital organs. These studies, largely unreported in the U.S. media, confirmed aspartame's carcinogenicity in laboratory rats.

In 2006, media reports spoke of a "new study" that countered Soffritti's research. This study was not new. It was actually conducted in the mid-1990s and reported that researchers could find no link between aspartame and cancer, according to Unhee Lim, PhD, a researcher at the National Institutes of Health (NIH). Lim and colleagues worked with 473,984 men and women between the ages of fifty and seventy-one who participated in this diet-and-health study. In 1995 and 1996, participants were asked how much they drank of three popular diet beverages—soda, fruit drinks, and iced tea. They were also asked if they added aspartame to their coffee and tea. From their answers, the researchers calculated how much aspartame they consumed on a daily basis. During the next five years, 1,972 of those studied developed lymphoma or leukemia, and 364 developed brain tumors. When the researchers looked at people who consumed an average of at least four hundred milligrams of aspartame a day—about the amount found in two cans of soda—they found no link between aspartame consumption and cancer.

Critics noted that this study was subject to "recall bias," since those in the study were being asked to remember what and how much they drank. "If their recollections weren't accurate, it compromises the findings," said Michael F. Jacobson, executive director of the Center for Science in the Public Interest, a consumer watchdog organization. There also was no consideration of the many other foods and additives that contained aspartame, which added to the daily intake. Yet, the few corporate mass media outlets that carried the story in 2006 introduced the ten-year-old study with headlines such as "Findings May Help to Alleviate Concerns Raised by Rat Study Last Year."

Why such aversion by the media to dealing with controversial health issues? According to the Center for Public Integrity (CPI), in the past seven years the pharmaceutical and health-products industry spent in excess of $800 million to lobby legislators and government officials at both the federal and state levels. Manufacturers of pharmaceuticals, medical devices, and other health products spent nearly $182 million on federal lobbying from January 2005 through June 2006. "No other industry has spent more money to sway public policy," stated a 2005 CPI special report titled "Drug Lobby Second to None." "Its combined political outlays on lobbying and campaign contributions is topped only by the insurance industry."

It should also be noted that the large pharmaceutical corporations annually spend nearly twice as much money on marketing as they do on research and development. In 2004, the CPI reported that pharmaceutical direct-to-consumer advertising has grown from $791 million in 1996 to more than $3.8 billion in 2004. Drug ads on television are now ubiquitous.

The cross-corporate ownership of both pharmaceutical houses, medical institutions, and the mass media, combined with the extraordinary amount of pharmaceutical advertising, might explain the media's hesitation in reporting the deleterious effects of drugs. According to Dr. Marcia Angell, former editor in chief of the *New England Journal of Medicine*, profit is the driving force behind medicine today. "In 2002 the combined profits for the ten drug companies in the Fortune 500 ($35.9 billion) were more than the profits for all the other 490 businesses put together ($33.7

billion)," she states. "Over the past two decades, the pharmaceutical industry has moved very far from its original high purpose of discovering
and producing useful new drugs. Now primarily a marketing machine to
sell drugs of dubious benefit, this industry uses its wealth and power to
co-opt every institution that might stand in its way, including the U.S.
Congress, the FDA, academic medical centers, and the medical profession
itself." Dr. Angell, also author of the 2004 book *The Truth About the
Drug Companies: How They Deceive Us and What to Do About It* brings
focus to the argument that the current power of the pharmaceutical industry can be directly traced to its phenomenal growth during the Reagan
years.

"The election of Ronald Reagan in 1980 was perhaps the fundamental
element in the rapid rise of Big Pharma—the collective name for the largest drug companies," wrote Angell. Dr. Angell and a number of others
took note of a strong pro-business attitude shift during the Reagan-Bush
years—not just in government but within American society.

There was a time in the not-so-distant past when educated persons of
class looked upon commercial businessmen only slightly more kindly
than they had once looked upon theater folk. They also had a slight disdain for enormous inherited wealth. Scientists, teachers, public servants
such as firemen and policemen chose their careers for service and
community-betterment rather than for lavish salaries and retirement benefits. But times and attitudes change. Today, the corporate mass media
portrays the race for wealth as practically virtuous. The wealthy are considered winners while everyone else is a loser. "The gap between the rich
and poor, which had been narrowing since World War II, suddenly began to widen again, until today it is a chasm," remarked Dr. Angell.

She went on to say that before 1980, pharmaceuticals was a good business, but afterward, it was a stupendous one. From 1960 to 1980, prescription drug sales were fairly static as a percentage of U.S. gross domestic
product, but from 1980 to 2000, they tripled. "They now stand at more
than $200 billion a year," said Dr. Angell. "Of the many events that con-
tributed to the industry's great and good fortune, none had to do with the
quality of the drugs the companies were selling."

The success of Big Pharma has more to do with marketing than with the efficiency of its drugs. Dr. Michael Wilkes described a recent process called "disease-mongering." This term is applied to large drug corporations' attempts to convince healthy people they are sick and need drugs. "This is all in an attempt to sell treatments," explained Dr. Wilkes. "When their profits don't match corporate expectations, they 'invent' new diseases to be cured by existing drugs." Dr. Wilkes cited these examples of medical conditions he considers disease-mongering: female sexual dysfunction syndrome, premenstrual dysphoric disorder, toenail fungus, baldness, and social anxiety disorder (formerly known as shyness). He said these are but a few areas "where the medical community has stepped in, thereby turning normal or mild conditions into diseases for which medication is the treatment."

Referring to the colossus that the pharmaceutical industry has become, Dr. Angell remarked, "It is used to doing pretty much what it wants to do." Beginning in the 1980s, important new laws were passed relaxing restrictions on pharmaceutical corporations. These included the Bayh-Dole Act, after its chief sponsors, Indiana Democratic senator Birch Bayh and Kansas Republican senator Robert Dole. The Bayh-Dole Act allowed universities and small businesses to patent discoveries from research underwritten by the National Institutes of Health (NIH), the major distributor of tax dollars for medical research. It also allowed taxpayer-financed discoveries formerly in public domain, to be granted to drug corporations through exclusive licenses. Dr. Angell said that today universities, where most NIH-sponsored work is carried out, can patent and license their discoveries and charge royalties. Subsequent but similar legislation allows the NIH itself to directly transfer NIH discoveries to industry. Today, "all parties cash in on the public investment in research," she noted.

Under this system, research paid for by public money becomes a commodity to be sold for profit by private concerns. Dr. Angell provides examples of the large consulting fees paid by pharmaceutical corporations to individual faculty members and to NIH scientists and directors, increasing the intrusion of the globalist pharmaceutical corporations into medical education and the almost complete domination of medical education, particularly when it comes to drugs. Recall that it was an NIH study that refuted peer-reviewed research linking cancer to the sweetener aspartame.

Approximately half of the largest pharmaceutical corporations are not American. About half of them are based in Europe. In 2002, the top ten were the American companies Pfizer, Merck, Johnson & Johnson, Bristol-Myers Squibb, and Wyeth (formerly American Home Products); the British companies GlaxoSmithKline and AstraZeneca; the Swiss companies Novartis and Roche; and the French company Aventis (which in 2004 merged with another French company, Sanafi Synthelabo, and that put it in third place). "All are much alike in their operations. All price their drugs much higher here than in other markets," stated Dr. Angell.

The lucrative connection between Big Pharma and medical schools and hospitals has brought about a definite corporate-friendly atmosphere. "One of the results has been a growing pro-industry bias in medical research—exactly where such bias doesn't belong," argues Dr. Angell.

She also blasted pharmaceutical corporations for their claims that high drug prices are necessary to fund research and development. "Drug industry expenditures for research and development, while large, were consistently far less than profits. For the top ten companies, they amounted to only 11 percent of sales in 1990, rising slightly to 14 percent in 2000. The biggest single item in the budget is neither R&D nor even profits but something usually called 'marketing and administration'—a name that varies slightly from company to company. In 1990, a staggering 36 percent of sales revenues went into this category, and that proportion remained about the same for over a decade. Note that this is two and a half times the expenditures for R&D."

Dr. Angell further noted what many people see as excessive salaries of pharmaceutical executives such as Charles A. Heimbold Jr., the former chairman and CEO of Bristol-Myers Squibb, who made $74,890,918 in 2001. This does not count his $76,095,611 worth of unexercised stock options. During this same time, John R. Stafford, chairman of Wyeth, made $40,521,011, not counting his $40,629,459 in stock options.

Congress expressly prohibited Medicare from negotiating lower drug prices through its bulk purchasing power and, in 1997, the FDA permitted the drug industry to do direct advertising, previously restricted to physicians, to the public, with no mention of side effects except for the most serious.

The excesses of the globalists' pharmaceutical corporations have prompted many Americans to seek price relief by traveling to Canada or Mexico to purchase drugs.

Dr. Angell concluded that only an aroused American public can rein in the power of the pharmaceutical monopoly. Noting that drug companies have the largest lobby in Washington, and they give copiously to political campaigns, Dr. Angell said legislators and the mass media corporations are now so dependent on the pharmaceutical industry for campaign contributions and advertising that it will be exceedingly difficult to break their power. "But the one thing legislators need more than campaign contributions is votes. That is why citizens should know what is really going on. . . . there will be no real reform without an aroused and determined public to make it happen," she said.

IF ASPARTAME IS not worry enough, a 2007 report by Peter Piper, a professor of molecular biology and biotechnology at Britain's Sheffield University, stated that sodium benzoate, a mold-prevention substance used routinely by the $160 billion soft-drink industry, creates the carcinogen benzene when mixed with vitamin C in drinks. Worse yet, according to Piper, "These chemicals have the ability to cause severe damage to DNA in the mitochondria to the point that they totally inactivate it: they knock it out altogether. . . . there is a whole array of diseases that are now being tied to damage to this DNA—Parkinson's and quite a lot of neuro-degenerative diseases, but above all the whole process of aging."

This report intensified the controversy over chemical food and drink additives that have been linked to hyperactivity in children. One British news report on sodium benzoate quoted the makers of Coca-Cola, Pepsi Max, and Diet Pepsi, which all contain sodium benzoate, as saying they entrusted the safety of additives to the government. Unfortunately, many government agencies are under the control of the giant pharmaceutical corporations.

Don't look for any real relief from the Democrats. Although two of the leading Democratic presidential hopefuls in 2008, New York senator Hillary Clinton and Illinois senator Barack Obama, both pledged to fight the huge pharmaceutical and insurance industries—promises similar to those Mrs. Clinton made during her husband's time in office—campaign contributions data released in April 2007 showed that, with the exception of Republican Mitt Romney, both Clinton and Obama were the largest recipients of Big Pharma largess in campaign funding.

And despite announced plans by Mrs. Clinton to pass laws to prevent insurers from charging higher rates to people in poor health, the insurance industry contributed a whopping $226,245 to her campaign.

WHILE THE FASCIST globalists took swift charge of Nazi drug technology after the war, it is most interesting that they neglected a little-known and little-publicized aspect of the Third Reich—the fight against cancer, tobacco, alcohol abuse, and occupational hazards.

In fact, the National Socialists' predilection for health foods and preventative medicine may have been yet another reason the globalists turned against Hitler and his regime. After all, most food additives, colorings, and preservatives are petrochemicals, and any decrease in human consumption would spell loss of profits to the globalists' corporations. Early on, the Nazi regime instituted policies designed to create healthier environments within the workplace. However, as the imperatives of wartime production grew, these measures lost priority.

One example of globalist neglect of Nazi science can be seen in the issue of asbestos. By the late 1930s, Nazi Germany had firmly documented the link between asbestos and lung cancer. This connection was flatly stated in a 1939 textbook, and by 1943 the Nazi government had recognized asbestos-induced cancer as a compensable occupational disease. This Nazi research would be used in later years to counter asbestos producers' claims that they were unaware of the danger of asbestos until modern studies.

"The net effect in the field of cancer research was to slow recognition of the asbestos hazard," noted author Robert N. Proctor, a professor of the history of science at Pennsylvania State University and author of *The Nazi War on Cancer*. "The consensus achieved in Germany in the early 1940s would not [be] obtain[ed] in Britain or the United States until more than two decades later. Science and political stigma [and commercial obstinacy] thus conspired—at least for a time—to confine the truth to the shadows."

In addition to confining occupational health hazards to the shadows in corporate America, the owners of the U.S. tobacco industry fought a successful, decades-long rearguard action against the claims that cigarettes are a leading cause of cancer. Utilizing one hired expert after another, they bought time while they diversified their ownership away from tobacco.

Contrary to the popular belief that the link between smoking and cancer was demonstrated in postwar Britain and America, "it was in Germany in the late 1930s that we first find a broad medical recognition of both the addictive nature of tobacco and the lung cancer hazard of smoking," according to Proctor. The Nazis were among the first to ban smoking in public places such as Nazi party offices, post offices, hospitals, rest homes, and waiting rooms—a restriction today becoming prevalent across America. In 1938, the Nazi Luftwaffe barred all smoking on its properties. As in modern America, limitations were placed on tobacco advertising.

Much of the attack on alcohol and tobacco stemmed from the Nazi ideals of racial hygiene and Aryan purity. But it also was well supported by German science.

Although the connection between smoking and cancer has been theorized for many years, it was the German physician Fritz Lickint who brought the connection to public knowledge with the publication of his 1939 opus *Tabak und Organismus,* or "Tobacco and the Organism." In 1940, Lickint, described as "most hated by the tobacco industry," escaped persecution by the Nazis for belonging to the Social Democratic Party, thanks to official Nazi support for his antitobacco work.

As in America from the 1960s to the 1980s, German tobacco interests formed organizations and hired various experts to counter the claims of antitobacco activists. As in America, they claimed that the medical evidence against tobacco was "unscientific" and the propaganda of health fanatics. But it was an uphill fight, considering the amount of scientific data then available, plus the fact that Hitler disdained tobacco and alcohol.

In the widespread National Socialist effort to stamp out both smoking and drinking, it was continually pointed out that Hitler neither smoked nor drank. Hitler would not permit his lover, Eva Braun, or his deputy, Martin Bormann, to smoke in his presence. Once, Hitler even suggested that tobacco was "the wrath of the Red Man against the White Man, vengeance for having been given hard liquor." It was also publicly noted in wartime propaganda that while Hitler, Mussolini, and Franco were nonsmokers, Churchill, Roosevelt, and Stalin all smoked cigarettes, cigars, or a pipe.

The attack on tobacco also has been traced to economic concerns. The Nazis, like modern American corporations, came to realize that tobacco-related illnesses could impact the hospital industry as well as the insurance industry. This concern, coupled with the increasing demands for healthy wartime workers, undoubtedly was an added stimulus for the antitobacco campaign.

It is ironic to learn that some of the Nazis' most ardent antismoking activists, such as Karl Astel, director of Jena University's Institute for Tobacco Hazards Research, who committed suicide in 1945, were also virulent anti-Semites and supporters of euthanasia. This is a fascinating example of how social idealism can be subverted for tyrannical purposes. As Proctor noted, "[T]here is the fact that many of Germany's leading antitobacco activists were also war criminals."

THE NAZIS' CONCERN over rising cancer rates also resulted in a bizarre confrontation over the use of X-rays. "The SS radiologist Hans Holfelder, who spearheaded an ambitious drive to X-ray hundreds of thousands of Germans, was trying to identify illness so steps could be taken to treat or isolate afflicted individuals," reported Proctor. "[Berlin's Kaiser Wilhelm Institute for Anthropology, Human Genetics and Eugenics founding director] Eugen Fischer's concern in warning against overexposure was the longer-term 'genetic health of the race.' Both were solid Nazis, but the two had very different conceptions of how to preserve the health of the favored race."

One of the only Jewish cancer researchers allowed to continue working during the Nazi regime was Nobel laureate, biochemist Dr. Otto Warburg, a relative of the banking family and director of Berlin's Kaiser Wilhelm Institute for Cell Physiology. Warburg's institute was founded in 1931, following a substantial donation by the Rockefeller Foundation to the Kaiser Wilhelm Gesellschaft, today known as the Max Planck Institute.

More than forty years ago, Dr. Warburg gave a lecture describing both the cause and cure for cancer. "Summarized in a few words, the prime cause of cancer is the replacement of the respiration of oxygen in normal body cells by a fermentation of sugar. All normal body cells meet their energy needs by respiration of oxygen, whereas cancer cells meet their energy needs in great part by fermentation. All normal body cells are thus obligate aerobes, whereas all cancer cells are partial anaerobes. . . . Oxygen gas, the donor of energy in plants and animals, is dethroned in the cancer cells and replaced by an energy-yielding reaction of the lowest living forms, namely, a fermentation of glucose."

In other words, while most living cells require oxygen to live, cancer cells can do well without oxygen, instead drawing energy from the fermentation of sugars. To maintain normal health, humans require a minimum of 22 percent oxygen in the air they breathe. Most American cities regularly fall below this minimum, and on so-called ozone-alert days, the oxygen level often drops to 18 percent or lower. And the less said about the amount of sugar in the American diet the better. Obesity is quickly becoming a major national health problem. If Dr. Warburg is correct, and he stated that "on the basis of anaerobiosis there is now a real chance to get rid of this terrible disease," it is astounding that nothing has been done to cure cancer in the intervening four decades. Perhaps this is because, as has been pointed out by suspicious researchers, more people are making a living off cancer than dying from it.

Meanwhile in America, Rockefeller executive Frank Howard, after convincing Alfred Sloan and Charles Kettering of General Motors to fund a cancer institute, was named chairman of the new Sloan-Kettering Institute. Howard chose Cornelius "Dusty" Rhoads, former chief of research for the medical division of the U.S. Chemical Warfare Service, to direct the institute's experimentation with chemotherapy. Later, Howard represented Rockefeller interests in the drug company Rohm and Haas.

IT IS INTRIGUING that in modern health-conscious America, very little has been done to educate citizens about the dangers of overly prescribed drugs, including the depressant alcohol.

During the Third Reich, there was even a substantial antialcohol movement in beer-loving Germany, including a sizeable German Antialcoholism Association with a membership numbering in the thousands. The Nazis outlawed alcohol advertising aimed at youth, as well as any that suggested alcohol was healthful. Alcohol-related hazards such as cirrhosis of the liver, cardiomyopathy, fetal abnormalities, and esophageal cancer were well known even prior to the rise of National Socialism but mostly ignored in modern America's popular media. Yet, like Prohibition in America, the Nazi antialcohol effort largely failed, due to the pressure for consumption from a thirsty population, coupled with the sizeable amounts of money gained by the government through taxes on alcohol.

Like Americans today, the Nazis introduced a variety of nonalcoholic beers and even produced some made from liquefied vegetables. In 1936, a certification system was instituted, designed to protect children from "unsuitable" drinks. Coca-Cola was declared one such beverage due to its sugar and additives.

In fact, today's "New Age" issues echo aspects of National Socialism in the Third Reich. In addition to high-ranking Nazis' fascination with the occult, organic foods, herbs, and healing plants were all encouraged in Nazi Germany, along with a "back to nature" idealism and respect for the rural life. "From 1934 to 1937, the amount of land devoted to herbs and healing plants . . . increased by more than a factor of ten—from 820 hectares to 3,896 hectares," noted Proctor. "Cultivation was especially strong in the forests of Thuringia and north-central Germany, but every part of the country was involved. Popular magazines celebrated the importance of natural foods and drugs, and professional apothecaries took steps to evaluate the efficiency of medicinal herbs."

Hitler advocated a vegetarian lifestyle. "One may regret living at a period when it's impossible to form an idea of the shape the world of the future will assume. But there's one thing I can predict to eaters of meat: the world of the future will be vegetarian." Many Germans followed Hitler's dream of a meatless society. About 83,000 voluntarily participated in his vegetarian-lifestyle program.

But Hitler was also a consummate politician. Although he disdained both hunting and eating meat, he did not attempt to force his ideals on his followers, for purely pragmatic reasons. "Personally, I cannot see what possible pleasure can be derived from shooting. . . . I have never fired at a hare in my life. I am neither poacher nor sportsman. . . . [But] if I excluded poachers from the Party, we should lose the support of entire districts."

In viewing the reality behind benign, if tyrannical, government efforts at social control, Robert N. Proctor correctly concluded, "The Nazi campaign against tobacco and the 'wholegrain-bread operation' are, in some sense, as fascist as the yellow stars [worn to identify Jews] and the death camps. Appreciating these complexities may open our eyes to new kinds of continuities binding the past to the present; it may also allow us better to see how fascism triumphed in the first place."

THE FASCIST GLOBALISTS, in addition to making unconscionable profits from tobacco and dangerous drugs such as aspartame, may be promoting a program of population reduction.

As previously noted, Third Reich Nazis, as well as their prominent American business partners, were greatly interested in the field of eugenics, the study of scientifically applied genetic selection to maintain and improve ideal human characteristics, which grew to include birth and population control. The concept grew from the writings of the Victorian scientist Sir Francis Galton, who after study reached the conclusion that prominent members of British society were such because they had "eminent" parents.

In 1925, after more than a decade in which at least sixty thousand "defectives" in the United States were legally sterilized, Justice Oliver Wendell Holmes, writing for the majority in a Supreme Court case, stated, "It is better for all the world, if instead of waiting to execute degenerate offspring for crime or to let them starve for their imbecility, society can prevent those who are manifestly unfit from continuing their kind."

Of course, to determine who was dirtying the gene pool required extensive population statistics. In 1910, the Eugenics Records Office was established as a branch of the Galton National Laboratory in London, endowed by Mrs. E. H. Harriman, the wife of railroad magnate Edward Harriman and mother of diplomat Averell Harriman. In 1912, Mrs. Harriman sold her substantial shares of New York's Guaranty Trust bank to J. P. Morgan, thus assuring his control over that institution. After 1900, the Harrimans, the family that gave the Prescott Bush family its start, along with the Rockefellers, provided more than $11 million to create a eugenics research laboratory at Cold Springs Harbor, New York, as well as eugenics studies at Harvard, Columbia, and Cornell. The first International Congress of Eugenics was convened in London in 1912, with Winston Churchill as a director. Obviously, the concept of "bloodlines" was significant to these people.

In 1932, when the Congress met in New York, it was the Hamburg-Amerika shipping line controlled by Harriman associates George Walker and Prescott Bush that brought prominent Germans to the meeting. One leader was psychiatry professor Dr. Ernst Rudin of the Kaiser Wilhelm Institute for Genealogy and Demography in Berlin. Rudin was unanimously elected president of the International Federation of Eugenics Societies for his work in founding the Deutschen Gesellschaft fur Rassenhygiene, or the German Society for Racial Hygiene, a forerunner of Hitler's racial institutions.

Honored personally by Hitler in 1939 and 1944, Rudin continued to be acknowledged as a leader in psychiatry. In 1992, the prestigious Max Planck Institute praised Rudin for "following his own convictions in 'racial hygiene' measures, cooperating with the Nazis as a psychiatrist and helping them legitimize their aims through pertinent legislation."

Despite much public renunciation of eugenics following the revelations of the Nazi racial extermination programs at the Nuremberg trials, work continues right up to today, under more politically correct names.

General William H. Draper Jr. was a "supporting member" of the International Eugenics Congress in 1932 and, despite or because of his ties to the Harriman and Bush families, was named head of the economic division of the U.S. Control Commission in Germany at the end of hostilities. According to Webster Tarpley and Anton Chaitkin, authors of *George Bush: The Unauthorized Biography*, "General Draper [in later years] founded 'Population Crisis Committee' and the 'Draper Fund,' joining with the Rockefeller and Du Pont families to promote eugenics as 'population control.' The administration of President Lyndon Johnson, advised by General Draper on the subject, began financing birth control in the tropical countries through the U.S. Agency for International Development (USAID)." Draper also served as a population consultant to President George H. W. Bush, and he and his son were in charge of Bush's campaign fund-raising in 1980. The younger Draper went on to work with population-control activities of the United Nations.

Early efforts at reducing the birth rate by sterilization met with resistance in the United States, so the rhetoric was softened and other means pursued. "Castration evidently hit a little too close to home for the average member of the public to stomach," wrote authors Jonathan Vankin and John Whalen, "so vasectomy became the preferred method for sterilizing males, and its equivalent, salpingectomy, became the preferred sterilization method for women."

The abortion controversy stems from the eugenics views of the globalist families and their belief that some form of population control must be allowed in a well-run society. The Human Genome Project has only elevated fears that human DNA can be manipulated and controlled.

Organizations such as the Planned Parenthood Federation of America, Incorporated, a tax-exempt corporation that has 860 centers nationwide and claims to prevent more than 617,000 unintended pregnancies a year, are subsidized by the plutocracy, usually through their foundations and think tanks. In 2006, more than a third of the group's contributions came from corporate and foundation grants.

Rudin's eugenics work was, in a large part, funded by Rockefeller money. "The plutocrats were in league with scientists, many with formidable reputations," noted Vankin and Whalen. "These scientists expended immeasurable energy trying to 'prove' that blacks were stupid, Jews were greedy, Mexicans were lazy, women were nutty, and so on—as well as the corollary: rich, white people with good table manners and glowing report cards were genetically superior."

General Maxwell Taylor, a former chairman of the Joint Chiefs of Staff and U.S. ambassador to South Vietnam, who began addressing the Council on Foreign Relations in 1952, reflected his fellow globalists' viewpoint in a 1981 interview with *Executive Intelligence Review*. Taylor stated that the underlying cause of world problems was overpopulation. He said it would be necessary by the beginning of the twenty-first century to reduce the human population, mostly in Third World countries, by disease, starvation, and regional conflicts. "I have already written off more than a billion people. These people are in places in Africa, Asia, Latin America. We can't save them. The population crisis and the food-supply question dictate that we should not even try. It's a waste of time," Taylor said.

When one considers the starvation that wracks so many poor countries, the AIDS epidemic sweeping Africa, and the ongoing strife in Afghanistan and Iraq, plus dozens of smaller conflicts all over the world, it would seem that Taylor's vision of the future has come to pass.

But the real fear is to be found in the desire to control others, not in voluntary population control or human genes. "[T]he twentieth century suffered two ideologies that led to genocides," said MIT cognitive scientist Steven Pinker, author of the 2002 book *The Blank Slate*. Referring to the wrongness of Nazi genetics beliefs, Pinker observed, "The other one, Marxism, had no use for race, didn't believe in genes, and denied that human nature was a meaningful concept. Clearly, it's not an emphasis on genes or evolution that is dangerous. It's the desire to remake humanity by coercive means (eugenics or social engineering) and the belief that humanity advances through a struggle in which superior groups (race or classes) triumph over inferior ones."

Before anyone looks for relief from social manipulation from heaven, it is instructive to see the parallels in the use of religion both in the Third Reich and in modern America.

CHAPTER 13

RELIGION

PRESIDENT GEORGE W. BUSH HAS BEEN ONLY THE MOST RECENT world leader who has used religious factions to gain support for his policies and objectives.

"National Socialism was a religion," noted Professor George Lachmann Mosse of the University of Wisconsin-Madison, whose wealthy Jewish family fled Germany in 1933. "The depth of the ideology, the liturgy, the element of hope, all helped to give the movement the character of a new faith. It has been shown that [Nazi propaganda minister Paul Joseph] Goebbels quite consciously used religious terminology in many of his speeches. Moreover, Nazism was a total worldview which by its very nature excluded all others. From this it followed that traditional Christianity was a rival, not a friend. But here Hitler at first went very slowly indeed, for he needed (and got) the support of the majority of the Christian churches."

Mosse concluded that "the Nazi future would have lain with the Evangelical Christians had the war been won."

In *Mein Kampf,* Hitler spoke condescendingly of religion, offering this rationalization for organized religion. "The great masses of people do not consist of philosophers, and it is just for them that faith is frequently the sole basis of a moral view of life."

He also saw in Christian fundamentalism a reflection of his own National Socialist zeal and ambition. "The greatness of Christianity was not rooted in its attempted negotiations of compromise with perhaps similarly constructed philosophical opinions of the old world," he wrote, "but in the inexorably fanatical preaching and representation of its own doctrine."

Despite this public support for religion, Hitler, who, as has been seen, was surrounded by occultism, privately expressed disdain for formal religions, as evidenced by this discourse related in *Hitler's Table Talk:*

> An educated man retains the sense of the mysteries of nature and bows before the unknowable. An uneducated man, on the other hand, runs the risk of going over to atheism (which is a return to the state of the animal) as soon as he perceives that the state, in sheer opportunism, is making use of false ideas in the matter of religion, while in other fields it bases everything on pure science. That's why I've always kept the Party aloof from religious questions. I've thus prevented my Catholic and Protestant supporters from forming groups against one another, and inadvertently knocking each other out with the Bible and the sprinkler. So we never became involved with these churches' forms of worship.... In any case, the main thing is to be clever in this matter and not to look for a struggle where it can be avoided.... So it's not opportune to hurl ourselves now into a struggle with the churches.... The dogma of Christianity gets worn away before the advances of science. Religion will have to make more and more concessions. Gradually the myths crumble. All that's left is to prove that in nature there is no frontier between the organic and the inorganic. When understanding of the universe has become widespread, when the majority of men know that the stars are not sources of light but worlds, perhaps inhabited worlds like ours, then the Christian doctrine will be convicted of absurdity. Originally, religion was merely a prop for human communities. It was a means, not an end in itself. It's only gradually that it became transformed in this direction, with the object of maintaining the rule of the priests, who can live only to the detriment of society collectively.

Hitler's thoughts were echoed by his deputy Martin Bormann, who flatly stated in a 1942 German Evangelical Church yearbook: "National Socialist and Christian concepts are incompatible."

Because of his private opposition to true Christianity, Hitler quickly took steps to subdue the church. On July 23, 1933, just six months after he came to power, a Nazi-dominated National Synod in Wittenberg named a former German Army chaplain and virulent anti-Semite, Ludwig Mueller, as Reich bishop. Six months later, Mueller issued what came to be known as the "Muzzling Order," a decree designed to bring control over the German Evangelical Church. Ministers were forbidden to speak about controversial or political matters; hence there could be no opposition to the Nazi regime. Mueller proclaimed that church services were "for the proclamation of the pure Gospel, and for this alone." This same no-involvement-with-politics message can be heard in many churches in America today.

Despite Nazi hostility to Christianity and thanks to Goebbels's propaganda, many Germans believed that Hitler was heaven-sent. A Cologne children's prayer began, "Fuehrer, my fuehrer, bequeathed to me by the Lord." And, with the notable exception of some anti-Nazi clerics such as Pastors Martin Niemoeller and the martyred Dietrich Bonhoeffer, the German congregations all fell into lockstep with the Nazi government. Many churchgoers were zealous Nazis, but many were simply hesitant or afraid to speak up against their noisy fellow members.

"We will discover that the Nazi era shouts its lessons to the church of America," concluded the Reverend Erwin W. Lutzer, senior pastor of Moody Church in Chicago, who made a detailed study of the National Socialists' seduction of German Christians. He stated: "It warns us, challenges us, and forecasts what might happen in the days ahead. Whether we heed its warnings, accept its challenges, and recognize its subtle deceptions is up to us."

GERMANY IN THE 1930s was a predominantly religious nation with the majority divided between Catholics and Lutherans. The fascist globalists realized that the multisectarian United States could not be brought under one religious control system. Through their corporate control over the large media outlets, these would-be global rulers have instituted a decades-long campaign to undermine and discredit organized religion, regardless of denomination. Some wayward TV evangelists and Catholic priests have only exacerbated this campaign.

There also appears to be a movement to control the church's message in the campaign for the 2008 election. According to a June 2007 CNN press release, the TV network "will serve as the exclusive broadcaster of a presidential candidate forum on faith, values and politics during the Sojourners 'Pentecost 2007' conference in Washington, D.C. . . . The Rev. Jim Wallis of Sojourners and author of the best-selling book *God's Politics: Why the Right Gets It Wrong and the Left Doesn't Get It,* has invited Democratic presidential candidates Sen. Hillary Clinton, Sen. John Edwards and Sen. Barack Obama to share their ideas and proposals about pressing social issues with a special emphasis on poverty." Soledad O'Brien, a CNN anchor and correspondent, was asked to moderate the forum.

Jim Wallis in 1971 founded Sojourners, an organization that wishes "to articulate the biblical call to social justice, inspiring hope and building a movement to transform individuals, communities, the church, and the world." Detractors accuse Wallis of attempting to divide evangelical Christians to the benefit of secular liberals. In an open letter, William J. Anderson, a teacher of economics at Maryland's Frostburg State University, accused Wallis of serving as a leftist political operative for the 2004 presidential campaign of John Kerry. Anderson wrote: "I am familiar enough with you [Wallis] and Sojourners to know that much of what you have written reeks of the worst kind of hypocrisy. . . . the central theme of Sojourners from day one . . . has been anticapitalism."

According to a special report by the Traditional Values Coalition, which claims to be the largest nondenominational, grassroots church lobby in America, "Throughout the history of Sojourners, Wallis has taken a consistently left-wing and anti-American stance. He was an antiwar activist against the Vietnam War. . . . Wallis is also a darling of the liberal media. He is often quoted in articles critical of conservative Christians or of President Bush's faith." The report goes on to accuse Wallis of supporting socialist programs, noting that while Wallis was in seminary, he founded a magazine he named *Post-American*. Within its pages, Wallis called for the redistribution of wealth and for government-managed economies, described as "social justice." Other critics saw Wallis as an example of the plutocracy's propensity for supporting—and thus controlling—both sides of an issue.

While Obama in early 2008 was criticized for intemperate remarks by his former pastor, Jeremiah Wright, little attention was given to Hillary Clinton's longtime active participation with a secretive Capitol Hill prayer and Bible study group known as "The Family" or "The Fellowship." According to an article by Barbara Ehrenreich posted on *The Nation* Web site, a former member of The Family—Jeff Sharlet—described the group's real work as "knitting together international networks of right-wing leaders, most of them ostensibly Christian." Quoting Sharlet, reporter Ehrenreich wrote that in the 1940s, The Family reached out to former and not-so-former Nazis, and its fascination with that exemplary leader, Adolf Hitler, has continued, along with ties to "a whole bestiary of murderous thugs."

Considering Hillary's ties to the secretive Bilderbergs, her husband's membership in the Council on Foreign Relations and the Trilateral Commission, as well as her work with the Nazi-connected group called The Family, it could be said that she provides a connective tissue between the globalists and their new Fourth Reich.

Pastor Lutzer described what he saw as attempts to suppress and denigrate Christianity in present-day America. "As the state expands its powers, it can initiate laws that limit the church's freedom," he noted. "Consider the phrase 'separation of church and state.' Interpreted in one way, it can mean that the church should be free to exercise its influence and practice religion without interference from the state. That kind of separation is exactly what the church in Germany so desperately needed.

"However, here in America the phrase 'separation of church and state' is given a sinister twist by civil libertarians. To them, it means that religious people should not be allowed to practice their religion in the realm that belongs to the state. Religion, we are told, should be practiced privately; the state must be 'cleansed' from every vestige of religious influence. By insisting that the state be 'free for all religions,' organizations such as the ACLU in effect makes it free for none!"

Some churches in America are already feeling the eyes of the government on them. In 2007, Pastor Mark Holick of the Spirit One Christian Center in Wichita, Kansas, urged the IRS to brush up on the constitutional freedoms guaranteed by the U.S. Constitution. This came after his church received a letter from IRS officials warning it against "political activity" and demanding answers to thirty-one questions regarding its beliefs. The IRS particularly cited church signs, such as one reading "[Kansas Governor Kathleen] Sebelius accepted $300,000 from abortionist [name withheld], price of 1,000 babies."

Holick notified the IRS that "the church cannot agree to not engage in any activity that *may* favor or oppose a candidate. Simply preaching the word of God on a moral issue to which a candidate is opposed, *may* be deemed to oppose a candidate. While it is the church's policy not to oppose or endorse a candidate for office, it will not stop preaching God's word."

Others have questioned the lack of public concern over a political candidate forum called "Pentecost 2007." "The Americans United for the Separation of Church and State have suddenly gone mute," noted Marsha West, founder and editor of the *E-Mail Brigade News Report,* an online news service for conservative Christians.

Evangelist Bill Keller, founder of the fifteen-year-old Bill Keller Ministries, which created the *Liveprayer with Bill Keller* television program and Liveprayer.com, reportedly the world's largest interactive Christian Web site, publicly complained that he too faced problems with his right to free speech. He specifically mentioned Americans United for Separation of Church and State, claiming this "liberal group . . . [would] try and silence churches and ministries by asking the IRS to investigate them for allegedly violating their 501(c)(3) status. Of course, this is designed to intimidate people into silence, even though in 76 previous attempts [they have] yet to be successful in getting anyone's tax exemption pulled. Our attorneys are confident that nothing I said violated our nonprofit status, but we are now going to be forced to defend ourselves from the IRS."

Keller also complained his freedom of speech was being curtailed by Internet corporations. "For the first seven years, we sent our Daily

Devotional every day to our subscribers around the world without any problems, including those who use Microsoft e-mail accounts," he said. "Last Thanksgiving [2006], Microsoft went to new filters many ISPs are now using to try and reduce spam. These new filters are 'content filters' and work off of a dictionary that can have any words added the operator wants. For six months, we have been getting our Daily Devotional blocked sporadically by Microsoft's servers based on the 'content' of my message. This is also happening to other Christian organizations as well as conservative political groups who rely heavily on e-mail. We have done all we can to get Microsoft to rectify this problem, but they have arrogantly failed to even respond. . . . Even though we could show considerable financial damages over these past six months, we aren't seeking any money from Microsoft, only that they stop blocking our Daily Devotional from going to our subscribers who use their e-mail accounts."

The issue here is not abortion or content but the right of free speech, whether it is a church or an individual. Large mainstream monied churches have long been used as platforms for politicians, both local and national. They seem to fare well but it is the fringe churches and religions where we find long-established freedoms being chipped away.

FOR THOSE UNAWARE of the tactics of the fascist globalists, it must seem strange that churches can be intimidated by the government much like in Nazi Germany, even with a professed Christian in the White House.

Some Christians have been less restrained in their comparison between professed fundamental Christianity on today's political scene and the use of religion in Nazi Germany. "I have been telling conservative Christians that who should be howling at the top of their lungs is not the Liberal Left, it is the Far Right Christian Conservatives, for they are being lied to, seduced, and misled even more so than the Liberal Left. They are being seduced into fascism and that is not Christianity," wrote Christian Republican Karl W. B. Schwarz, who, probably without knowing of the GOP's fascist past, nevertheless styled the Bush-dominated Republican Party a "fascist cult."

An *Online Journal* contributing writer, Schwarz stated, "In fact, if you look real close at Bush-Cheney and understand the fundamental dynamics of what brought Hitler to power, how he controlled the masses, how he sold the Great Lie, it is very easy to see that Bush-Cheney 'compassionate Conservatism' and Fascism are one and the same. Many hear the term 'neocon' and do not recognize that in its current operative sense, it is a term meaning 'New World Order Fascist.'"

Whole books have been written about the rise to power in America of the "Religious Right," a critical support base for the Republican Party. But most people appeared not to notice the parallels between the fascism of Nazi Germany and the conservative Christian movement in America today, both with deep roots in the conservative faction of the population. In America, this faction tends to be pro-business, which makes it a prime target of the fascist globalists, who largely control the corporate life of the nation.

This faith-based political movement began in the late 1970s with the formation of the Moral Majority, a coalition of Christian conservative groups who were seeking to defeat President Jimmy Carter in the 1980 election. One of its founders was Southern Baptist preacher Pat Robertson, who in 1988 severed all connections with the church in order to run for president on the Republican ticket. Defeated in the primaries, Robertson urged his followers to vote for George H. W. Bush. Robertson went on to become an influential TV evangelist, primarily thanks to the Christian Broadcasting Network he founded in 1961. In 2005, he was forced to apologize for comments interpreted by many as advocating the assassination of Venezuelan president Hugo Chavez. "I don't know about this doctrine of assassination, but if [Chavez] thinks we're trying to assassinate him, I think that we really ought to go ahead and do it. It's a whole lot cheaper than starting a war, and I don't think any oil shipments will stop," Robertson told his audience.

Another Moral Majority founder was Jerry Falwell, a televangelist who became a firm supporter of George W. Bush's Faith-based Initiative. Following the attacks of September 11, 2001, Falwell, on Pat Robertson's *700 Club* TV show, said pagans, abortionists, feminists, gays, lesbians, the ACLU, and everyone else trying to secularize America "helped this happen." He was found dead in his office of heart failure on May 15, 2007.

Erik Prince, a former Navy SEAL, is the founder of Blackwater USA, a private security contracting firm that has grown into one of the largest private armies in the world. In 2007, Blackwater came under criticism and congressional scrutiny following more than two hundred shooting reports in Iraq, one in September of that year that left seventeen Iraqis dead and more than two dozen wounded.

Prince's father, Edgar, a self-made millionaire from selling auto parts, supported the Family Research Council (FRC), a right-wing fundamentalist Christian group close to the Bush administration. Both men were significant contributions to the elections campaigns of George W. Bush. Edgar's widow served on the boards of FRC and another heavyweight Christian right organization, Dobson's Focus on the Family. She runs the Edgar and Elsa Prince Foundation, of which Erik is a vice president. The foundation gave more than $1 million to the Christian right from July 2003 to 2006.

Author Jeremy Scahill compared Prince's private army to Hitler's "Brownshirt" storm troopers.

It has been noted that prewar Germany and the United States both had Christian roots, a widespread acceptance of biblical social values, and a basic commitment to private virtue. Pastor Lutzer pointed out that America has differed from Germany in that it has benefited from a constitutional guarantee of the separation of church and state, as well as its history of democracy. But he warned: "Despite the differences, the American church, like that of Nazi Germany, is in danger of wrapping the cross of Christ in some alien flag."

Like so many in modern America, most Germans of the 1930s offered no resistance to the ever-encroaching fascism of National Socialism. "Many welcomed the abolition of individual responsibility for one's actions; for some it is easier to obey than to accept the dangers of freedom," wrote Gerald Suster in his 1981 book *Hitler: The Occult Messiah*.

No one in the area of religion seems able to get a clear picture of what is happening in modern America. The push-pull between liberty and security, scripture and social consciousness appears to have created a stultifying tension. The globalists have found that such ongoing controversies coming from many different directions is an effective mechanism to keep Americans arguing with each other, off balance, and ineffective in uniting to learn the truth behind their New World Order agenda.

It might be wise to consider the words of the New Testament. On three separate occasions—Matthew 12:31–31, Mark 3:28–29, and Luke 12:10—Jesus specifically stated that all sins can be forgiven, even from those who choose not to believe in Him or have blasphemed against Him. But he stated the one sin that can never be forgiven is to speak against the Holy Spirit. In John 15:26, it states, "But I will send you the Comforter— the Holy Spirit, the source of all truth [*Living Bible*]." God is truth. Nothing in God's universe is a lie; it simply exists. Only humans, with our free will, can choose to be deceitful. So to speak against the Holy Spirit is to distort, deform, or deny truth, an insight many politically conservative Christians seem to have missed.

Truth must be gleaned through study and contemplation. No one, no matter how smart, can make correct decisions without truthful knowledge and understanding, the very cornerstones of wisdom. And knowledge begins with education.

CHAPTER 14

EDUCATION

"EDUCATION IS CRUCIAL..." STATED PROFESSOR GEORGE MOSSE IN his book *Nazi Culture,* "for if an ideology can be institutionalized through the education establishment, it has won a major battle. The Nazis realized this only too well."

Upon close inspection, it will be found that the American education establishment has been created and guided for many years by the progenitors of the globalists who created both communism and National Socialism.

The oil magnate John D. Rockefeller created the General Education Board (GEB) in 1903 to dispense Rockefeller donations to education. By 1960, it had ceased operating as a separate entity and its programs were rolled into the Rockefeller Foundation. In 1917, the GEB made a $6 million grant to Columbia University to create the New Lincoln School, a private experimental coeducational school in New York City. According to current school literature, the facility engages in "enrolling students from kindergarten through grade 12. Its predecessor was founded as Lincoln School in 1917 by the Rockefeller-funded General Education Board as 'a pioneer experimental school for newer educational methods,' under the aegis of Columbia University's Teachers College."

According to Eustace Mullins: "From this school descended the national network of progressive educators and social scientists, whose pernicious influence closely paralleled the goals of the Communist Party, another favorite recipient of the Rockefeller millions. From its outset, the Lincoln School was described frankly as a revolutionary school for the primary and secondary schools of the entire United States. It immediately discarded all theories of education that were based on formal and well-established disciplines, that is the McGuffey Reader type of education, which worked by teaching such subjects as Latin and algebra, thus teaching children to think logically about problems."

Other Rockefeller-connected entities that still shape the United States include the Brookings Institution, the National Bureau of Economic Research, the Public Administration Clearing House, the Council of State Governments, and the Institute of Pacific Relations. During the Carter administration, a former Rockefeller assistant, Paul Volcker, was named chairman of the U.S. Central Bank, the Federal Reserve System.

John D. Rockefeller, who claimed to be a devout Baptist, began his attempt to influence American education in the late 1800s, with a $600,000 donation to the American Baptist Education Society for an endowment for the University of Chicago. Between 1890 and 1914, Rockefeller, through the society, handed out close to a million dollars to many different schools. But the lynchpin of his attempt to guide American education was his formation of the National Education Board and the continuing Rockefeller support to the University of Chicago.

The mammoth university now encompasses an undergraduate college, four graduate divisions, six professional schools as well as libraries, laboratories, museums, clinics, and other institutions; nursery and K-12 schools; a continuing-studies program; and an academic press. In 2003, a university center opened in Paris to accommodate the school's program of European studies. Closely connected to Rockefeller's University of Chicago is the English world's most accepted authority on everything, Encyclopaedia Britannica. Formerly a privately held not-for-profit company, Encyclopaedia Britannica, headquartered in Chicago, also owns Merriam-Webster Incorporation, one of the world's leading publishers of dictionaries and thesauri.

Until recently, the encyclopedia firm was owned by the William Benton

Foundation, whose sole beneficiary was the University of Chicago, in accordance with the wishes of its namesake, Senator William Benton, a former vice president of the university.

A former senator from Connecticut, Benton graduated from Yale University in 1921, then worked in advertising until becoming a part-time vice president of the University of Chicago in 1932. During World War II, he served as assistant secretary of state in Washington, D.C., until 1947, during which time he was active in organizing the United Nations. He served as a member of and delegate to numerous United Nations and international conferences and commissions. A declared Democrat, he was U.S. senator from 1949 until 1953. From 1943 until his death in 1973, Benton was chairman of the board and publisher of *Encyclopaedia Britannica*.

It was announced in 1996 that 100 percent of the company's stock was purchased for an undisclosed amount by an investment group led by Jacob Safra, who headed the Swiss bank of his name, which in 1955 opened a branch in Brazil. Bank Jacob Safra Switzerland is part of the Safra Group, a far-flung chain of financial institutions. Some conspiracy researchers see the private ownership of the world's premier encyclopedia as the perfect means for controlling public knowledge, particularly in the areas of history and science.

"The creation and funding of the University of Chicago had done much to enhance Rockefeller's public relations profile among Baptists and educators. . . . The only difficulty was that education, on the whole, wasn't in bad shape," explained Paolo Lionni, author of *The Leipzig Connection*, his 1993 book that traced the deleterious effects of experimental psychology on the education system back to German professor of philosophy Wilhelm Max Wundt, the founder of experimental psychology. "The indigenous American educational system was deeply rooted in the beliefs and practices of the Puritan Fathers, the Quakers, the early American patriots and philosophers. Jefferson had maintained that in order to preserve liberty in the new nation, it was essential that its citizenry be educated, whatever their income. Throughout the country, schools were established almost immediately after the colonization of new areas." These school systems included ones established by the Quakers in Pennsylvania and the

Midwest, the free school movement in New York, and a large number of "normal schools," so named for helping to set the norms for education. By the start of the twentieth century, the United States was home to many major universities, which turned out thousands of well-trained teachers each year.

Lionni noted: "Educational results far exceeded those of modern schools. One has only to read old debates in the Congressional Record or scan the books published in the 1800s to realize that our ancestors of a century ago commanded a use of the language far superior to our own. Students learned how to read not comic books but the essays of Burke, Webster, Lincoln, Horace, Cicero. Their difficulties with grammar were overcome long before they graduated from school, and any review of a typical elementary school arithmetic textbook printed before 1910 shows dramatically that students were learning mathematical skills that few of our current high school graduates know anything about. The high school graduate of 1900 was an educated person, fluent in his language, history, and culture, possessing the skills he needed in order to succeed."

The agenda behind Rockefeller's creation of the GEB may have been revealed in correspondence from Frederick T. Gates, Rockefeller's choice to head the board.

Gates wrote, "In our dreams, we have limitless resources and the people yield themselves with perfect docility to our molding hands. The present education conventions fade from their minds, and unhampered by tradition, we work our own good will upon a grateful and responsive rural folk. We shall not try to make these people or any of their children into philosophers or men of learning, or men of science. We have not to raise up from among them authors, editors, poets or men of letters. We shall not search for embryo great artists, painters, musicians nor lawyers, doctors, preachers, politicians, statesmen, of whom we have an ample supply.

"The task we set before ourselves is very simple as well as a very beautiful one, to train these people as we find them to a perfectly ideal life just where they are. So we will organize our children and teach them to do in a perfect way the things their fathers and mothers are doing in an imperfect way, in the homes, in the shops and on the farm."

Paolo Lionni wrote: "It would be false to say [John D. Rockefeller] was the mastermind of international intrigue and deception. But it wouldn't be false to say the Rockefeller money has been used in various ways to forward social and global control through economics, foundations, the United Nations, universities, banking, industry, medicine, and, of course, education, psychology, and psychiatry . . .

"It is not incorrect to say that a major segment of today's modern institutions exist not because of honest study and concern for the truth in the respective fields but because Rockefeller's money was available at their inception to fund incredible PR campaigns, establish 'professional' publications and societies, steamroller over any competition (regardless of their legitimacy or value), and continue selling the ideas until accepted and institutionalized within the basic fabric of society.

". . . That's a tremendous amount of control and involvement for one group! What if the theories and practices they funded and continue to fund are fundamentally flawed and don't lead to the best possible situations in the various fields mentioned? Well, the views in most of those areas *are* fundamentally flawed and they *don't* lead to the best solutions in 'mental health,' education, medicine, sanity, and happiness. But, most likely, despite all 'humanitarian' posturing, they were never intended to."

NORMAN DODD, DIRECTOR of research for the House Select Committee to Investigate Foundations and Comparable Organizations, reported that in 1952 the president of the Ford Foundation told him bluntly that "operating under directive from the White House," his foundation was to "use our grant-making power so as to alter our life in the United States that we can be comfortably merged with the Soviet Union." Now, with the collapse of communism, the advent of the United Nations and NATO, along with various economic treaties now in place, it would appear that this globalist goal is close to becoming realized.

Dodd also stated that the congressional investigation found that the Guggenheim, Ford, and Rockefeller Foundations and the Carnegie Endowment were "working in harmony to control education in the United States." He added that these entities had been subverted from the original goals of their creators by subsequent directors—another example of wealth taking control of existing organizations.

Some of the past and current organizations and foundations linked by membership or funding to the plutocracy that once supported the Nazis include the Agency of International Development, American Civil Liberties Union, American Council of Race Relations, American Press Institute, Anti-Defamation League, Arab Bureau, Aspen Institute, Association of Humanistic Psychology, Battelle Memorial Institute, Center for Advanced Studies in the Behavioral Sciences, Center for Constitutional Rights, Center for Cuban Studies, Center for Democratic Institutions, Christian Socialist League, Communist League, Environmental Fund, Fabian Society, Ford Foundation, Foundation for National Progress, German Marshall Fund, Hudson Institute, Institute for Pacific Relations, Institute on Drugs, Crime and Justice, International Institute for Strategic Studies, Mellon Institute, Metaphysical Society, Milner Group, Mont Pelerin Society, National Association for the Advancement of Colored People, National Council of Churches, New World Foundation, Rand Institute, Stanford Research Institute, Tavistock Institute of Human Relations, Union of Concerned Scientists, International Red Cross, and the YMCA.

David N. Gibbs, associate professor of political science at the University of Arizona, noted that the intelligence community has long taken advantage of academia to propagate their views and philosophies—as always, through the distribution of money. He wrote: "While pundits never tire of the cliché that American universities are dominated by leftist faculty who are hostile toward the objectives of established foreign policies, the reality is altogether different: The CIA has become 'a growing force on campus,' according to a recent article in the *Wall Street Journal*. The 'Agency finds it needs experts from academia, and colleges pressed for cash like the revenue.' Longstanding academic inhibitions about being publicly associated with the CIA have largely disappeared: in 2002, former CIA director Robert Gates became president of Texas A & M University, while the new president of Arizona State University, Michael Crow, was vice chairman of the agency's venture-capital arm, In-Q-Tel

Inc. . . . The CIA has created a special scholarship program, for graduate students able and willing to obtain security clearances. According to the London *Guardian,* 'the primary purpose of the program is to promote disciplines that would be of use to intelligence agencies.' And throughout the country, academics in several disciplines are undertaking research (often secret) for the CIA."

AMERICAN YOUTH WERE educated in the principles of National Socialism before World War II.

In 1935, Ernst Mueller, head of the German-American Settlement League, acquired lakeside property in Yaphank, Long Island, and invited Americans of German descent to visit and relax. He also formed a group called the German-American Youth. The youth in Yaphank constructed tents on platforms, called Camp Siegfried. Gustave Neuss, the grandson of German immigrants, whose father served as a judge in Yaphank, recalled: "Some of the parents complained about the harsh conditions and at least one removed her daughter from the camp because of this. The regimen included education in pro-Nazi doctrine to ensure a new generation having the pure Aryan philosophy." Neuss's father initially was friendly with the German organization but soon turned against them because of their anti-Semitism and un-American speeches.

The youth organization was part of the German-American Bund, an anti-Semitic fraternal group formed in the 1930s by a merger of the National Socialist German Workers Party and the Free Society of Teutonia. German-American Bund activity was not limited to Yaphank and the New York City area. Neuss wrote: "Groups of the pro-Nazis were located throughout the United States. Hitler's claim was that after he had conquered Europe he would then take over the USA. The Bund and other pro-German groups located throughout the country provided a cadre of subversives to assist in such a takeover."

Although many German-American organizations existed in the prewar United States, the Bund was among the only ones to express support for Nazi ideals. In February 1939, Bund leader Fritz Kuhn addressed a crowd of about twenty thousand in Madison Square Garden and railed against President "Frank D. Rosenfeld" and the "Bolshevik-Jewish conspiracy" threatening America. Once the nation entered the war, some Bund members fled the country while others were placed in internment camps.

Although the American Nazi Bund never developed into a serious threat—even Hitler tried to distance himself from it—the subversion of the educational process under which it prospered continued.

THE NATIONAL TEACHERS Association, now known as the National Education Association, was founded in 1850 as a professional teacher organization, but today has become the largest labor union in the United States, representing almost 3 million educators. In 1966, the NEA merged with the American Teachers Association, an organization primarily concerned with education in the black community, yet another special interest of the Rockefellers, who funneled money into black education programs and organizations. By 1964, Rockefeller's General Education Board had spent more than $3.2 million in gifts to support black education, criticized by some as merely a means to instill white values and worldviews in black students.

According to William H. Watkins, author of *The White Architects of Black Education: Ideology and Power in America, 1865–1954,* John D. Rockefeller Sr. was more concerned with shaping a new industrial social order than providing a useful education. "The Rockefeller group demonstrated how gift-giving could shape education and public policy," commented Watkins.

Conservative Republican Pat Buchanan decried the efforts of the NEA as a negative influence on American education. In a 1999 interview, he stated that "ever since the judges have gotten heavily into education, and the National Education Association has gotten into control of the Department of Education, test scores go down, there's violence in classroom, things are going wrong." Criticism also has been leveled at the NEA Ex-Gay Educators Caucus, whose literature states the purpose of the caucus is to "work within the NEA to make policy changes to ensure that the Ex-Gay voice is heard." Opposing the Ex-Gay Educators Caucus at the 2004 NEA National Convention was Kevin Jennings, a former private-school teacher in Massachusetts and founder of the Gay Lesbian Straight Education Network, who has partnered with the NEA in promoting acceptance of homosexuality using curriculum materials in the nation's schools, beginning as early as kindergarten and the elementary grades. At the convention, Jennings was presented the NEA's human rights "creative leadership" award.

Education today is mixing drugs with student control. In years past, if a child was acting up or caught staring out the window, he or she received a rap on the knuckles with a ruler and was told to stay with the rest of the class. Today, the child is sent to the school nurse, who oftentimes tells the parents the student has attention-deficit hyperactivity disorder (ADHD) and recommends the administration of Prozac (94% sodium fluoride) or Ritalin, psychotropic drugs that have been shown to produce psychosis in lab rats. At least one state has put a stop to this practice. In 2001, the Connecticut House voted 141–0 on a law prohibiting school personnel from recommending to parents that their children take Ritalin or other mood-altering drugs. Republican State Representative Lenny Winkler, one of the bill's primary sponsors, quoted studies showing the number of children taking Ritalin nationally jumped from 500,000 in 1987 to more than 6 million by 2001. The bill also prohibited the state's Department of Children and Families from taking children away from parents who decline to put their children on mood-altering drugs.

A 1999 study at the Human Development Center at the University of Wisconsin in Eau Claire found that thirteen "ADHD" children on medication, over four years, performed progressively worse on standardized tests than a group of thirteen normal children with similar IQs and other characteristics. Another study, by Dr. Gretchen LeFever, an assistant professor of pediatrics and psychiatry at Eastern Virginia Medical School, revealed that while children in her community used the drug Ritalin two to three times more than the national rate, their academic performance in relation to their peers showed no improvement. For her persistence in questioning the rising incidence of drug use in schoolchildren, Dr. LeFever was fired in 2005.

Psychiatrist Peter Breggin, in his 1991 book *Toxic Psychiatry,* wrote: "Hyperactivity is the most frequent justification for drugging children.

The difficult-to-control male child is certainly not a new phenomenon, but attempts to give him a medical diagnosis are the product of modern psychology and psychiatry. At first psychiatrists called hyperactivity a brain disease. When no brain disease could be found, they changed it to 'minimal brain disease' (MBD). When no minimal brain disease could be found, the profession transformed the concept into 'minimal brain dysfunction.' When no minimal brain dysfunction could be demonstrated, the label became 'attention deficit disorder.' Now it's just assumed to be a real disease, regardless of the failure to prove it so. 'Biochemical imbalance' is the code word, but there's no more evidence for that than there is for actual brain disease."

Alan Larson, a former secretary of the Oregon Federation of Independent Schools, criticized the expanding diagnosis of attention deficit disorder (ADD) and was an outspoken critic of the indiscriminate use of drugs, proclaiming, "[T]he labeling of children with ADD is not because of a problem the kids have; it is because of a problem teachers who cannot tolerate active children have."

According to John Cornwell, author of *Hitler's Scientists*, the center of psychoanalysis shifted from Germany to the United States after the war. "[M]any of the homegrown analysts were German trained," he noted. This charge was echoed by Dr. Thomas Roeder, Volker Kubillus, and Anthony Burwell in their book *Psychiatrists—the Men Behind Hitler*. "In the period since 1971," they wrote, "child and adolescent psychiatry developed completely along the theoretical and methodological lines developed by its Nazi-era founders."

The influx of Nazi-trained psychiatrists after World War II, particularly in the military and intelligence fields, has produced a blossoming of psychological disorders. The American Psychiatric Association, in its 1952 *Diagnostic and Statistical Manual for Mental Disorders* (DSM), defined only 112 mental disorders. By the publication of DSM-IV in 1994, the number had grown to 374.

"Today, though psychiatry may still be suspect among the public, it has won over both government and the media. The profession and its treatments inundate talk shows, magazines, and the front pages of our newspapers," wrote Bruce Wiseman, the U.S. national president of the Citizens

Commission on Human Rights and former chairman of the history department at John F. Kennedy University.

In the 1970s, when the drug companies tried to find substitutes for LSD because of its serious side effects, they developed the antidepressant Prozac (fluoxetine) followed by Zoloft (sertraline), Effexor (venlafaxine), and Paxil (paroxetine). Dr. Helmut Remschmidt, who directed the Society for Child and Adolescent Psychiatry until 1984, was a leader in research into hyperactivity in children. He studied under Dr. Hermann Stutte, who was associated with Nazi psychiatrists involved in the German euthanasia program and had received his doctorate from Robert Sommer, director of the Deutscher Verband fur psychische Hygiene, or the German Association for Mental Hygiene. Dr. Remschmidt, long after the war, still pointed to "a genetic answer" to hyperactivity and was a leading proponent of the use of drugs such as Ritalin.

"It is by no means shocking for Remschmidt to be a prominent advocate of horrible things," commented Roeder, Kubillus, and Burwell. "After all, he is a disciple and protégé of Nazis. . . . What is equally frightening and obvious is that the racist and elitist theories of the original child psychiatrists—as documented in 1940 at the First Congress—have not only survived but flourished. It has been a natural passage of poison from teacher to student, from the Nazis to subsequent generations of child and adolescent psychiatrists. The efforts 'to spot and screen'—that is, to distinguish between the valuable and less valuable human beings—are more than dubious—they are indefensible." Yet, the number of child psychologists in U.S. schools grew from a mere 500 in 1940 to more than 22,000 by 1990.

According to Kelly Patricia O'Meara, a former chief of staff in the U.S. House of Representatives, whose investigative reports on child vaccines and mood-altering drugs prompted congressional hearings, "Thirty years ago, the World Health Organization (WHO) concluded that Ritalin was pharmacologically similar to cocaine in the pattern of abuse it fostered, and cited it as a Schedule II drug—the most addictive in medical use. The Department of Justice also cited Ritalin as a Schedule II drug under the Controlled Substances Act, and the Drug Enforcement Agency (DEA) warned that 'Ritalin substitutes for cocaine and d-amphetamine in a number of behavioral paradigms.'" O'Meara pointed to a 2001 study at the Brookhaven National Laboratory that confirmed the similarities between cocaine and Ritalin, but found that Ritalin is more potent than cocaine in its effect on the dopamine system, an area of the brain many doctors believe is most affected by these drugs.

Drugs such as Ritalin, now used to treat questionable mental afflictions, are taken by tens of millions of American youngsters. A 1986 edition of *The International Journal of the Addictions* listed 105 adverse reactions to Ritalin, including suicidal tendencies.

Americans wonder why there has been a rash of school shootings and teen suicides in recent years, yet virtually all of these killings have involved a student on mood-altering drugs or just coming off them. In five cases of school shootings between March 1998 and May 1999—including the tragedy at Columbine High School—at least seven of the students involved were being medicated. It was downplayed but reported that Seung-Hui Cho, the gunman in the Virginia Tech shootings in April 2007, had been undergoing psychological counseling and had prescription psychoactive drugs in his possession.

In his book *Reclaiming Our Children,* psychiatrist and drug critic Dr. Peter Breggin analyzed the clinical and scientific reasons for asserting that Eric Harris's violence at Columbine was caused by the prescription drug Luvox. "I've also testified to the same under oath in depositions in a case related to Columbine," Breggin wrote, adding, "I also warned that stopping antidepressants can be as dangerous as starting them, since they can cause very disturbing and painful withdrawal reactions."

A Web site called TeenScreenTruth is dedicated to gathering information off the Internet to help teens "connect the dots to see the revealing connections" between mood-altering drugs and teen violence. The Web site states: "Here's a statistic that is rarely mentioned in news reports: in nearly every school-shooting incident, the children and teens involved were already taking one or more psychiatric drugs or had just recently come off them, and had been under the care of a psychiatrist or mental-health practitioner. The same is true for the majority of child and teen suicides—they were already on some type of psychiatric drug program that was supposed to be treating their 'mental illness' yet they killed themselves anyway."

This assessment was echoed in a 1999 article in *Health and Healing* by Dr. Julian Whitaker, who stated, "[V]irtually all of the gun-related massacres that have made headlines over the past decade have had one thing in common: they were perpetrated by people taking Prozac, Zoloft, Luvox, Paxil, or a related antidepressant drug."

In 1998, GlaxoSmithKline, maker of Paxil, was ordered to pay $6.4 million to surviving family members after Donald Schnell, sixty, flew into a rage and killed his wife, daughter, and granddaughter just forty-eight hours after taking Paxil.

The TeenScreenTruth site and the *Indianapolis Star* compiled a list of violent episodes dating as far back as 1985, when Steven W. Brownlee, an Atlanta postal worker on psychotropic drugs, killed two coworkers. The list includes:

- In 1986, fourteen-year-old Rod Mathews of Canton, Massachusetts, beat a classmate to death with a baseball bat while on Ritalin.
- In 1988, thirty-one-year-old Laurie Dann, who had been taking Anafranil and lithium, walked into a second-grade classroom in Winnetka, Illinois, and began shooting. One child was killed and six wounded.
- Later that same year, nineteen-year-old James Wilson went on a shooting rampage at the Greenwood, South Carolina, elementary school and killed two eight-year-old girls and wounded seven others. He had been on Xanax, Valium, and five other drugs.
- In 1989, Patrick Purdy, twenty-five, opened fire on a schoolyard filled with children in Stockton, California. Five kids were killed and thirty wounded. He had been treated with Thorazine and Amitriptyline.
- In 1993, Steve Lieth of Chelsea, Michigan, walked into a school meeting and shot and killed the school superintendent, wounding two others, while on Prozac.
- In 1996, ten-year-old Tommy Becton grabbed his three-year-old niece as a shield and aimed a shotgun at a sheriff's deputy who

had accompanied a truant officer to his Florida home. He had been on Prozac.

- In 1997, Michael Carneal, fourteen, opened fire on students at a high school prayer meeting in Heath High in West Paducah, Kentucky. Three died and one was paralyzed. Carneal reportedly was on Ritalin.
- In 1998, Kip Kinkel, a fifteen-year-old in Springfield, Oregon, murdered his parents and proceeded to his high school, where he went on a rampage, killing two students and wounding twenty-two others. Kinkel had been prescribed both Prozac and Ritalin.
- In 1998, eleven-year-old Andrew Golden and fourteen-year-old Mitchell Johnson apparently faked a fire alarm at Westside Middle School in Jonesboro, Arkansas, and shot at students as they left the building. Four students and a teacher were killed. The boys were said to be on Ritalin.
- In 1999, Shawn Cooper, fifteen, of Notus, Idaho, took a shotgun to school and injured one student. He had been taking Ritalin.
- On April 20, 1999, Eric Harris, eighteen, and Dylan Klebold, seventeen, shot and killed twelve classmates and a teacher and wounded twenty-four others at Columbine High School in Colorado. Harris had been taking Luvox.
- In 1999, Todd Cameron Smith walked into a high school in Taber, Alberta, Canada, with a rifle and killed one student and injured another. He had been given a drug after a five-minute phone consultation with a psychiatrist.
- In 1999, Steven Abrams drove his car into a preschool playground in Costa Mesa, California, killing two. He was on probation with a requirement to take lithium.
- In 2000, T. J. Solomon, fifteen, opened fire at Heritage High School in Conyers, Georgia, while on a mix of antidepressants. Six were wounded.
- The same year, Seth Trickey of Gibson, Oklahoma, thirteen, was on a variety of prescription drugs when he opened fire on his middle-school class, injuring five.

- In 2001, Elizabeth Bush, fourteen, was on Prozac. She shot and wounded another student at Bishop Neumann High in Williamsport, Pennsylvania.
- Also in 2001, Jason Hoffman, eighteen, was on Effexor and Celexa, both antidepressants, when he wounded two teachers at California's Granite Hills High School.
- Same year, in Wahluke, Washington, Cory Baadsgaard, sixteen, took a rifle to his high school and held twenty-three classmates hostage. He had been taking Paxil and Effexor.
- In Osaka, Japan, also in 2001, Mamoru Takuma, thirty-seven, went into a second-grade classroom and started stabbing students. He killed eight. He had taken ten times his normal dose of an antidepressant.
- In 2005, sixteen-year-old Native American Jeff Weise, on the Red Lake Indian Reservation in Minnesota, was under the influence of the antidepressant Prozac when he shot and killed nine people and wounded five before committing suicide.
- In 2006, Duane Morrison, fifty-three, shot and killed a girl at Platte Canyon High School in Colorado. Antidepressants later were found in his vehicle.

Other incidents cited, but not apparently related to schools, included:

- In 1987, William Cruse was charged with killing six people in Palm Bay, Florida, after taking psychiatric drugs for "several years."
- The same year, Bartley James Dobben killed his two young sons by throwing them into a 1,300-degree foundry ladle. He had been on a "regimen" of psychiatric drugs.
- In 1989, Joseph T. Wesbecker, forty-seven, just a month after he began taking Prozac, shot twenty workers at Standard Gravure Corporation in Louisville, Kentucky, killing nine. Eli Lilly, which makes Prozac, later settled a lawsuit brought by survivors.
- In 1991, sixty-one-year-old Barbara Mortenson was arrested by

San Jose, California, police, who said she had "cannibalized" her eighty-seven-year-old mother while on Prozac.

+ In 1992, Lynnwood Drake III shot and killed six in San Luis Obispo and Morro Bay, California. Prozac and Valium were found in his system.

+ In 1993, sixteen-year-old Victor Brancaccio attacked and killed an eighty-one-year-old woman and covered her corpse with red spray-paint. He was two months into a Zoloft regimen.

+ In 1995, while on four medications, including Prozac, Dr. Debora Green set her Prairie Village, Missouri, home on fire, killing her children, ages six and thirteen.

+ In 1996, Kurt Danysh, eighteen, shot and killed his father seventeen days after his first dose of Prozac. He told authorities, "I didn't realize I did it until after it was done.... This might sound weird, but it felt like I had no control of what I was doing, like I was left there just holding a gun."

It would appear that German drug science and German psychiatry have provided the foundation for action toward today's schoolchildren who are being increasingly steered to drugs for any complaint, from true antisocial behavior to mere daydreaming.

And why hasn't the "watchdog" media put these stories together and presented it to the public? Direct-to-consumer pharmaceutical advertising grew from $791 million in 1996 to more than $3.8 billion by 2004. Those familiar with this subject claim media executives fear the loss of advertising revenues from the giant pharmaceutical corporations, some of the largest advertisers in the nation. Why haven't physicians spoken out about this? Many have, but they don't receive significant coverage in the corporate mass media, and many more fear reprisals from both the drug corporations and the federal government. Additionally, corporate drugs are heavily promoted to physicians. "The pharmaceutical companies send representatives to physicians' offices talking about their drugs, giving free samples of their drugs, those types of things. That effort dwarfs the [pharmaceutical corporations'] advertising expenditures," explained

Alan Mathios, a dean at the College of Human Ecology at Cornell University.

BUT THE EDUCATION issue that has drawn the greatest recent controversy is Public Law 107-110, better known as the No Child Left Behind Act of 2001 (NCLB), a prized legacy of the Bush administration. According to the act, its purpose is "to ensure that all children have a fair, equal, and significant opportunity to obtain a high-quality education and reach, at a minimum, proficiency on challenging state academic achievement standards and state academic assessments."

Signed into law by President George W. Bush on January 8, 2002, the act nevertheless brought immediate criticism from educators, state authorities, and libertarians alike. They questioned the act's sweeping proposals, which range from forcing teachers to conform to federally mandated curricula to inflicting monetary punishments on school districts that do not live up to federal expectation, and even taking state control or turning them over to private management companies. They also questioned the $410 million apportioned for the education of migratory children, most of whom come from the families of illegal aliens. Standardized testing has proven a handicap to children who speak English as a second language.

While no caring person would want to be caught leaving some poor child "behind," there was nevertheless the irksome feeling by many that the act was a thinly disguised attempt to force conformity on students and standardize the minds of American youth. It smacks of the same uniformity of education sought by the National Socialists under Hitler. Some conservatives and libertarians even claim that the act is an usurpation of state authority by the federal government. In 2007, the new Congress began taking steps to protect states from the controls and punishments of NCLB. For example, in 2005, when Utah passed a state law allowing school districts to ignore portions of NCLB, the Department of Education threatened to withhold federal education funds.

The backbone of NCLB is federally mandated standardized testing, which long has been accused of cultural bias. In fact, the entire practice of testing as a determinant of educational quality has been called into question, because the emphasis on tests forces teachers to teach only material that will get students to pass the tests, leaving a deficiency in grasping greater understanding and thinking critically. It should also be realized that both the textbook publishers and the standardized testing firms are, for the most part, controlled by the same globalist corporations under discussion.

The act rewards districts with better test scores, so critics claim that schools lower their standards to show improvement on test scores. The kids are not learning more, just being assessed differently. A 2007 report from the Center on Education Policy (CEP) indicated increased test scores in reading and math, but it was unclear if this reflected enhanced learning or lower standards on student tests. "Look at any state that has a 90 percent proficiency level with lots of students in poverty," commented Jack Jennings, president and CEO of CEP. "That doesn't happen without either an extraordinary effort to raise the quality of education for all students or setting lower standards."

Another portion of the NCLB Act that rankled libertarians was Section 9528, which requires schools to give military recruiters the name, home phone number, and address of every enrolled student. Schools are not required to tell the students or parents that their information has been passed on, but students can ask to not have their contact information shared. But filing the required form often means student information is withheld from universities and job recruiters as well as the military.

Unlike the Nazis, who placed great emphasis on athletics and physical fitness, the NCLB Act narrowly focuses on two main skills: reading and math. As a result, there are claims that other areas of schooling have been neglected, especially physical education.

This idea was reinforced by recent data from the Centers for Disease Control, showing kids between the ages of six and nineteen, some nine million youngsters, suffer from obesity. "With the obesity rates going up and it's in our face, why are we cutting PE time? I don't get it," questioned PE teacher Garrett Lydic, Delaware's 2006 Teacher of the Year. "The focus right now is on testing," he said, referring to a series of academic tests now mandated by federal law. "The result is that there's less time to get kids more active."

314 THE RISE OF THE FOURTH REICH

"It's a stretch even to call the law 'well-intentioned' given that its creators, including the Bush administration and the right-wing Heritage Foundation [Paul Weyrich, its founder, has been accused of ties to Nazi collaborators] want to privatize public education. Hence NCLB's merciless testing, absurd timetables and reliance on threats," commented *USA Today* education writer Alfie Kohn. "No wonder 129 education and civil rights organizations have endorsed a letter to Congress deploring the law's overemphasis on standardized testing and punitive sanctions. No wonder 30,000 people [mid-2007] have signed a petition at educatorroundtable .org calling the law 'too destructive to salvage.'"

Like Hitler, the globalist creators of a new empire carry an innate distrust of education that might explain why their education programs appear to savage true learning. "I do not wish any intellectual upbringing whatsoever, knowledge may only demoralize youth," Adolf Hitler once said. He echoed the statement of John D. Rockefeller, founder of the National Education Board, who said, "I don't want a nation of thinkers. I want a nation of workers."

Hitler also felt that intellectuals might not only present a rival to Nazi ideology but could form a group separate from the common man through a feeling of superiority due to their knowledge and education. "What we suffer from today is an excess of education," he stated in 1938. "What we require is instinct and will."

Hitler's sentiment was echoed recently by President George W. Bush. Journalist Ron Suskind, writing in the *New York Times Magazine,* reported an incident in Washington: "Forty democratic senators were gathered for a lunch in March [2004] just off the Senate floor. I was there as a guest speaker. Joe Biden was telling a story, a story about the president. 'I was in the Oval Office a few months after we swept into Baghdad,' he began, 'and I was telling the president of my many concerns.' . . . Bush, Biden recalled, just looked at him, unflappably sure that the United States was on the right course and that all was well. 'Mr. President,' I finally said, 'How can you be so sure when you know you don't know the facts?' Biden said that Bush stood up and put his hand on the senator's shoulder. 'My instincts,' he said. 'My instincts.'"

The Fourth Reich in America, it seems, is guided by "instincts," just as during Hitler's Reich.

ONE MAJOR DIFFERENCE between the Third Reich and the Fourth is the lack of emphasis on flag ceremonies and repetitious pledges. In Nazi Germany, a school day did not pass without these ceremonies of the state.

But in multicultural and globalized America, although schoolchildren still recite the Pledge of Allegiance and raise the U.S. flag, any formal ceremony has dropped away. Today, hardly any American—child or adult (with the possible exception of some Girl and Boy Scouts)—knows or observes proper flag protocols.

If the Pledge of Allegiance is used in schools, it is generally spoken over the loudspeaker. Students can recite along or not, as they will. If they can obtain a note from their parents, they are not even required to stand. Patriotic allegiance to one's nation is not conducive to the globalists' agenda of borderless countries under the control of multinational corporations.

Today, such nationalistic trappings have been replaced by ubiquitous corporate logos and slogans. More and more educational programs are being underwritten—and guided—by corporate officials. President Bush's secretary of education Margaret Spellings hosted the 2006–07 Siemens Competition in Math, Science, and Technology. In May 2006, she told attendees of the first National Summit on the Advancement of Girls in Math and Science in Washington, D.C., "I recently met with George Nolen, president and CEO of Siemens Corporation, and I look forward to working with him on President Bush's American Competitiveness Initiative."

Business leaders began to realize that education had failed to keep up with the corporatism of America. Prudential CEO Art Ryan complained that hiring high school graduates today is a high risk. "They can't do many of the things you would like them to do. But have high schools changed to reflect that economy? I would argue not enough."

Another rising concern in education that can be traced back to corporate intrusion is the rise of advertising in schools. According to a debate posted on the official NEA Web site, teachers have complained that getting kids to buy products, feel good about a corporation, or adopt the viewpoints of an industry on an important issue is not the purpose of education and that promotional sponsored education materials blur the line between education and propaganda and lead to distorted lessons.

Manny Lopez, a fourth-grade teacher at the International Community School in Oakland, California, warned, "Upon entering our schools, advertisers would dictate the placement of their billboards, banners, and lightboxes. The highly visible areas normally reserved for students' artwork, bulletin boards, and school/community message centers would be taken over. It would only be a matter of time before advertisers would attempt to have a hand in the curriculum, molding a school into a corporate image." Retired Georgia teacher Elizabeth Gould wrote, "[Advertising] gives students a warped sense of the world. They think everything is up for sale—ethics, morals, children."

Arguing in favor of school advertising, Kathleen McMahon, a fourth-grade teacher at Alice Costello School in Brooklawn, New Jersey, said, "In an ideal world, schools would have all the money they need to fund programs and buy equipment, but we're not living in Utopia. Public and private colleges accept money from plutocrats every day."

Gary Ruskin, writing in *Advertising Age* magazine, urged advertisers to exercise self-restraint or face future legal restrictions. He noted, "Advertisers are being expelled from schools in droves. Channel One, the in-school advertising service, was removed last school year from Nashville, and will soon be kicked out of Seattle. New restrictions on the marketing or sale of soda or junk food in school have been approved in places such as California, Texas, New York City and Philadelphia."

CORPORATE INVOLVEMENT WITH education is not a new idea. In the late 1800s, a group of Chicago businessmen became concerned that education in Germany and Austria was moving ahead of American education. "Unlike American schools, German classrooms divided kids into two tracks: one for those destined to become managers and the other for those destined to be their employees," explained Elizabeth Weiss Green writing in *U.S. News & World Report.*

But even in the 1800s, thoughtful persons questioned the propriety of business guiding education. "Education would then become an instrument of perpetuating unchanged the existing industrial order of society. Who, then, shall conduct education so that humanity may improve?" questioned John Dewey, the American philosopher, psychologist, and education reformer.

In mid-2007, billionaires Bill Gates and Eli Broad announced they had pooled $60 million to finance a campaign to emphasize education in the 2008 elections. However, the pair's ideas for improving American education were immediately criticized, because the three tenets of their campaign were: making teachers' salary increases dependent on student test scores, keeping students in class with more days and longer hours, and setting federal curriculum standards based on input from corporate leaders.

NATIONAL SOCIALISM IN the Germany of the 1930s made a strong appeal to its youth. The Nazis fully realized that if the younger generations could be brought to their worldview, the future of National Socialism would be assured. "Fascism in all countries made a fetish of youthfulness," commented George L. Mosse, author of *Nazi Culture.* "What a contrast this offered to the elderly politicians haggling in parliaments, or to the fossilized bureaucracies which ran the nations (and the political parties) of Europe."

Modern America also has witnessed conflict between the younger and older generations. The genesis of this generational conflict began with the rock-and-roll music of the 1950s and grew full-blown with the Vietnam War, when families were split along lines of age. The young embraced the antiwar movement while the older generations, tempered by the propaganda of World War II, supported the war policies of Lyndon Johnson and Richard Nixon.

Here again was a reflection of Nazi Germany, where Hitler's promises of a more prosperous future held considerable sway with the younger generations. "The young were set off against the old," observed Mosse, "and the distinction that was made between the old and young nations was operative within the *Volk* itself. When Hitler damned the bourgeoisie, he was inveighing against the older generation, brought up under the [First and Second] Empire."

This generational conflict, caused by control over popular culture such as music, films, art, etc., was merely another use of the divide-and-conquer tactic.

American journalist and educator Milton Mayer was both Jewish and of German ancestry. He traveled in Europe before the war and tried unsuccessfully to gain an audience with Hitler in 1935. Seven years after the war ended, Mayer traveled in Germany, searching to understand what had made the average German blindly follow National Socialism.

"I never found the average German," he recounted in a 1955 book, "because there is no average German. But I found ten Germans sufficiently different from one another in background, character, intellect, and temperament to represent, among them, some millions or tens of millions of Germans and sufficiently like unto one another to have been Nazis.

"I found—and find—it hard to judge my Nazi friends," he wrote. "I liked them. I couldn't help it. . . . I was overcome by the same sensation that had got in the way of my newspaper reporting in Chicago years before. I *liked* Al Capone. I liked the way he treated his mother. He treated her better than I treated mine."

Mayer recounted the story of one unnamed German academic, a language teacher, and his experience as Hitler's Third Reich grew in prominence. In light of modern America, it is worth repeating.

This teacher said that after 1933 no one seemed to notice the ever-widening gap between the government and the people. "This separation of government from people, this widening of the gap, took place so gradually and so insensibly, each step disguised (perhaps not even intentionally) as a temporary emergency measure or associated with true patriotic allegiance or with real social purposes. And all the crises and reforms (real reforms, too) so occupied the people that they did not see the slow motion underneath, of the whole process of government growing remoter and remoter."

As a scholar, this man was consumed with "meetings, conferences, interviews, ceremonies, and, above all, papers to be filled out, reports, bibliographies, lists, questionnaires.... It was all rigmarole, of course, but it consumed all one's energies, coming on top of the work one really wanted to do. You can see how easy it was, then, not to think about fundamental things. One had no time. The dictatorship, and the whole process of its coming into being, was above all diverting. It provided an excuse not to think for people who did not want to think anyway. Unconsciously, I suppose, we were grateful. Who wants to think?"

He said to live in this process required not noticing it. "Each step was so small, so inconsequential, so well explained or, on occasion, 'regretted,' that, unless one were detached from the whole process from the beginning, unless one understood what the whole thing was in principle, what all these 'little measures' that no 'patriotic German' could resent must some day lead to, one no more saw it developing from day to day than a farmer in his field sees the corn growing. One day it is over his head."

Pastor Martin Niemoeller spoke for men like this academic when he said: "First they came for the Communists, but I was not a Communist so I did not speak out. Then they came for the Socialists and the Trade Unionists, but I was neither, so I did not speak out. Then they came for the Jews, but I was not a Jew so I did not speak out. And when they came for me, there was no one left to speak out for me." A great part of this hesitancy to resist encroaching fascism was not due to fear, according to the academic Milton Mayer, but to genuine uncertainty that anyone else was seeing the things as he did. In Nazi Germany, people who questioned the motives behind government policies were deemed alarmists. In America today, they are called conspiracy theorists.

"[I]n small gatherings of your oldest friends, you feel that you are talking to yourselves, that you are isolated from the reality of things. This weakens your confidence still further and serves as a further deterrent.... It is clearer all the time that, if you are going to do anything, you must make an occasion to do it, and then you are obviously a troublemaker. So you wait, and you wait," explained an unnamed teacher quoted by Mayer. "But the one great shocking occasion, when tens or hundreds or thousands will join with you, never comes. That's the difficulty. If the last and worst act of the whole regime had come immediately after the first and smallest, thousands, yes, millions would have been sufficiently shocked. . . . But of course this isn't the way it happens. In between, come all the hundreds of little steps, some of them imperceptible, each of them preparing you not to be shocked by the next." He said that the person who is aware suddenly sees the world in a new way. "The world you live in—your nation, your people—is not the world you were born in at all. The forms are all there, all untouched, all reassuring, the houses, the shops, the jobs, the mealtimes, the visits, the concerts, the cinema, the holidays. But the spirit, which you never noticed because you made the lifelong mistake of identifying it with the forms, is changed. Now you live in a world of hate and fear, and the people who hate and fear do not even know it themselves; when everyone is transformed, no one is transformed. Suddenly it all comes down, all at once. You see what you are, what you have done, or, more accurately, what you haven't done—for that was all that was required of most of us: that we do nothing. . . . And the people in Germany who, once the war had begun, still thought of complaining, protesting, resisting, were betting on Germany's losing the war. It was a long bet. Not many made it."

Of particular concern to Americans who look beyond the advancing fascist agenda is the question of who will come to their rescue? Nazi-occupied Europe, and even many Germans toward the end of the war, looked hopefully to the Allied nations for their liberation. If America, today the world's foremost empire—a new Reich—falls under fascist domination, where can Americans look for deliverance?

CHAPTER 15

PSYCHOLOGY AND PUBLIC CONTROL

WHILE NOT DIRECTLY RELATED TO EDUCATION, IT SHOULD ALSO be noted that in Nazi Germany, while labor was extolled and glamorized, the power of labor unions was all but abolished. All labor matters were combined into the Deutsche Arbeitsfront, or German Labor Front, a monolithic Nazi organization created by Hitler to replace the old labor union system.

Again, the agenda of the corporate globalists was at work to curtail any meaningful power within the working class. However, despite early labor support based on pro-labor Nazi slogans and propaganda, the workers soon realized that the promised labor-professional equality was a myth, and labor problems continued to plague the Nazi government right through the end of the war.

In light of the continuing problems in Nazi Germany, the globalists declined to create a labor-controlling mechanism with modern America. Instead, they began a successful program of buying corrupt labor leaders, making deals with the crime syndicates that controlled certain unions, and crippling labor through federal legislation pushed through Congress with corporate money. The Reagan and both Bush administrations were particularly antiunion.

One example of this occurred in the 1980s, when the Reagan administration crushed collective bargaining by air traffic controllers, who desired increased air safety through better working conditions rather than just higher wages. Another example came in mid-2002, when a labor dispute between the International Longshore and Warehouse Union (ILWU) and the Pacific Maritime Association (PMA) caused a stack-up of cargo ships along the West Coast from San Diego to Seattle, which threatened to cut deeply into the 2002–03 holiday season profits. The strike was broken in October, when President Bush invoked the Taft-Hartley Act, a controversial 1947 union-busting law that was passed over President Truman's veto. Under this law, an eighty-day "cooling off" period can be ordered during a "national emergency."

Although Bush's action received scant attention in a media focused on the impending invasion of Iraq, one official of ILWU, Jack Heyman, termed Bush's intervention "a historic juncture in the labor movement." Heyman added, "By invoking Taft-Hartley against the longshore workers, Bush is effectively declaring war on the working class here and the Iraqi people simultaneously."

It has already been demonstrated how many U.S. government and corporate policies have been instituted by firms and organizations created and controlled by America's wealthy elite globalists.

The globalists also created the myth that the Rockefellers represent the apex of America's wealthy elite, a description the Rockefellers have done little to discourage. The long-standing idea that American oil magnate John D. Rockefeller was driven solely by greed obscured the fact that he was financed by outside sources. According to Eustace Mullins, the Rockefeller combine has never been an independent power. It was this Rockefeller myth of a homegrown elite that distracted attention from the international globalists and allowed the American public to accept the family as the nation's preeminent power. "[T]he Rockefeller oil trust [became] the 'military-industrial complex,' which assumed political control of the nation; the Rockefeller medical monopoly attained control of the health care of the nation; and the Rockefeller Foundation, a web of affiliated tax-exempt creations, effectively controlled the religious and educational life of the nation," wrote Mullins.

"The Rockefeller Syndicate operates under the control of the world financial structure, which means that on any given day, all of its assets could be rendered close to worthless by adroit financial manipulation," noted Mullins, who observed that patriarch John D. Rockefeller was able to gain almost total control over U.S. oil with financing from the National City Bank of Cleveland, named in congressional reports as a branch of the Rothschild banking empire. According to Mullins, "This is the final control, which ensures that no one can quit the organization. Not only would he be stripped of all assets, but he would be under contract for immediate assassination. Our Department of Justice is well aware that the only 'terrorists' operating in the United States are the agents of the World Order, but they prudently avoid any mention of this fact."

In pointing to the awareness of the Justice Department, Mullins, like authors John Loftus, Gary Allen, Mark Aarons, and others, was fully aware of the lack of detection or prosecution of Nazi war criminals who came to America, or of any serious prosecution of prominent corporate miscreants. Examples of the machinations of foreign financial powers in the United States, including the secret societies behind the War Between the States, may be found in my book *Rule by Secrecy*.

While it is true that the families that originated the wealthy elite in America—the Rockefellers, the Morgans, the Schiffs, and the Warburgs—have in recent years lost much of their previous influence, their giant global enterprises today remain as powerful as ever, maintaining all the functions for which they were first organized. This mechanism today has been brought under the control of the fascist globalists, who created both communism and National Socialism using the wealth brought from Europe by the Nazi ratlines.

"Since he set up the Trilateral Commission, David Rockefeller has functioned as a sort of international courier for the World Order, principally concerned with delivering working instructions to the Communist bloc [now the Russian Federation], either directly, in New York, or by traveling to the area," Mullins argued.

AMAZINGLY, IN JANUARY 1945, with their cities in ruins from round-the-clock Allied bombings, German war production was actually higher than in 1940, the year of spectacular military successes. Germans were still heading off to work each day, and production facilities, many moved underground, were producing at record capacity.

Historians have faithfully recorded the reasons for this—propaganda and hidden terror. The official, controlled government news bombarded the public with assurances that there was light at the end of the tunnel. New secret weapons were coming on line, they were told, and this, as has been noted, was not a total lie. According to their media, God was on their side and all would end well.

There also was the knowledge, only heard in whispered rumors, that anyone who spoke out about the true state of Nazi Germany would disappear into the offices of the Gestapo, likely never to be seen again.

The globalists have learned from the Nazis of the Third Reich how to employ these methods to ensure obedience from an intimidated public. From the constantly changing terrorist alerts and color codes in public places, since September 11, 2001, Americans have learned to live with fear. Unspoken fear is rampant in many levels of contemporary American society, and it is not all caused by unknown foreign terrorists. Fear of government harassment and surveillance is widespread.

Beginning with the still-controversial assassinations of President John F. Kennedy, Robert F. Kennedy, and Dr. Martin Luther King Jr., on through the 2002 fatal plane crash of Minnesota Democratic senator Paul Wellstone eleven days before an election—whose seat was then taken by a Republican, which created a Republican majority in the Senate—the trail of dead dissidents, witnesses, accusers, and whistleblowers has grown longer with each passing year. Just as in the days of the Third Reich, if any individual threatens to become too popular or brings too much attention to the fascist activities, they seem to disappear from the scene quickly. One such person was Republican Texas senator John Tower, whose Tower Commission was highly critical of the Reagan-Bush handling of the Iran-contra scandal, and Tower had confided to friends that he planned to write a tell-all book. Tower, who had chaired the powerful Senate Armed Services Committee and the Republican Policy Committee, was killed in a plane crash at Brunswick, Georgia, on April 5, 1991. Other persons who died under suspicious circumstances included Clinton

White House counsel Vince Foster, whose July 1993 death was ruled a suicide, and James McDougal, a convicted partner of Bill Clinton in the Whitewater scandal, who was a source of insider information to prosecutor Kenneth Starr. Foster's body was found in a park with a pistol still in his hand. McDougal died of an apparent sudden heart attack while being held in solitary confinement in a Fort Worth federal institution. Others, while not actually killed, have been shot or intimidated from running for office, such as former Alabama governor George Wallace and 1992 presidential candidate Ross Perot, who publicly stated his reason for dropping out was concern for the safety of his family. With the protection of the FBI, CIA, Secret Service and U.S. military, who did Perot have to fear?

Thanks to the modern surveillance state, many members of Congress find themselves susceptible to blackmail by damaging information from any number of government and corporate databases. It is reminiscent of the many allegations that former FBI director J. Edgar Hoover—perhaps with files from Interpol, as previously discussed—blackmailed government employees and congressmen into supporting his agendas. All too often, this type of coercion is more effective than campaign contributions.

BUT IT IS not only Congress members who find themselves at the mercy of increasing surveillance and control.

Beginning in December 2009, Americans will face the prospect of carrying their "papers" to conduct daily life, similar to the identity papers demanded by the Nazi Gestapo. Under provisions of the Emergency Supplemental Appropriations Act for Defense, the Global War on Terror, and Tsunami Relief (2005), popularly known as the Real ID Act, for all practical purposes, a national identity card will be required of every citizen. Today it has become unlawful not to provide identification documents to a police officer upon demand. The sheer act of failing to properly identify oneself can today result in arrest and jail.

Under the pretext of combating terrorism, this law requires national standards for state-issued driver's licenses as well as nondriver identification cards. It specifically states that no federal agency can accept any state ID card or license unless it meets the requirements as stated in the Real

ID Act. Since the Transportation Security Administration provides security at airports, anyone without identification compliant with Real ID Act may be unable to fly on commercial aircraft. And as the federal Social Security Administration requires states to maintain a new-hire directory, employers would no longer be able to hire anyone without a Real ID Act–compliant document. Of particular concern to libertarians is the requirement that all financial institutions would be required to accept only Act-compliant IDs. Customers without such federally approved documentation could be denied financial and banking services.

This thinly disguised nationwide citizen-registration law languished in a hesitant Congress until it was attached as a rider to a military spending bill and signed into law on May 11, 2005. In 2007, perhaps in light of several states passing legislation opposing the Real ID Act, it was announced that enforcement of the law would be delayed until December 2009. In 2007, states that opted out of the Real ID Act were told their citizens might not be able to travel freely around the country.

Author Steven Yates is a teaching fellow at the Ludwig von Mises Institute in Auburn, Alabama, a research and education center of classical liberalism, libertarian political theory, and economics. Following the intellectual tradition of Ludwig von Mises, a renowned economist who has been called the "uncontested dean of the Austrian school of economics," the institute supports publications, programs, and fellowships. Yates noted, "It is a testimony to how much this country has changed since 9/11 that no one has visibly challenged [national IDs] as unconstitutional and incompatible with the principles of a free society." He and many others see the slow encirclement of law-abiding U.S. citizens with national ID technology advancing a globalist agenda while doing little if anything to safeguard us against terrorism.

However, many legislators, such as Representative Jane Harman of California, seemed agreeable to citizen registration. She stated: "I think this issue must be looked at. We don't automatically have to call it a national ID card, that's a radioactive term, but we can certainly think about smart cards [such as driver's licenses with chips] for essential functions, but we need the database to support that."

This need for a national database, so necessary for Hitler's euthanasia and extermination programs, was addressed in the USA PATRIOT Act, which authorized $150 million in tax money for the "expansion of the Regional Information Sharing System [to] facilitate federal-state-local law enforcement response related to terrorist acts." Asked if she thought the public was ready for such measures, Harman replied, "I think most people are really there. Keep in mind that if we have a second wave of attacks, the folks who are raising objections will probably lose totally."

What disturbs many thinking people is a vision of the near future in which, should the feds decide to stifle dissent, they could "freeze" the dissident's assets by reprogramming his database information. Scanners would not recognize him and he would become officially invisible, unable to drive or work legally, have a bank account, buy anything on credit, or even see a doctor. "Do we want to trust *anyone* with that kind of power?" Yates asked.

Lest anyone think this is naive or even paranoid nonsense, consider that in late October 2002, Applied Digital Solutions, Incorporation, a high-tech development company headquartered in Palm Beach, Florida, announced the launching of a national promotion for its new subdermal personal verification microchip. The "Get Chipped" promotion was describing a device that can be implanted under a person's skin to transmit data to various locations. The "VeriChip," according to company literature, is "an implantable, 12mm by 2.1mm radio frequency device . . . about the size of the point of a typical ballpoint pen. It contains a unique verification number. Utilizing an external scanner, radio frequency energy passes through the skin, energizing the dormant VeriChip, which then emits a radio frequency signal containing the verification number. The number is displayed by the scanner and transmitted to a secure data storage site by authorized personnel via telephone or Internet."

In addition to "VeriChip Centers" in Arizona, Texas, and Florida, the firm also fields the "ChipMobile," a motorized marketing and "chipping" vehicle. The new "Get Chipped" campaign was launched just days after the Food and Drug Administration ruled that the chip is not a regulated medical device and stated it found "reasonable assurance" that the chip was safe. However, neither the manufacturer nor the FDA mentioned a series of veterinary and toxicology studies conducted in the mid-1990s, which found that chip implants "induced" malignant tumors in some lab mice and rats. According to an Associated Press report in September 2007, Keith Johnson, who led a study in 1996 at Dow Chemical Company, said, "The transponders were the cause of the tumors." Several leading cancer experts contacted by the AP cautioned that while animal tests do not necessarily apply to humans, they were "troubled" by the findings and urged further study before the chips were implanted in people. Some stated they would not allow their family members to receive such implants. The head of the federal Department of Health and Human Services when the VeriChip was approved was Tommy Thompson, who after leaving his government post joined VeriChip Corporation as a director. He resigned from the company in early 2007 to run an unsuccessful campaign as a Republican presidential candidate. The law firm in which Thompson was partner—Akin, Gump, Strauss, Hauer & Feld LLP—was paid $1.2 million to represent VeriChip, according to the SEC.

Uses for the chip include controlling access to nonpublic facilities such as government buildings and installations, nuclear power plants, national research laboratories, correctional institutions, and transportation hubs— either using the chip by itself or in conjunction with existing security technologies such as retina scanners, thumbprint scanners, or face- recognition devices. Company officials envision the chip will come to be used in a wide range of consumer products, including PC and laptop computers, personal vehicles, cell phones, homes, and apartments. They said the implanted chip will help stop identity theft and aid in the war against terrorists.

By early 2006, fears of the chip became reality when a Cincinnati video surveillance firm, CityWatcher.com, began to place the VeriChip in the arms of some of its employees who worked in sensitive areas. While the firm did not require employees to receive the chip to keep their jobs, some saw the company as establishing an unsettling precedent.

A NATIONAL ID card or chip may be the least of a citizen's worries. Today, authorities are availing themselves of technologies the Nazis of the Third Reich could only have dreamed about. Satellite surveillance and the increasingly ubiquitous cameras in public places have curtailed privacy to a large extent in the industrialized world. Sophisticated miniature cameras now can read license plates and track vehicles even traveling at speeds of more than sixty miles per hour.

The U.S. federal government utilizes an electronic eavesdropping satellite and computer system called Echelon. This system tracks international telephone calls, faxes, and e-mail messages all around the world. It was so secret that the government would neither confirm nor deny its existence until 2001. According to a study by the European Union, Echelon accumulates electronic transmissions like a vacuum cleaner, using keyword-search software in conjunction with massive computer data banks.

The Echelon system, housed within the National Security Agency at Fort Meade, Maryland, has caused protests in several nations—excluding the United States, whose population rarely sees any news concerning this powerful global wiretapping system.

As technology continues to advance, so does the means of manipulating, even controlling, whole groups of individuals. The means to control the human mind has come a long way since the days of Nazi concentration camps and the subsequent CIA drug experimentation. Today, beamed electromagnetic frequencies can alter perceptions, instill emotions, and even cloud normal reasoning.

All organic life consists of living cells controlled by the DNA within them. The chemical action within the cells is driven by electromagnetic frequencies that pulse, oscillate, and vibrate. Collectively, the energy within a living organism creates a surrounding, albeit weak, electromagnetic field.

Dr. Nick Begich Jr., executive director of the Lay Institute on Technology and author of *Angels Don't Play This HAARP,* stated scientists today "have succeeded in isolating many of the healing frequency codes of the human body and, importantly, are adding to a growing body of remarkably practical medical advancement toward the diagnosis and treatment of numerous disease states and conditions."

In the 1930s, Dr. Royal Raymond Rife demonstrated the ability of precise electrical frequencies to disrupt viral and bacteria cells. A Special

Research Committee of the University of Southern California confirmed that Rife frequencies were reversing many ailments, including cancer. Opposition immediately came from Dr. Thomas Rivers of the Rockefeller Institute, who had not even seen Rife's equipment in operation. By 1934, Rife had isolated a virus that bred cancer and stopped it by bombarding it with electromagnetic frequencies. He was successful in killing both carcinoma and sarcoma cancers in more than four hundred tests on animals. It has been widely reported that in the summer of 1934, Rife, along with doctors Milbank Johnson and Alvin G. Foord, succeeded in using his frequencies to cure sixteen cancer patients diagnosed as terminal by conventional medicine.

Rife described the operation of his frequency machine thus: "With the frequency instrument treatment, no tissue is destroyed, no pain is felt, no noise is audible, and no sensation is noticed. A tube lights up and three minutes later the treatment is completed. The virus or bacteria is destroyed and the body then recovers itself naturally from the toxic effect of the virus or bacteria. Several diseases may be treated simultaneously." A general analogy to this effect is glass shattering when a singer's high note is sounded.

It did not take long for the medical establishment to realize that such a device not only would wreck the pharmaceutical industry but damage medicine in general, since cures meant fewer visits to the doctor. Overworked and underfunded, Rife and his associates were easy targets for attack. False claims were made against him, test procedures were altered, causing his demonstrations to fail, and impossible and diverting demands were made on Rife's research. Barry Lyne, who chronicled Rife's story in his book *The Cancer Cure That Worked,* elaborated: "[Rife] was curing cancer while the [International Cancer Research Foundation] broke their agreements, insisted on procedures with inexperienced people, which were doomed from the outset, and ignored the larger goal which Rife was achieving—the cure of cancer in human beings."

After he declined an offer to partner with Morris Fishbein, then head of the American Medical Association, Rife's troubles turned more serious, with lawsuits and health authorities coming at him from all sides. The university's Special Research Committee's work was ended, Rife was marginalized, and his device today is available only as a costly research instrument employed by a few doctors and private citizens. Rife died a broken man in 1971.

But while electromagnetic energy manipulation was seeking to make humans healthier, such technology also brought horrific possibilities for mind control. "The early attempts used chemicals and hallucinogenics to achieve some measure of control," wrote Begich, the son of Alaskan Democrat senator Nick Begich Sr., who disappeared along with congressman and member of the Warren Commission Hale Boggs when their plane was lost over Alaska in 1972. "Then in the early 1960s the interest changed to nonchemical means for affecting behavior. By the early 1970s, within certain military and academic circles, it became clear that human behavior could be modified by the use of subtle energetic manipulations. By 2006, the state of the technology had been perfected to the point where emotions, thoughts, memory, and thinking could be manipulated by external means."

According to reports from the military in both the United States and Russia, psychotronic generators are being developed, which can create an infrasonic oscillation in the 10–20 Hertz range. It is destructive to living organisms; can cause behavior modification by transmitting frequencies through normal telephone, TV, and radio networks; and can produce frequencies to paralyze the central nervous system.

"In recent years, the Defense Advanced Research Projects Agency (DARPA) has pursued research into brain decoding and the development of electronic micro and nanocircuits that will directly interact with the brain," stated Dr. Begich. "New microchip technology could be used for direct interaction between the brains of people and computers."

Several methods have been found to allow persons to hear sounds and speech without the use of the ears or normal auditory pathways—such as the Neurophone. This device uses a vibrational technology developed by Patrick Flanagan in 1958, which allows the transmission of sound vibrations through the skin, much like the vibration of a speaker. Bypassing the ear, a completely different part of the brain processes the sound, creating new neural pathways. Such technology is used to increase concentration while studying or learning languages, and it helps with meditation, relaxation, and healing.

Unfortunately, it was also found that amplified brainwave frequencies can be imposed on others. One U.S. patent even described how very low or very high audio frequencies can be used to transmit subliminal messages to the human brain. For example, U.S. Patent No. 3,170,993, accepted in 1965, is called the "Means for Aiding Hearing by Electrical Stimulation of the Facial Nerve System."

According to Dr. Begich, all Americans are already under constant bombardment from microwave frequencies "one thousand times higher than the level considered safe in the former Soviet Union [a leader in mind-affecting energy technology]. The reason that the Soviets set their safety standards as low as they did was because they detected biological effects at levels ignored by the West," he explained.

But while microwaves may be affecting large segments of the population in the United States, this issue pales when one considers the possibility of a program to consciously target whole populations. Such a program exists in the High Frequency Active Auroral Research Program (HAARP), a vast array of powerful transmitting dishes located near Gakona, Alaska. Officially, HAARP is designed to study the uppermost portion of the Earth's atmosphere: the ionosphere. However, critics of the program, including Dr. Begich and other scientists, claim this powerful array can be used as a weapon, to deliver energy blasts equal to an atomic bomb, destroy communications across the planet, and even influence human behavior.

Critics argue that HAARP has the capability of stimulating the ionosphere to return a pulsed electromagnetic signal, which, at the proper frequency, can override normal brain functions.

In 2002, the Russians, who experimented with both psychic and mind-altering technologies as far back as the 1970s, expressed their concern with HAARP. A complaint letter from the Russian State Duma to President Vladimir Putin stated that HAARP was influencing near-Earth atmosphere with high-frequency radio waves. "The significance of this qualitative leap [in science] could be compared to the transition from cold steel to firearms, or from conventional weapons to nuclear weapons. This new type of weapon differs from previous types in that the near-Earth medium becomes at once an object of direct influence," stated the letter. Despite this objection, no discernible action has been taken by the United States, and the HAARP system continues to operate.

Dr. Begich warned that allowing the military-industrial complex to solely guide such mind-control research not only runs the risk of creating an Orwellian thought-controlled society but would prevent the "enabling enhancement of human potentials in ways only reserved in the past to mystics, religious figures, and those who sought to change people."

INCREDIBLY, COMPUTER SYSTEMS are under development today to anticipate criminal or antisocial behavior. They are designed to read human body language that might indicate potential "criminal" activity and summon authorities, reminiscent of prewar Nazi plans to preempt crime and dissent.

In Vienna, during September 1940, at the First Conference of the German Society for Child Psychiatry and Therapeutic Education—later changed to the German Society for Child and Adolescent Psychiatry—Dr. Anna Leiter, a genetic researcher from Dresden, who studied three thousand youngsters for possible antisocial traits, stated, "We demand that as soon as a careful and responsible analysis shows an extremely unusual lack of emotions in connection with other criminogenic reactive tendencies, we detain these children as early as possible, since they represent an unbearable burden and danger for the entire country."

Nazi educators pointed to these symptoms of "bad student material" that would qualify for such detention: actual and potential repeaters of lower public school grades, students recommended for special school, "borderline and questionable cases," uneducable children, those with special educational difficulties, and schoolchildren whose siblings and families are or have been in special schools. It was stated that "genetic and national health considerations recommend their preventative registration." Of course, such registration led to the euthanasia centers.

The trend to identify and detain potential troublemakers before they have actually committed a crime is being perpetuated today. The British government, in May 2007, responding to news accounts, acknowledged it had secretly established a new national antiterrorist unit to protect VIPs by first profiling, then arresting persons considered to be potentially dangerous. Amazingly, this power to detain suspects even before they actually committed a crime was based on mental health laws.

"The Fixated Threat Assessment Centre (FTAC) was quietly established last October [2006] and is set to reignite controversy over the detention of suspects without trial," wrote *The Times* reporter Joanna Bale. "Until now it has been up to mental health professionals to determine if someone should be forcibly detained, but the new unit uses the police to identify suspects, increasing fears that distinctions are being blurred between criminal investigations and doctors' clinical decisions."

The FTAC unit will be staffed by four police officers, two civilian researchers, a psychiatrist, a psychologist, and a community mental health nurse. It was hailed as the first joint mental health–police unit in the United Kingdom and a "prototype for future joint services" in other areas. Even as this police unit was assuming the power to arrest and hold potential "suspects," Scotland Yard refused to discuss how many suspects have been forcibly hospitalized by the team, because of "patient confidentiality." Meanwhile, the British government was introducing legislation to broaden the definition of mental disorders to give doctors—and now police—more power to detain people.

"There is a grave danger of this being used to deal with people where there is insufficient evidence for a criminal prosecution," said Gareth Crossman, policy director for Britain's National Council for Civil Liberties. "This blurs the line between medical decisions and police actions. If you are going to allow doctors to take people's liberty away, they have to be independent. That credibility is undermined when the doctors are part of the same team as the police. This raises serious concerns. First, that you have a unit that allows police investigation to lead directly to people being sectioned without any kind of criminal proceedings. Secondly, it is being done under the umbrella of antiterrorism at a time when the government is looking at ways to detain terrorists without putting them on trial."

Is this coming to America soon? Libertarians fear that such measures might be slipped into legislation such as funding for the military or

Homeland Security. Many conspiracy researchers suspect that the globalists try out new policies and methods in the United Kingdom first, to see if they are accepted by the public.

Once offenders are picked up, today they face new types of unconstitutional trials. Some researchers saw Bush's Military Commissions Act with its secret tribunals as an echo of the Third Reich "special courts," which were designed to prosecute political resistance to the Nazi administration. "The main duty of the special courts was to criminally prosecute the political resistance to the Nazi regime," wrote Dr. Thomas Roeder, Volker Kubillus, and Anthony Burwell in their book *Psychiatrists—the Men Behind Hitler.* "During the Second World War, they gradually took over the duties of ordinary justice and from 1942 on, most of the sentencing. Experts estimate that the special court of Hannover alone, one of several in today's Saxony, sentenced 4,000 defendants, about 170 of them to death."

Another eerie parallel between the Bush administration and the Third Reich involved the *Fuehrerprinzip,* or leader principle, which was outlined in Hitler's book *Mein Kampf.* This principle stated the leader embodied National Socialism and therefore the people. All decision-making rested with him. Such thinking evolves from the lack of trust in the people. National Socialism, like our democracy, was supposedly a movement of the masses. But its leadership had little faith in its followers.

"[Nazi] ideology denounced civilian methods of elections, negotiation, and compromise as horse trading and called for authority of command, discipline and obedience," noted Professor Louis L. Snyder in his *Encyclopedia of the Third Reich.* This top-down leadership principle sounds eerily similar to calls by President George W. Bush for a "unified presidency" as well as his comment to reporters in April 2006, "I hear the voices, and I read the front page, and I know the speculation. But I'm the decider, and I decide what is best." "Perhaps the most unique feature of the Bush administration is its protracted period of unified party control of the government, a stark contrast to the divided governments of George H. W. Bush and Ronald Reagan," noted Kathryn Dunn Tenpas, a senior fellow with the Brookings Institution.

Civil libertarians historically have heeded the statement of patriot Thomas Paine, who wrote in *Common Sense,* "In America, the law is king. For as in absolute governments the king is law, so in free countries the law ought to be king; and there ought to be no other." In other words, no one is above the law.

Yet, Bush has argued that actions allowing him to ride roughshod over the Congress, the courts, and the Constitution were somehow necessary to preserve the presidency. "I have an obligation to make sure that the presidency remains robust and that the legislative branch doesn't end up running the executive branch," Bush argued in mid-2002. In launching the invasion of Iraq in 2003, Bush preempted the power of Congress, as the U.S. Constitution under Section 8 clearly states that only Congress has the power to declare war. When he and his appointees rammed the USA PATRIOT Act through a cowed Congress, with little or no input, he likewise took powers from the representatives of the people.

A panicky House of Representatives, still in shock over 9/11 and the subsequent anthrax attacks, rushed the PATRIOT Act into law by a vote of 339–79. The act was 342 pages long and made changes, both great and small, to more than fifteen different U.S. laws, many of them enacted following revelations about the misuse of surveillance powers by the FBI and CIA. It was hurriedly and enthusiastically signed into law by President Bush on October 26, 2001. The speed with which this legislation was presented to Congress left little doubt in many minds that it had long been prepared and simply needed some provocation as an impetus for action.

According to some congressmen, many lawmakers had not even read the entire document when it was passed. The ACLU also reported that some members of Congress had less than one hour to read the extensive changes of law contained within the act. Congressman Dennis Kucinich, a Democrat from Ohio, described the atmosphere in which the PATRIOT Act was passed: "There was great fear in our great Capitol. . . . The great fear began when we had to evacuate the Capitol on September 11. It continued when we had to leave the Capitol again when a bomb scare occurred as members were pressing the CIA during a secret briefing. It continued when we abandoned Washington when anthrax, possibly from a government lab, arrived in the mail. . . . It is present in the camouflaged armed national guardsmen who greet members of Congress each day we enter the Capitol campus. It is present in the labyrinth of concrete barriers through which we must pass each time we go to vote." Texas congressman Ron Paul, one of only three Republicans to vote against the House bill, said he objected to how opponents were stigmatized by the name alone. "The insult is to call this a 'patriot bill' and suggest I'm not patriotic because I insisted upon finding out what was in it and voting no. I thought it was undermining the Constitution, so I didn't vote for it—therefore I'm somehow not a patriot. That's insulting."

Paul confirmed rumors that the bill was not read by most members of the House prior to their vote. "It's my understanding the bill wasn't printed before the vote—at least I couldn't get it," he told *Insight* magazine. "They played all kinds of games, kept the House in session all night, and it was a very complicated bill. Maybe a handful of staffers actually read it, but the bill definitely was not available to members before the vote." Paul's view of the PATRIOT Bill was echoed by the only independent in the House, Congressman Bernie Sanders of Vermont, who said, "I took an oath to support and defend the Constitution of the United States, and I'm concerned that voting for this legislation fundamentally violates that oath. And the contents of the legislation have not been subjected to serious hearings or searching examination."

Most Americans would be surprised to learn that since March 9, 1933, the United States has been in a state of declared national emergency. In fact, until 1976, the USA operated under four presidentially proclaimed states of national emergency—the one declared by President Roosevelt in 1933; a national emergency proclaimed by President Truman on December 16, 1950, during the Korean conflict; and two states of national emergency declared by President Nixon, on March 23, 1970, and August 15, 1971. Years of debate over the need for such emergency powers resulted in the creation in 1973 of the U.S. Senate's Special Committee on the Termination of the National Emergency. By 1976, the committee had consolidated the emergency declarations and produced the National Emergencies Act (50 U.S.C. 1601–1651), which limits such emergencies to two years.

The Senate committee found that the weight of all these often-underreported proclamations "give[s] force to 470 provisions of federal law. These hundreds of statutes delegate to the president extraordinary powers, ordinarily exercised by the Congress, which affect the lives of American citizens in a host of all-encompassing manners. This vast range of powers, taken together, confer enough authority to rule the country without reference to normal constitutional processes. Under the powers delegated by these statutes, the president may: seize property; organize and control the means of production; seize commodities; assign military forces abroad; institute martial law; seize and control all transportation and communication; regulate the operation of private enterprise; restrict travel; and, in a plethora of particular ways, control the lives of all American citizens."

Few people in the United States today have been informed that they have been living under a state of emergency since September 11, 2001. This was quietly but officially declared by President George W. Bush three days later, when he issued a proclamation stating, "A national emergency exists by reason of the terrorist attacks at the World Trade Center, New York, NY, and the Pentagon, and the continuing and immediate threat of further attacks on the United States.

"Now, therefore, I, George W. Bush, president of the United States of America, by virtue of the authority vested in me as president by the Constitution and the laws of the United States, hereby declare that the national emergency has existed since September 11, 2001, and pursuant to the National Emergencies Act (50 U.S.C. 1601 et seq)."

With this proclamation Bush activated what the media called the "shadow government," those unelected officials and appointees who, under the guidance of his father, Donald Rumsfeld, and Dick Cheney, had years earlier began to alter the form of this former republic, including the use of warrantless electronic surveillance by the National Security Agency.

In mid-2007, Bush codified the "shadow government" with the ominously worded National Security Presidential Directive/NSPD-51 and Homeland Security Presidential Directive/HSPD-20, innocently titled "National Continuity Policy." In the interest of "continuity of government," this directive stated, "the president shall lead the activities of the federal government for ensuring constitutional government." The implication was that he would lead the entire government, not just the executive branch.

This takeover of the federal government was contingent on a "catastrophic emergency," defined as "any incident, regardless of location, that results in extraordinary levels of mass casualties, damage, or disruption severely affecting the U.S. population, infrastructure, environment, economy, or government functions."

Sharon Bradford Franklin, senior counsel at the Constitution Project, a bipartisan think tank that promotes constitutional safeguards, said the policy's definition "is so broad that it raises serious concerns about when and how this might be used to authorize unchecked executive action."

ALSO TROUBLESOME IS Bush's contention that he must defend his office from the loss of power. In January 2006, Vice President Cheney stated on NBC, "For thirty-five years that I've been in [Washington], there's been a constant, steady erosion of the prerogatives and the powers of the president of the United States. And I don't want to be a part of that." Bush has stated, "I have an obligation to make sure that the presidency remains robust and that the legislative branch doesn't end up running the executive branch."

This is blatantly untrue. The American president today carries far more power than ever imagined by our Founding Fathers or even more modern chief executives, like Franklin D. Roosevelt, as expounded in Richard Loss's 1990 book *The Modern Theory of Presidential Power*. John W. Dean, the White House counsel to Nixon, who served jail time after Watergate, said "the institutional powers of the presidency all but overwhelm those of Congress. They are, in fact, stronger today than thirty years ago." Dean added, "To claim a need for secrecy to restore presidential power is disingenuous at best, and a deliberate falsehood at worst. Secrecy is the way of dictatorships, not democracies."

A June 29, 2006, Supreme Court ruling bolstered efforts of those in Congress who had been trying to curtail overreaching presidential power that claims unilateral authority to determine not only how terrorism suspects are tried, but also to set rules for domestic wiretapping, to interrogate prisoners, and to pursue other wartime powers. It was this 5–3 decision to overrule the president's actions that required the Bush administration to draw up the Military Commissions Act of 2006.

Volumes have been written about the Bush White House's penchant for secrecy. It has almost become policy. Even conservative columnist Phyllis Schlafly has attacked the Bush policy of unnecessary secrecy, writing, "The American people do not and should not tolerate government by secrecy." She added that no one is "going to buy the sanctimonious argument that the Bush administration has some sort of duty to protect the power of the presidency." "What the president is claiming is legally and historically absurd and politically stupid," declared former Justice Department official Bruce Fein.

Bush's secretive manner of drawing ever more power unto himself came in the form of "signing statements"—written responses by the president issued upon the signing of a bill into law. Such statements have drawn severe criticism from credible legal sources. Jennifer Van Bergen holds a law degree from Benjamin N. Cardozo School of Law, is an adjunct faculty member at the New School for Social Research in New York, and is a member of the board of the ACLU Broward County (Florida) chapter. She criticized this activity by noting that from 1817 until the end of the Carter administration in 1981, only 75 "signing statements" were issued. From the Reagan administration until the end of the Clinton administration, this number had grown to 322. But in the first term alone, Bush issued at least 435 signing statements, many noting his concept of a "unitary executive."

Such signing statements convey a president's view toward the law and his own power. Bush's use of the term "unitary executive," according to Van Bergen, is merely a code word for a doctrine "that favors nearly unlimited executive power."

"In [Bush's] view, and the view of his administration, that doctrine gives him license to overrule and bypass Congress or the courts, based on his own interpretations of the Constitution—even where that violates long-established laws and treaties, counters recent legislation that he has himself signed, or (as shown by recent developments in the Padilla case) involves offering a federal court contradictory justifications for a detention," Van Bergen wrote.

Charlie Savage, writing in the *Boston Globe,* said Bush is the first president in modern history who has never vetoed a bill, thus giving Congress no chance to override his judgments. (In late 2007, Bush's veto of a $23 billion water resources bill critics claimed was laden with pork projects was overridden by both the House and the Senate, marking the first time in a decade that Congress passed legislation over a presidential veto.) Bush often invites the bills' sponsors to signing ceremonies, at which he lavishes praise upon their work. But Savage noted: "Then, after the media and the lawmakers have left the White House, Bush quietly files 'signing statements'—official documents in which a president lays out his legal interpretation of a bill for the federal bureaucracy to follow when implementing the new law. The statements are recorded in the federal register. In his signing statements, Bush has repeatedly asserted that the Constitution gives him the right to ignore numerous sections of the bills—sometimes including provisions that were the subject of negotiations with Congress in order to get lawmakers to pass the bill."

Van Bergen took particular note of Bush's signing statement while he was signing into law legislation curtailing torture on prisoners. "When President Bush signed the new law, sponsored by Senator [John] McCain, restricting the use of torture when interrogating detainees, he also issued a presidential signing statement," said Van Bergen. "That statement asserted that his power as commander in chief gives him the authority to bypass the very law he had just signed."

Portland State University law professor Phillip Cooper told newsmen Bush and his legal team spent the past five years quietly working to concentrate ever more governmental power in the White House. "There is no question that this administration has been involved in a very carefully thought out, systematic process of expanding presidential power at the expense of the other branches of government. This is really big, very expansive, and very significant," Cooper said.

Little is said of such things, because the Bush White House is more closed-mouthed than any previous administration. In the Nixon, Reagan, Clinton, and both Bush administrations, loyalty has been seen as a requisite for those serving the White House. Many people working with the highest levels of power in the United States see their superiors as public servants, only looking after the best interests of America. Such unquestioning loyalty and allegiance was a hallmark of the Third Reich.

Many loyalists around Nixon, Reagan, Clinton, and the two Bushes truly saw their leader as a paragon. They appeared blinded to their leader's actions, as happened to so many in the Third Reich. "I admit, I was fascinated by Adolf Hitler," recalled Traudl Junge, the last surviving occupant of the *Fuehrerbunker,* shortly before her death in 2002. She was twenty-two years old when, in 1942, she was selected to become a secretary to Hitler. "He was a pleasant boss and a fatherly friend," she recalled. "I deliberately ignored all the warning voices inside me and enjoyed the time by his side almost until the bitter end. It wasn't what he said, but the way he said things and how he did things."

Soon after the War on Terror got underway, the Germans, who should know, could see the parallels between their Nazi era and modern America. The attacks of 9/11 and the Reichstag fire, Bush's PATRIOT Act and Hitler's Enabling Act, the use of German Army reserves to attack Poland and Bush's use of reserves in Afghanistan and Iraq to avoid a military draft must have seemed quite familiar to them. German justice minister Herta Daeubler-Gmelin brought heated criticism from President George W. Bush in September of 2002, when, in criticizing Bush's policy in Iraq, she stated publicly, "Bush wants to distract attention from his domestic problems. That's a popular method. Even Hitler did that." For her remark, Daeubler-Gmelin was asked to resign by German chancellor Gerhard Schroeder, an attendee of Bilderberg meetings.

And it is not just politicians who must be careful of voicing their opinions in the new American empire. Demonstrators are routinely herded into fenced "free speech zones," or arrested for speaking out. Thanks to the Internet, millions witnessed twenty-one-year-old Florida university student Andrew Meyer screaming in pain after being tasered by campus police on September 17, 2007, after simply asking John Kerry why he had not challenged the 2006 presidential election.

On top of the ubiquitous nature of surveillance technology and the intimidation of dissenters, there is very real control through the print and electronic matrix we know as the mass media.

CHAPTER 16

PROPAGANDA

ONE TOOL OF PUBLIC CONTROL QUICKLY UNDERSTOOD AND WELL
utilized by the Nazis of the Third Reich was propaganda. And even the
definition of commonly used words can change. Many still remember
when "gay" meant "happy."

The 1952 edition of *The New Webster Encyclopedic Dictionary of the
English Language* defined propaganda as "The dissemination and the defense of beliefs, opinions, or actions deemed salutary to the program of a
particular group; the propagation of doctrines and tenets of special interests, as an effort to give credence to information partially or wholly falla-
cious." A more diluted definition was given in the 1996 edition of *The
Reader's Digest Oxford Complete Wordfinder,* which merely stated it was
"an organized program of publicity, selected information, etc., used to
propagate a doctrine, practice, etc." So, where the definition of propaganda once included allegations and false information, today it is just "organized" and "selected" information, of the type seen daily in the mass
media.

A noteworthy component of propaganda is not just false or spun information, but the omission of critical material that is essential to enable the
reader or viewer to place the presented information in a meaningful context. Critics see the single greatest failing of the modern corporate mass

media as that while it daily bombards its audience with facts, figures, and statistics, it rarely attempts to bring coherence or meaning to this data.

"The media may not always be able to tell us what to think, but they are strikingly successful in telling us what to think about," stated media critic Michael Parenti.

Many people complain that the major media are superficial, conformist, and subjective in their selection of news. A Pew Research Center poll showed respondents who thought news reporting unfair and inaccurate ranged into the 60 percentiles. A survey by the news industry publication *Editor & Publisher* showed that journalists themselves do not disagree. Nearly half of its participants indicated their belief that news coverage is shallow and inadequate. "Much of what is reported as 'news' is little more than the uncritical transmission of official opinions to an unsuspecting public," wrote Parenti. Fox news commentator Brit Hume stated, "What [the mass media] pass off as objectivity, is just a mindless kind of neutrality." Hume added that reporters "shouldn't try to be objective, they should try to be honest."

Hitler understood propaganda intuitively. He wrote in *Mein Kampf:* "[O]ne started out with the very correct assumption that in the size of the lie there is always contained a certain factor of credibility, since the great masses of people may fall victim to a great lie rather than a small one, since they themselves also lie sometimes in little things but would certainly be too ashamed to tell a great lie. Thus, it would not enter their heads to tell a great lie, and they would not believe that others could tell such an infamous lie; indeed, they will doubt and hesitate; even after learning the truth, they will continue to think there must be some other explanation; therefore, just for this reason, some part of the most impudent lie will remain and stick, a fact which all great lying artists and societies of this world know only too well and therefore villainously employ." Hitler was stating that most people are honest and would never tell a great lie; therefore, it is hard for them to believe that anyone—especially an admired leader—would tell one. This presents a great weapon to unprincipled officials.

For example, during the Nuremberg trials, Reichsmarschall Hermann Goering explained how to persuade the public to go to war. He told one of his interrogators, "Why of course the people don't want war. . . . That is understood. But after all it is the leaders of the country who determine the policy, and it is always a simple matter to drag the people along, whether it is a democracy, or a fascist dictatorship, or a parliament, or a communist dictatorship. . . . Voice or no voice, the people can always be brought to the bidding of the leaders. That is easy. All you have to do is to tell them they are being attacked, and denounce the pacifists for lack of patriotism and exposing the country to danger." Conspiracy researchers believe such methodology has been used on the American public.

Ubiquitous propaganda remains a prime method for controlling public thought and worldviews. The use of the public media to frame the issues of the day and bolster support for the government was quickly understood by Hitler and his top lieutenants. "It is no coincidence that the Ministry of Propaganda and Enlightenment became so important, or that Hitler was personally closest to Joseph Goebbels, the expert manipulator of mass opinion," explained Professor George Mosse, author of *Nazi Culture*.

In America today, there are thousands of persons working in the advertising agencies, public relations firms, and programming departments of the TV networks and cable/satellite channels—all under the command of their corporate owners—toiling unceasingly to present the corporate worldview to their viewers. They weave an electromagnetic "matrix" around our society, which is difficult to break with even for the most sophisticated and educated.

By 2008, SIX multinational corporations controlled almost everything the average American reads, sees, or hears. The Internet has done a marvelous job of bringing alternative news and information but only to those who own and can use a computer. Everyone else is at the mercy of the corporate-controlled mass media.

A cursory look at the six major media corporations reveals:

◆ *Time Warner Inc.* made history in 2000 by its $112 billion merger with America Online. *Time* magazine, one of the corporation's media assets, noted, "The combination will be, for better or worse, the world's biggest media conglomerate. (Of which *Time* will be a part.) It's a vast empire of broadcasting, music, movies and publishing assets, complemented by AOL's dominant Internet presence, all fed to consumers, ultimately, through Time Warner's cable network. Think of it as AOL Time Warner Anywhere, Anytime, Anyhow.... Time Warner is in the traditional media business; AOL is an Internet company. Because the two didn't overlap, antitrust lawyers saw no need for concern. But the more people looked, the more they thought this was not just a marriage of two companies in different arenas. It was potentially game changing." The AOL–Time Warner conglomerate involved twelve film companies, including Warner Bros.; multiplex theaters; Hanna-Barbera cartoons; the CNN network; the HBO cable system; twenty-four book brands, including Time-Life Books and Little, Brown and Company; fifty-two record labels and the Turner Entertainment Corporation, which includes the Turner Broadcasting System and four national sports teams; theme parks; and Warner Bros. studio stores in thirty countries. It is the leading magazine publisher in the world, with approximately thirty periodicals, including *Time* and *Fortune* magazines. AOL–Time Warner president and CEO, Richard Dean Parsons, and at least one other director are members of the Council on Foreign Relations.

♦ *The Walt Disney Company,* in a far cry from its Mickey Mouse origins, has grown into one of the largest media and entertainment corporations in the world. Disney now owns ABC TV network as well as the film studios of Walt Disney Pictures, Touchstone Pictures, Hollywood Pictures, MGM, Miramax, Buena Vista, and Pixar Animation Studio; magazines and newspapers; Disneyland and Disney World theme parks; almost twenty online ventures; six music labels; Disney Book; the ESPN, Soap, and Lifetime cable networks; Jim Henson's Muppets; and several baseball and hockey teams. "In 1998, ABC News discarded an investigative report that raised embarrassing questions about hiring and safety practices at Disney World," noted Leo Bogart, illustrating the dangers of cross-media ownership. Bogart is a Media Studies Center fellow and former general manager of the Newspaper Advertising Bureau.

• **Viacom** is a media conglomerate created from the remerging of Viacom and CBS Corporation in 1999. The corporation's properties include CBS with its two hundred affiliates, Blockbuster, Paramount Pictures, MTV Films, Nickelodeon Movies, Dreamworks, United International Pictures (a joint venture with Universal Studies), Republic Pictures, MTV and Nickelodeon cable networks, TV Land, CMT, Spike TV, VH1, BET, Comedy Central, book publishers, almost two hundred radio stations; and Infinity Outdoor, the largest advertising company in the world. According to the Center for Public Integrity, "Viacom broke U.S. rules controlling media ownership when it bought TV network CBS. Within a week, Senator [John] McCain had proposed a change to those rules. Viacom is McCain's fourth biggest 'career patron.'"

• **Vivendi Universal,** best known as one of the largest owners of privatized water in the world, also is one of the giant media/entertainment corporations. Through subsidiaries, Vivendi supplied water-related services to 110 million people in more than one hundred countries. Vivendi Universal Entertainment was created by the 2000 merger of Vivendi media with the French-based Canal+television and the purchase of Universal Studios from the Seagram Company of Canada. By 2006, with the sale of Vivendi Universal Entertainment to General Electric, it was reformed as NBC Universal. The corporation owns Universal Music Group, which holds 22 percent of the world's music market with labels like Polygram and Motown; TV series; movie theater chains; five Universal Studio theme parks; two mobile phone companies; Connex, a U.K. rail line; and Havas, the world's sixth largest advertising and communications group, which was renamed

Vivendi Universal Publishing with its acquisition in 1998 (now known as Editis, it owns sixty publishing houses selling 80 million books and 40 million CD-ROMS yearly).

- *News Corporation,* a creation of Australian news magnate Rupert Murdoch, originally was incorporated in Australia, but in late 2004, it was reincorporated in the United States under the laws of the State of Delaware. News Corporation owns Fox TV, Fox News, and seven other U.S. news networks; Sky TV in the U.K.; Foxtel in Australia; Star TV, which reaches some 300 million Asian viewers; more than a third ownership in HughesNet, the largest American satellite TV system; 20th Century Fox movie studio; 20th Century Fox Television; Fox Searchlight Pictures; Internet properties including MySpace, Photobucket, Flektor, Grab.com, and IGN Entertainment; thirty-four magazines, including *TV Guide*; HarperCollins Publishers and seven other publishers; a global network of newspapers, including the *New York Post*, the *Times* in the United Kingdom, and the *Daily Telegraph* and the *Australian* in Australia; as well as several sports teams, including the Los Angeles Dodgers. With his media conglomerate reaching tens of millions each day, Murdoch remarked in an older News Corporation annual report, "Our reach is unmatched around the world. We're reaching people from the moment they wake up until they fall asleep."

- *Bertelsmann AG,* the world's largest publisher in the English language, is a multinational media corporation based in Gutersloh, Germany, which employs more than 88,000 employees in sixty-three countries. Bertelsmann controls BMG Music Publishing, the world's third largest music publisher; in 2002, it bought the Internet music-sharing firm Napster; fourteen music labels, including Arista Records, BMG, RCA Records, RCA Victor Group, and Windham Hill; ten television networks; seven radio stations; five production houses, five newspapers; twenty-one magazines, including *Family Circle, Parents, Geo,* and *Stern.* Bertelsmann's real power comes from their ownership of publishing houses, which include Random House, Ballantine Books, Del Rey, Fawcett, Bantam, Delacorte Press, Dell, Delta, Spectra, The Dial Press, Bell Tower, Clarkson Potter, Crown Publishers, Harmony Books, Broadway Books, Doubleday, Doubleday Religious Publishing, Main Street Books, Harlem Moon, Alfred A. Knopf, Anchor, Everyman's Library, Pantheon Books, Vintage, The Modern Library, Random House Children's Books, Fodor's Travel Publications, Living Language, RH Puzzles and Games, RH Reference Publishing, and The Princeton Review, to name just a few. It also owns YES Solutions, a large provider of creative, media, and operational services. A partnership between Bertelsmann and Time Warner in 2002 produced Bookspan, the parent organization of Doubleday Entertainment, which directs the world's largest book clubs, including Arvato, Book-of-the-Month Club, Black Expressions, Children's Book-of-the-Month Club, Behavioral Science Book Club, Computer Books Direct, Architects and Designers Book Service, Country Homes & Gardens, Crafter's Choice, Discovery Channel Book Club, Doubleday Book Club, History Book Club, Intermediate & Middle Grades Book Club, The Literary Guild, The Military Book Club, Mystery Guild, Primary Teachers' Book Club, Scientific American Book Club, Science Fiction Book Club, and various foreign book clubs. In April 2007, Directgroup Bertelsmann acquired a 50 percent share of the Bookspan partnership.

Not publicly listed, Bertie, as Bertelsmann is sometimes called, is 76.9 percent owned by the Bertelsmann Foundation created by the German Bertelsmann and Mohn families. The remaining 23.1 percent ownership resides with the Mohns.

Heinrich Mohn, chief of Bertelsmann house from 1921 onward, was a member of a patrons group sponsoring the Nazi SS. During wartime, Bertelsmann had close ties to Goebbels's Reich Ministry of Public Enlightenment and Propaganda and published 19 million books, which made it the largest publisher for the German Army, according to the Independent

Historical Commission for Investigating the History of the Bertelsmann House During the Third Reich (IHC), a group created in Britain in the late 1990s. In announcing the group's findings in 2002, chairman Saul Friedlaender, an Israeli historian, said, "Bertelsmann published a variety of papers and books that clearly had anti-Jewish bias." The group also found that Bertelsmann made indirect use of "slave labor" in some occupied countries. The IHC found that Bertelsmann particularly targeted wartime propaganda toward youngsters with series like Exciting Stories and the Christmas Book of the Hitler Youth.

Bertelsmann's history came under scrutiny in 1998, when the company took over Random House, the largest book publisher in the United States. In public announcements, the firm said it was closed down by the Nazis. The IHC, however, reported that this story was a "legend" and that the firm's closure in 1944 was more likely due to shortages of material than any subversive activities.

The IHC's report forced Bertelsmann chairman Gunther Thielen to publicly state, "I would like to express our sincere regret for the inaccuracies the commission has uncovered in our previous corporate history of the World War II era as well as for the wartime activities that have been brought to light."

Although Bertelsmann spokesman Tim Arnold said, "The values of Bertelsmann then are irreconcilable with the company today. The company is now a global player in the media industry," many Americans, especially World War II veterans, might wonder about the propriety of a German corporation once tied so closely to the Nazis controlling so much of their media. As has been noted here previously, many of today's wealthiest multinational corporations might deserve the same scrutiny as Bertelsmann.

IT HAS BEEN said that freedom of the press belongs to whomever owns the press, and when it comes to media ownership, the name of the game is money.

Like other major monopolies, money is the purchaser of influence for the communication corporations in Washington, D.C. A Center for Public Integrity (CPI) investigation of campaign contributions through November 2006 showed the communications industry has spent $486 million since 1997 to affect election outcomes and influence legislation before Congress and the White House.

Whose money is this and what has the American public gotten in return?

"Since the landmark 1996 Telecommunications Act, in which the cable industry promised that deregulation would stimulate market competition and lower monthly cable bills for all Americans, rates shot up 45 percent, nearly three times as fast as inflation. That same law relaxed the rules on ownership of radio, and since then the two largest companies have greatly enriched themselves, increasing their number of stations owned from 130 to 1,400. In 1997, broadcasters lobbied and received portions of the digital broadcast spectrum worth, according to some estimates, upward of $70 billion—for free.... With the past relaxed ownership and control rules, I have not seen any evidence to credibly suggest that the quality of information provided to the American people has improved, or that the values and commitment to serious journalism in this country have changed for the better," said Charles Lewis, former executive director of the CPI, in a 2003 talk at the Columbia University School of Law in New York City.

Lewis also explained how media companies win influence with lawmakers and regulators. "They do it the old-fashioned way, by using the time-honored techniques with which business interests routinely reap billions of dollars worth of subsidies, tax breaks, contracts, and other favors. The media lobby vigorously. They give large donations to political campaigns. They take politicians and their staffs on junkets.... Not only does the media aggressively lobby and contribute to the two political parties and politicians at the federal level, they *also* decide whose face and voice make it onto the airwaves. Such raw power provokes fear and trepidation in the political realm."

Lewis illustrated this power by noting that when President Clinton requested that the Federal Communications Commission (FCC) provide free or reduced-cost television time to candidates, within days "the powerful broadcast corporations and their Capitol Hill allies managed to halt this historic initiative. In the Senate, incoming Commerce Committee chairman John McCain, the Arizona Republican, and Conrad Burns, a Republican from Montana and the chairman of that panel's communications subcommittee, announced that they would legislatively block the FCC's free air-time initiative. 'The FCC is clearly overstepping its authority here,' McCain declared."

Lewis concluded: "So this is where we are. A regulated industry has a stranglehold over the regulator and its congressional overseers. Not a new story in Washington, I'm afraid, but one of the reasons Americans frequently distrust government, its officials, and its policies."

And has this tightly controlled news media brought us better coverage of the events and issues crucial to national well-being? Not according to studies by the nonpartisan Project for Excellence in Journalism. They found that "hard" news stories, which constituted more than half of all news reports thirty years ago, today have fallen to less than one-third. Furthermore, story content today has moved away from discussions of the political process, war and peace, and policy, to people, human interest, and news you can use. There is also increasing emphasis on scandal, the bizarre, and fear of the future.

Declining viewership and readership has become commonplace in the mass media and does not seem to overly concern corporate managers. Arthur Ochs Sulzberger, publisher and board chairman of the New York Times Company, recently stated: "I really don't know whether we'll be printing the *Times* in five years, and you know what? I don't care."

Increasingly, the attempt to present news objectively has given way to advocacy, which borders on sheer propaganda. According to a State of the News Media 2007 report by the Project for Excellence in Journalism, "A growing pattern has news outlets, programs, and journalists offering up solutions, crusades, certainty, and the impression of putting all the blur of information in clear order for people." Comparing programming today with "shock jock" radio talk-show hosts, the report stated, "The tone may be just as extreme as before, but now the other side is not given equal play."

Distraction, ignorance, and fear are integral parts of the fascist globalists' plan for a new American empire. Like Hitler before him, President

George W. Bush has used both nationalism and middle-class moral values to gain the unquestioning allegiance of a large swath of the public.

Georges Eugene Sorel, a French conservative Socialist whose philosophical work provided a foundation for fascism, noted at the start of the twentieth century that all great political movements are generated by "myths," defined as the belief held by a group who believe themselves to be an army of truth fighting an army of evil.

The Nazis used the myth of Aryan brotherhood and superiority to create their National Socialist empire. In the United States, it has become the myth of America's God-ordained right to create an empire which has launched a worldwide war for American-style freedom and democracy. Today, numerous books and articles have been written about a new American empire—or Reich, in German. "No one can argue credibly that America today is not an empire. Militarily, economically, and culturally, the United States wields a hegemonic influence unparalleled in world history," stated Jim Garrison, founder of the State of the World Forum and author of *America as Empire: Global Leader or Rogue Power?*

The creation of this empire has been facilitated by the power of the corporate mass media, increasingly falling into fewer and fewer hands. The ownership of the corporations that today control the information available to the broadest portion of the public can be traced back to the same families and companies who backed Hitler.

Like Hitler, who sought to bring individuals alienated by the Industrial Revolution and Depression into a hive-like German *Volk,* or united people, the globalists who control America's mass media have attempted to bring citizens into one common worldview by the unremitting dissemination of homogeneous news and information. Networks and cable channels routinely accept reports from "pool" reporters, which results in many channels presenting the same version of the news. Furthermore, news content has increasingly been simplified down to the lowest common denominator; *60 Minutes* correspondent Morley Safer argued: "I challenge any viewer to make the distinction between [TV talk-show host] Jerry Springer and the three evening newses and CNN." Only those who consciously tune in to alternative media or surf the Internet gain access to information differing from the party line.

Hitler achieved this conformity of thought among his followers through the use of radio, mass rallies, and meetings. Today, it is the corporate-controlled mass media that determines the worldview of most Americans. The corporate media owners, many of them members or close associates of the fascist globalists' secret societies, have learned the lessons of Nazi media manipulation well—i.e., simplistic catchwords repeated constantly with no real opposing viewpoints allowed. For example, at the time of the Iraqi invasion, the corporate media called the enemy "insurgents," defined as anyone who opposed the established authority. This, of course, meant Iraqis opposed to the U.S. occupation, but the term "Iraqis" did not fit well with government pronouncements about U.S. troops being well received by that nation's population. By 2007, the term "insurgent" was being superseded by the term "al-Qaeda" in an effort to connect fighting in Iraq to the attacks of September 11, 2001. This attempt came in the wake of President Bush's admission that neither Saddam Hussein nor Iraq played any role in the attacks. Likewise, the mass media long and loudly has trumpeted the term "War on Terror," with its attendant warnings to be watchful for terrorists trying to slip weapons of mass destruction into the United States, yet noticeably failed to report that even seven years after 9/11, no serious attempt has been made to secure the nation's borders. Official government pronouncements are merely broadcast uncritically, with very little effort to check their reliability.

Also, like Hitler, the directors of modern American viewpoints speak of a brighter and better tomorrow, yet constantly regale the public with images and evocations of great moments in history. The attacks of 9/11 initially were compared to Pearl Harbor, and President Bush early on garnered criticism for calling his War on Terror a "crusade," a term with ugly historical connotations. The sacrifices of wartime America during World War II were pointed to as models for the war against terrorism.

The Nazis brought complex social and economic issues down to one single concept—the Aryan German in a death struggle with the International Jew. Nearly the same concept is widespread in America—the freedom-loving American in a "war on terror" with Muslim fanatics. Such "us against them" mentality has been used by despots for centuries to rally populations behind them.

"Media manipulation in the United States today is more efficient than it was in Nazi Germany, because here we have the pretense that we are getting all the information we want. That misconception prevents people from even looking for the truth," said Professor Mark Crispin Miller of New York University, who specializes in propaganda and the media.

The arrogance of the Nazi mindset along with their accepted method for hoodwinking the public was voiced by an unnamed aide to President George W. Bush in 2002. In a *New York Times Magazine* article by Ron Suskind, this person referred to Suskind, and apparently all media reporters, as living "in what we [the Bush administration] call the reality-based community." Such persons, stated this presidential aide, "believe that solutions emerge from your judicious study of discernible reality.... That's not the way the world really works anymore. We're an empire now, and when we act, we create our own reality. And while you are studying that reality—judiciously, as you will—we'll act again, creating other new realities, which you can study too, and that's how things will sort out. We're history's actors . . . and you, all of you, will be left to just study what we do." This clearly echoes Hitler's stated belief that big lies repeated loudly and often will sway the masses of people more effectively than well-articulated arguments based on reason and facts. It also reflects the arrogance of the fascist globalists, those who feel they are somehow superior to others and that they alone are fit to rule.

This reach for empire—the "Future Belongs to Us" Nazi mentality—cannot be separated from its cultural context, for a nation's culture clarifies its worldview and sensibilities.

In describing Nazi culture under Hitler, Professor George L. Mosse might well have been describing popular tastes and prejudices reflected in America today. "The mass of people (and not just in Germany) do not like 'problem art.' Do not care for the distorted pictures of expressionists; they do not understand the searchings of such art. The same can be said about literature, indeed all cultural endeavor. People like their pictures simple and easily understandable and their novels should have gripping plots and large amounts of sentiment. The lowest common denominator of popular taste has a sameness about it which does not vary from the nineteenth to the twentieth century, or indeed from country to country."

As America's economic wheels slow in the New World Order's global economy, the poverty numbers have continued to rise, with an estimated 36.5 million impoverished persons counted in 2005, according to the U.S. Census Bureau. The bureau did note a three-tenths of a percent decline for the first time in a decade in mid-2006 but termed it "not statistically different from 2005." Meanwhile, America's middle class is being squeezed out. The percentage of middle-income neighborhoods in metropolitan areas like New York, Los Angeles, Chicago, and Washington, D.C., has continued to drop since 1970, according to a recent report from the Brookings Institution, as cited in the *New York Times*.

It would be wise to reflect on the words of psychiatric researchers Roeder, Kubillus, and Burwell as the U.S. economy threatens to topple, like that of post–World War I Germany: "In time of virtually universal economic ruin, it was far easier to sell an ideology which supported the extermination of the social and political—and, inevitably, economic—'deadweight.' "

NO ONE QUESTIONS that the attacks of 9/11 were the result of a large conspiracy. The question has become: whose conspiracy was it?

A cursory count showed more than four hundred Web-site citations on Google, a major Internet search engine, in which Americans have drawn parallels between the attacks of September 11, 2001, and the Reichstag fire that launched Hitler's Third Reich—the destruction of a prominent national structure with the immediate affixing of blame on terrorists, the rapid passage of restrictive laws and the curtailment of civil liberties. But because of the tight corporate control over news and information in America, many citizens still have not been exposed to the controversies, contradictions, and unexplained events of 9/11.

Despite attempts by both government and corporate leaders to deny that the attacks were anything but what was claimed, separate national polls show a growing number of Americans have declined to fully accept the official explanation, which, when carefully examined, is nothing more than a conspiracy theory. In mid-June 2006, when questioned as to why Osama bin Laden's Most Wanted poster did not include any mention of the

9/11 attacks, Rex Tomb, FBI chief of Investigative Publicity, stated, "The reason why 9/11 is not mentioned on Osama bin Laden's Most Wanted page is because the FBI has no *hard evidence* connecting bin Laden to 9/11" (emphasis added).

The conspiracy theory accepted by the Bush administration and the corporate mass media contends that nineteen fanatical Muslims somehow overcame the $40 billion U.S. defense system, simultaneously hijacked four commercial airliners, managing to disconnect their transponders at approximately the same time, and crashed them into the World Trade Center towers and the Pentagon, with the fourth crashing in western Pennsylvania as the result of a revolt by the passengers—the entire operation conducted under the command of a Muslim cleric using a computer in a cave in Afghanistan.

Zogby International poll results released in September 2007 showed that 90 percent of persons on the East Coast and 75 percent of those in the western United States believe the attacks of 9/11 have been the most significant historical event of their lifetime. However, 65 percent of those polled gave President Bush's attempts to combat terrorism negative marks, a dramatic drop from those who supported his efforts just after 9/11. And despite the globalist control over the mass media, national polls, such as a September 2006 poll of more than 64,000 persons, have shown more than half the country has failed to buy into the official story of the attacks. Only slight mass media coverage was given to a Zogby poll conducted between August 24 and 26, 2004, on the eve of the Republican National Convention. The poll showed that almost one half of New York City residents (49.3 percent) and 41 percent of New York state residents believed that some national leaders "knew in advance that attacks were planned on or around September 11, 2001, and that they consciously failed to act." Despite the political implications of such an accusation, nearly 30 percent of registered Republicans and more than 38 percent of those who described themselves as "very conservative" supported this proposition. A September 11, 2006, MSNBC Question of the Day poll asked, "Do you believe any 9/11 conspiracy theories that indicate the U.S. government was involved?" An astounding 58 percent answered, "Yes, I believe there is evidence," with only 30 percent voting "No, that's ridiculous," and 11 percent stating, "I'm not sure." This means a whopping 68 percent were at least open to the suggestion that 9/11 was contrived within the U.S. government.

Much of this attitude was due to unanswered questions concerning the attacks, such as:

◆ Who was truly behind the creation of al-Qaeda, terrorists whose origins are connected to a Nazi intelligence organization and later the CIA? Recall it has been previously shown that al-Qaeda was taken over by Nazi intelligence before World War II and, after the war, passed by British intelligence to the American CIA.

◆ Who authorized the war-game exercises for the morning of September 11, 2001? Various exercises with names like Vigilant Warrior, Vigilant Guardian, Northern Vigilance, and Tripod II were being conducted by the North American Aerospace Defense Command (NORAD) and other government agencies that morning. According to Sergeant Lauro Chavez of the U.S. Central Command, these exercises included a scenario of hijacked commercial airliners being used to crash into prominent U.S. buildings, specifically naming the World Trade Center towers. Chavez, who worked with the command center's computers, also noted that false images called "inputs" representing hijacked planes were placed on military radar screens, causing much initial confusion. Chavez said, much to the surprise of the military, it was announced shortly before 9/11 that command over NORAD, normally a military position, had been assumed by Vice President Cheney.

◆ Who sent the Federal Emergency Management Agency's emergency response team and military rescue units to New York City as part of the Tripod II exercises the evening before 9/11?

◆ Who authorized the flight of more than two dozen members of the bin Laden family across the United States during the "no fly" period when Americans were not permitted to fly?

◆ If a Boeing-757 with a wingspan of 124 feet and a height of 44 feet hit the Pentagon, why did photos taken the day prior to the collapse of the west wall show only one hole, approximately 15×20 feet, in the ground floor, with no evidence of wings, engines, or wheel assemblies?

- Who told New York City mayor Rudy Giuliani to evacuate his temporary command center because the Twin Towers were about to collapse, as he stated to Peter Jennings of ABC News that afternoon?

- Why has no other steel-reinforced building in the history of the world collapsed due to only fire, and why did the firefighting industry publication *Fire Engineering* describe the FEMA investigation of the towers collapse as "half-baked farce"?

- Why has there been such controversy over what caused the symmetrical collapse of the forty-seven-story World Trade Center Building 7 at 5:25 P.M. on 9/11? Although it was not hit by airplanes, it fell neatly between the Verizon Building and the U.S. Post Office, neither of which suffered critical damage. Why did the official government study of the collapse of Building 7 conclude their "best hypothesis has only a low probability of occurrence"?

- Why did the short-selling of stock in American and United Airlines suggest foreknowledge of the 9/11 attacks, and why was it not properly investigated once it was learned these sales were made by persons connected to the CIA?

These comprise only a short list of the more pertinent questions that have not been addressed or adequately explained. Evidently, the "watchdog" mass media has been cowed by its corporate masters. Nor have these and many other questions been answered by any of the official government investigations, including President Bush's handpicked panel, the National Commission on Terrorist Attacks upon the United States, popularly known as the 9/11 Commission. Bush initially tried to appoint globalist Henry Kissinger to head his commission but was thwarted by public outcry.

The growing 9/11 Truth Movement, joined by many of the victims' families, remains divided in their view of the real conspiracy, sometimes referred to as LIHOP (Let It Happen on Purpose) and MIHOP (Made It Happen on Purpose). The LIHOP argument is that certain individuals within the federal government had foreknowledge of the attacks—more than a dozen nations, including Israel, Cuba, and even the Taliban in Afghanistan attempted to warn Washington of the coming attacks—yet did nothing to prevent them, as the attacks furthered their political agenda, while the MIHOP supporters took notice of the close relationships between the Bush and bin Laden families, as reported in the Texas media, as well as the role of Saudi Arabia and the CIA in the creation of al-Qaeda. They argue that the attacks were actually precipitated by elements within the U.S. government.

The controversy over the truth of the 9/11 attacks undoubtedly will continue for many years, but how long it will take before the corporate mass media are allowed to objectively investigate and report on this festering scandal is anybody's guess.

The major corporations that control the mass media, like the other corporations previously discussed, are for the most part filled with honest and honorable people. They, like most of their corporate superiors, are men and women with credible backgrounds. They are capable managers, but they have little inkling of who truly controls their corporate structure—i.e., the fascist globalists' banks and foundations that seek to control every major aspect of modern life, including energy, transportation, communications, education, religion, pharmaceuticals, and health.

EPILOGUE

AMERICA TODAY IS A NATIONAL SOCIALIST'S DREAM COME true.

Individuals are computerized, databased, logged, and categorized. Video cameras, motion sensors, metal detectors, and spy satellites monitor our movements, while think tanks and foundations study our every habit. We are constantly bombarded with "official" pronouncements and advertising. Television is everywhere—in bars, waiting rooms, airports, and usually constantly on in our very living rooms. In our fast-paced society, no one has time to think, much less read deeply.

Business, especially corporate business, is king. Giant corporations, governed by faceless directors answering to shadowy owners, control everything, from water to wing nuts. Even the time-honored profession of soldiering has been usurped by private corporate armies like Blackwater, in 2007 already being accused of becoming America's version of the Nazi Brownshirts.

Meanwhile, the American taxpayer is footing the bill, even though, as convincingly shown in Aaron Russo's 2006 documentary *America: Freedom to Fascism,* there is no law requiring Americans to pay an income tax. Of course, the IRS, through its myriad rules and regulations, can drag into court and even jail those who fail to fulfill "voluntary compliance."

"[F]ascism's principles are wafting in the air today, surreptitiously masquerading as something else, challenging everything we stand for. The cliché that people and nations learn from history is not only overused, but also overestimated; often we fail to learn from history, or draw the wrong conclusions. Sadly, historical amnesia is the norm," stated author Dr. Laurence W. Britt, in an article for *Free Inquiry,* a long-standing publication of the Council for Secular Humanism, which promotes secular humanist principles.

Following a careful study of the regimes of Nazi Germany, Fascist Italy, Franco's Spain, Salazar's Portugal, Papadopoulos's Greece, Pinochet's Chile, and Suharto's Indonesia, Britt concluded that these fascist governments had observable similarities. "Analysis of these seven regimes reveals fourteen common threads that link them in recognizable patterns of national behavior and abuse of power," he noted. "These basic characteristics are more prevalent and intense in some regimes than in others, but they all share at least some level of similarity."

Britt's fourteen characteristics of a fascist regime, many sounding ominously close to what's happening today in the United States, include:

+ Powerful and continuing expressions of nationalism.

   From the prominent displays of flags and bunting to the ubiquitous lapel pins, the fervor to show patriotic nationalism, both on the part of the regime itself and of citizens caught up in its frenzy, was always obvious. Catchy slogans, pride in the military, and demands for unity were common themes in expressing this nationalism. It was usually coupled with a suspicion of things foreign that often bordered on xenophobia. Examples of such patriotic zeal may be found in the ever-present yellow ribbons showing support for U.S. troops to the plethora of American flags and bunting at large public events such as the Super Bowl.
+ Disdain for the importance of human rights.

   The regimes themselves viewed human rights as of little value and a hindrance to realizing the objectives of the ruling elite. Through clever use of propaganda, the population was brought to accept these human rights abuses by marginalizing, even demonizing, those being targeted. When abuse was egregious, the tactic was to use secrecy, denial, and disinformation. In November 2007, former federal judge Michael B. Mukasey was sworn in as attorney general of the United States, despite contentious confirmation hearings focused on the issue of torturing prisoners. He replaced Alberto R. Gonzales, who was criticized for his part in crafting the Bush administration's secretive legal arguments permitting the torture of suspects. Mukasey, who served eighteen years as judge of U.S. district court for the Southern District of New York, presided over the trials of Omar Abdel Rahman and El Sayyid Nosair, the convicted bombers of the World Trade Center in 1993; the trial of José Padilla, the man declared an "enemy combatant" by President Bush and the only person convicted in connection with the 9/11 attacks; and the lawsuits between World Trade Center leaser Larry Silverstein and several insurance companies over damages stemming from the 9/11 attacks.

• Identification of enemies/scapegoats as a unifying cause.

The most significant common thread among these regimes was the use of scapegoating as a means to divert the people's attention from other problems, to shift blame for failures, and to channel frustration in controlled directions. The methods of choice—relentless propaganda and disinformation—were usually effective. Often the regimes would incite "spontaneous" acts against the target scapegoats, usually communists, socialists, liberals, Jews, ethnic and racial minorities, traditional national enemies, members of other religions, secularists, homosexuals, and "terrorists." Active opponents of these regimes were inevitably labeled as terrorists and dealt with accordingly. Examples of such tactics can be heard from the mouths of those who constantly use racial slurs. Afghanistan's former "freedom fighters" have semantically changed into "insurgents" then into "al-Qaeda terrorists" in the news columns, while such epitaphs as "rag head" and "sand nigger" are commonly used in the general population.

• The supremacy of the military and avid militarism.

Ruling elites always identified closely with the military and the industrial infrastructure that supported it. A disproportionate share of national resources was allocated to the military, even when domestic needs were acute. The military was seen as an expression of nationalism, and was used whenever possible to assert national goals, intimidate other nations, and increase the power and prestige of the ruling elite. The U.S. military budget for many years has consumed the bulk of the national spending. President Bush's 2008 budget provides $439.3 billion for the Department of Defense's base budget—a 7 percent increase over 2006 and a whopping 48 percent increase over 2001. This figure does not include military-related expenditure such as nuclear weapons research or the wars in Afghanistan and Iraq. Neither does it count trust funds, anticipated costs of Social Security, and Veterans Administration costs of services to veterans. "The government practice of combining trust and federal funds began during the Vietnam War, thus making the human-needs portion of the budget seem larger and the military portion smaller," according to literature from the War Resisters League (WRL), an antiwar organization founded in 1923. By totaling all government figures relating to the military, the WRL estimated that more than half (51 percent) of all federal spending goes to the military.

- Rampant sexism.

Beyond the simple fact that the political elite and the national culture were male-dominated, these regimes inevitably viewed women as second-class citizens. They were adamantly anti-abortion and also homophobic. These attitudes were usually codified in draconian laws that enjoyed strong support by the orthodox religion of the country, thus lending the regime cover for its abuses. This practice is less prevalent in the United States today, although many women still find it difficult to break through what has been termed the "glass ceiling," in which they can see higher positions in the workplace but never seem to get there. Modern America also differs from Nazi Germany and other cultures in that women are beginning to fill the corporate chairs formerly held by men. Many seem agreeable to advancing fascist and globalist philosophy.

- A controlled mass media.

    Under some of the regimes, the mass media were under strict direct control and could be relied upon never to stray from the party line. Other regimes exercised more subtle power to ensure media orthodoxy. Methods included the control of licensing and access to resources, economic pressure, appeals to patriotism, and implied threats. The leaders of the mass media were often politically compatible with the power elite. The result was usually successful in keeping the general public unaware of the regimes' excesses. As previously detailed, the American corporate mass media today is essentially in the hands of six giant multinational communications corporations. The owners of these corporations are proponents of "free trade" in business policies, yet coverage of alternative news and views is mostly ignored. "One of our best-kept secrets is the degree to which a handful of huge corporations control the flow of information in the United States. Whether it is television, radio, newspapers, magazines, books, or the Internet, a few giant conglomerates are determining what we see, hear, and read. And the situation is likely to become much worse as a result of radical deregulation efforts by the Bush administration and some horrendous court decisions," warned Congressman Bernie Sanders, adding, "This is an issue that Congress can no longer ignore."

- Obsession with national security.

    Inevitably, a national security apparatus was under direct control of the ruling elite. It was usually an instrument of oppression, operating in secret and beyond any constraints. Its actions were justified under the rubric of protecting "national security," and questioning its activities was labeled unpatriotic or even treasonous. While all Americans should be concerned about national security, many see it as a pretext to strip away constitutional rights. Thoughtful persons also worry about a man like Michael Chertoff, son of a Jewish rabbi, who has been accused of having dual citizenship (American and Israeli) and was a major architect of Bush administration policies, being named secretary of the Homeland Security Department.

+ Religion and ruling elite tied together.

Unlike communist regimes, the fascist and protofascist regimes were never proclaimed godless by their opponents. In fact, most of the regimes attached themselves to the predominant religion of the country and chose to portray themselves as militant defenders of that religion. The fact that the ruling elite's behavior was incompatible with the precepts of the religion was generally swept under the rug. Propaganda kept up the illusion that the ruling elites were defenders of the faith and opponents of the "godless." A perception was manufactured that opposing the power elite was tantamount to an attack on religion. Earlier in this work, the obvious parallels have been drawn between the use of religion in Nazi Germany and modern America to support government policies.

+ Power of corporations protected.

Although the personal life of ordinary citizens was under strict control, the ability of large corporations to operate in relative freedom was not compromised. The ruling elite saw the corporate structure as a way to not only ensure military production (in developed states) but also as an additional means of social control. Members of the economic elite were often pampered by the political elite to ensure a continued mutuality of interests, especially in the repression of "have-not" citizens. According to the Federalism Project of the American Enterprise Institute, a group that conducts and sponsors original research on American federalism, "Consumer advocates, plaintiffs' attorneys, and state officials argue that broad federal preemption claims—often by federal regulatory agencies, without a clear congressional mandate—interfere with the states' historic role in protecting citizens against corporate misconduct. Corporations and federal agencies respond that preemption is often the only viable safeguard against unwarranted state interferences with the national economy." In a 2006 article in the *Los Angeles Times,* Alan C. Miller and Myron Levin noted how a series of steps by federal agencies were meant to "shield leading industries from state regulation and civil lawsuits on the grounds that they conflict with federal authority."

- Power of labor suppressed or eliminated.

  Since organized labor was seen as the one power center that could challenge the political hegemony of the ruling elite and its corporate allies, it was inevitably crushed or made powerless. The poor formed an underclass that was viewed with suspicion or outright contempt. Under some regimes, being poor was considered akin to a vice. As previously noted, antilabor actions of the Bush administration prompted Jack Heyman, an official of the International Longshore and Warehouse Union, to state that "Bush is effectively declaring war on the working class here." Those with long memories know that labor news has largely dropped from the mainstream media's radar screen.

- Disdain and suppression of intellectuals and the arts.

  Intellectuals, and the inherent freedom of ideas and expression associated with them, were anathema to these regimes. Intellectual and academic freedom were considered subversive to national security and the patriotic ideal. Universities were tightly controlled, politically unreliable faculty harassed or eliminated. Unorthodox ideas or expressions of dissent were strongly attacked, silenced, or crushed. To these regimes, art and literature had to serve the national interest or they had no right to exist. In the wake of the 9/11 attacks, many conservative groups on college campuses denounced academic freedom, according to a report by John K. Wilson, coordinator of the Independent Press Association's Campus Journalism Project. Other academics were fired or reprimanded for merely speaking out on the issues of war or questioning the official story of 9/11.

- Obsession with crime and punishment.

  Most of these regimes maintained draconian systems of criminal justice, with huge prison populations. The police were often glorified and had almost unchecked power, leading to rampant abuse. "Normal" and political crime were often merged into trumped-up criminal charges and sometimes used against political opponents of the regime. Fear and hatred of criminals or "traitors" was often promoted among the population as an excuse for more police power. The United States today has a higher incarcerated population than all European jails combined, and in certain areas, such as Washington, D.C., police presence is at an all-time high. One visitor to Washington in the summer of 2007 asked a police officer why there were so many cops around. He replied, "People would rather have security than freedom."

- Rampant cronyism and corruption.

Those in business circles and close to the power elite often used their position to enrich themselves. This corruption worked both ways: the power elite would receive financial gifts and property from the economic elite, who in turn would gain the benefit of government favoritism. Members of the power elite were in a position to obtain vast wealth from other sources as well: for example, by stealing national resources. With the national security apparatus under control and the media muzzled, this corruption was largely unconstrained and not well understood by the general population. The cronyism and outright nepotism of the Bush administration has been well documented. Elizabeth Cheney, the vice president's daughter, was named as a deputy secretary of state in late February 2002, and within about a week, her husband, Philip Perry, became chief counsel for the Office of Management and Budget, where he joined director Mitchell Daniels, whose sister Deborah is an assistant attorney general. "That's just the beginning," noted *Washington Post* reporter Dana Milbank. "Among Deborah Daniels' colleagues at Justice is young Chuck James, whose mother, Kay Coles James, is the director of the Office of Personnel Management, and whose father, Charles Sr., is a top Labor Department official. Charles James Sr.'s boss, Labor Secretary Elaine L. Chao, knows about having family members in government: Her husband is [Kentucky] Sen. Mitch McConnell and her department's top lawyer, Labor Solicitor Eugene Scalia, is the son of Supreme Court Justice Antonin Scalia.... Ken Mehlman, the White House political director, regularly calls his younger brother Bruce, an assistant commerce secretary, to get his input." Former secretary of state Colin L. Powell is the father of Michael Powell, who chaired the Federal Communications Commission. An informal survey of 415 historians conducted by George Mason University's History News Network found that eight in ten, or 81 percent, of the responding historians rated Bush's presidency as an overall failure. One respondent to the survey wrote that Bush "ranks with U. S. Grant as the worst. His oil interests and Cheney's corporate Halliburton contracts smack of the same corruption found under Grant." Central to this belief were the numerous Bush administration scandals, including the deceit that preceded the invasion of Iraq; the Abu Ghraib mistreatment of prisoners; pre-9/11 intelligence failures; the $2.3 trillion missing from the Pentagon, announced by Donald Rumsfeld the day before 9/11; the mishandling of the Katrina disaster, which resulted in the resignation of Bush's appointee Michael D. Brown as director of FEMA; Bush's Medicare prescription drug plan that shifted 6.2 million low-income seniors whose medications had been covered by Medicare over to private insurers; the noncompetition government contracts to Halliburton, Dick Cheney's former employer; and the substitution of political ideals for science. In 2004, the Union of Concerned Scientists issued a statement blasting the administration's politicization of science. Ultimately, this statement was signed by 4,062 scientists, including 51 Nobel laureates, 63 National Medal of Science recipients, and 195 members of the National Academies. Buzzflash.com, which styles itself as marketplace for progressives, after listing several debacles and scandals of the Bush administration, said it operated in a "culture of cronyism and corruption."

• Fraudulent elections.

Elections in the form of plebiscites or public opinion polls were usually bogus. When actual elections with candidates were held, they would, as a rule, be perverted by the power elite to get the desired result. Common methods included maintaining control of the election machinery, intimidating and disenfranchising opposition voters, destroying or disallowing legal votes, and, as a last resort, turning to a judiciary beholden to the power elite.

Americans are well aware of the controversies concerning the presidential elections of 2000 and 2004. George W. Bush's first term was decided by the Supreme Court, not the voters. And it was just as bad in 2004. Robert F. Kennedy Jr., writing in *Rolling Stone* magazine, stated, "Republicans prevented more than 350,000 voters in Ohio from casting ballots or having their votes counted—enough to have put John Kerry in the White House." Controversy over both elections continues today and in 2008 charges of vote fraud were already being voiced in the state primary elections, primarily over computer voting machines.

Many Americans noticed the similarities between George W. Bush's unprovoked attacks on Afghanistan and Iraq and Hitler's unprovoked attacks on Poland, the Low Countries, and France. In both cases, the pretext for invasion proved false and reservists were used rather than the option to resort to a military draft.

In early 2008, a study by the nonpartisan Center for Public Integrity documented 935 "false statements" by the Bush administration in the months leading up to the 2003 invasion of Iraq. "Nearly five years after the U.S. invasion of Iraq, an exhaustive examination of the record shows that the statements were part of an orchestrated campaign that effectively galvanized public opinion and, in the process, led the nation to war under decidedly false pretenses," stated the CPI report. Most people would term this telling lies.

"DOES ANY OF this ring alarm bells?" asked Britt. "Of course not. After all, this is America, officially a democracy with the rule of law, a constitution, a free press, honest elections, and a well-informed public constantly being put on guard against evils. Historical comparisons like these are just exercises in verbal gymnastics. Maybe, maybe not."

It seems that by comparing Britt's characteristics of fascism to current events, the argument can definitely be made that globalist fascists are turning the once free and independent United States into a not-so-profitable subsidiary of their global corporate structure—their empire of the rich.

You are free to accept this idea or not. But when secular humanists, conservative Christians, Jews, liberal Democrats, bedrock Republicans, and moderates, not to mention the activist fringe elements, all start issuing the same warning against fascism, perhaps it is time we start paying serious attention. Commentators like Noam Chomsky and Gore Vidal have spoken out against the "national security state" from the left. The late Senator Barry Goldwater and evangelist Pat Robertson have spoken out from the right. Even mainstream centrists, like commentator Bill Moyers and attorney Gerry Spence, have warned of the abuses of a "secret government." When historical figures along with concerned citizens from opposite ends of the political spectrum all say the same thing, it is time to consider the true state of the American union. And perhaps time to stand up and be counted for true freedom—freedom from the corporate state.

The Reverend Erwin W. Lutzer, senior pastor of Moody Church in Chicago, wrote: "We must support our government, but we must be ready to criticize it or even defy it when necessary. Patriotism is commendable when it is for a just cause. Every nation has the right to defend itself, the right to expect the government to do what is best for its citizens. However, if the German church has taught us the dangers of blind obedience to government, we must eschew the mindless philosophy 'My country, right or wrong.'"

Media critic Michael Parenti observes, "To oppose the policies of a government does not mean you are against the country or the people that the government supposedly represents. Such opposition should be called what it really is: democracy, or democratic dissent, or having a critical perspective about what your leaders are doing. Either we have the right to democratic dissent and criticism of these policies or we all lie down and let the leader, the fuehrer, do what is best, while we follow uncritically, and obey whatever he commands. That's just what the Germans did with Hitler, and look where it got them."

There are those who would argue that it is perhaps unpatriotic or at least not politically correct to speak out on issues involving taxation, immigration, political beliefs, race, eugenics, or criticism of the military-industrial complex.

The term "political correctness," which has entered today's discourse, is defined by the *Merriam-Webster Dictionary* as "conforming to a belief that language and practices which could offend political sensibilities (as in matters of sex or race) should be eliminated." Today many believe that definition has grown to include the perceived need to conform to restrictions on speech and behavior set by politicians, corporate leaders, and other self-appointed authorities. This is the same self-imposed restriction that was adopted by too many Germans during the Third Reich. Not only was the man on the street afraid to speak out against the Nazi regime but free speech was denied the intelligentsia. Nazi academic Walter Schultze in 1939 stated that "the reorganization of the entire university system must begin with people who understand that freedom has limits and conform to National Socialist thinking." Germans in the Third Reich did not know the term "political correctness," but they well understood the penalties for freely voicing their opinions.

Recent legislation targeting so-called hate speech can easily slip into official punitive action against any speech that arouses the ire of politicians, police, or judges. Jonathan Rauch writing in *Harper's* magazine noted that equating verbal violence with physical violence is a "treacherous, mischievous business." Rauch quoted author Salman Rushdie, who was sentenced to death in absentia by Muslim ayatollahs after writing a book they claimed slandered the beliefs of millions of Muslims. "What is freedom of expression?" asked Salman Rushdie. "Without the freedom to offend, it ceases to exist." Rauch wrote that the public should learn a lesson from Rushdie's experience. Rauch proclaimed: "The campaigns to eradicate prejudice—all of them, the speech codes and workplace restrictions and mandatory therapy for accused bigots and all the rest—should stop, now. The whole objective of eradicating prejudice, as opposed to correcting and criticizing it, should be repudiated as a fool's errand."

Even though the German Nazis preached the unity of the *Volk* and spoke out against the old divisions of class and education, the leaders operated in an entirely different manner. "In reality, the Third Reich was a network of rival leaders, each with his own followers and his own patronage," noted George Mosse in his book *Nazi Culture*. "Hitler kept them competing against one another and in this way was able to control the whole leadership structure."

Likewise, the globalist rulers of America pit bureaucrats, politicians, academics, corporate leaders, and the public against one another in an agenda of divide and conquer. They maintain control in a society fragmented by combative ideologies and philosophies as well as competing corporate interests. In today's America it seems the only common denominator is consumerism and debt.

Because of their loss of control over Hitler, the globalists learned well the dangers of allowing any one individual to gain the power over masses of people. Consequently, there has not been one prominent figure in recent American history who has commanded the popular respect and esteem of a majority of the population. Even the assassinated President John F. Kennedy, beloved by so many, never held popular goodwill to the extent of Franklin D. Roosevelt. Since World War II, no national leader has gained the stature of Roosevelt, Churchill, or Hitler.

"Hitler's world has gone forever. But many of the basic attitudes and prejudices which went into his worldview are still with us, waiting to be actualized, to be directed into a new mass consciousness," prophesied Professor Mosse from the relatively naive year of 1966.

Ladislas Farago, author of *Aftermath: Martin Bormann and the Fourth Reich,* wrote: "The despicable forces loosed by the Third Reich are not expunged, although, like some virulent virus, they may have changed to other forms and be difficult to identify. They remain malignant and as potentially dangerous as before." In his 1997 book *The Beast Reawakens,* Martin Lee wrote, "Fascism is on the march again. . . . unchecked corporate power has, to a significant degree, stultified the democratic process, and fascist groups in Europe and the United States feed upon this malaise." These sentiments came from writers unaware of the fascist globalists' plan being woven around them. Yet, they could sense that Americans could easily fall sway to the pernicious ideology of National Socialism.

THE BIGGEST STUMBLING block to the plans of the globalists has always been the United States, with its tradition of individual freedom, its Constitution that guarantees that freedom, and the fact that so many

Americans own firearms to protect their freedoms. But true freedom is a transient quality.

National politicians no longer refer to the "republic," because modern America has ceased to be one. It is now an empire—a new Reich.

Obviously, there are dissimilarities between Hitler's Third Reich and the new American Reich. After all, the United States today is a very different time and culture. But it has been demonstrated how the same philosophies and methodologies employed by the same families, corporations, and organizations that at one time supported Hitler's Third Reich, have now found roots in modern America.

It has been necessary for these fascist globalists to break up the United States into divisions of race, sex, age, generation and culture. This has been accomplished through a degrading of popular culture, downgrading the education process, permitting a steady flow of illegal immigrants, and the fragmentation of the population over issues such as abortion, immigration, nonheterosexual relationships, and foreign policy. Control over a diminished national economy and corporate downsizing has brought undue stress on workers, resulting in the gradual destruction of the nuclear family.

None of this construction of the new American empire has come about suddenly.

The global National Socialists—Nazis—are in it for the long haul. The owners of the multinational corporations, with their membership in secret societies, know their goals will not be achieved overnight, although since 9/11 they seem to have redoubled their efforts, speeding up the timetable. While businessmen deal with yearly quarters, and the average worker lives for his weekly paycheck, these people look ahead fifty years or a hundred, if that's what it takes. They realize that their program of a global fascist socialism is the only means of maintaining their power and control, the only way—in their view—to maintain the purity of their race and class. They laugh at the concepts of true individual freedom and multiculturalism, for they have no faith in the innate goodness of humankind or its ability for self-government. They have no real faith in God and use religious ideals and concepts merely as another tool for social control.

The struggle against such steadfast will to power and its attendant control will not be easy. Sacrifices and change will have to be made in all areas of society. Lifestyles will have to be altered. But it can be done—hopefully before the United States falls into depression, anarchy, and then a police state. New energy sources and technologies are lurking in the wings. Technological breakthroughs await only the change of attitude on the part of conventional politics, commerce, and finance.

A sea change in the public consciousness is well under way, although it is not reflected in the corporate-controlled mass media. Yet it is happening. Informed consumers are beginning to realize they can vote with their spending. If enough people refuse to buy a certain product—whether it's a brand of car, gasoline, or something else—or even reject a federal policy proposal, it can force a change of direction in the controllers.

We may do well to recall the words of President Franklin D. Roosevelt, who had to deal with a previous "New World Order." In a 1940 address, he stated, "The history of recent years proves that the shootings and the chains and the concentration camps are not simply the transient tools but the very altars of modern dictatorships. They may talk of a 'new order' in the world, but what they have in mind is only a revival of the oldest and the worst tyranny. In that there is no liberty, no religion, no hope. The proposed 'new order' is the very opposite of a United States of Europe or a United States of Asia. It is not a government based upon the consent of the governed. It is not a union of ordinary, self-respecting men and women to protect themselves and their freedom and their dignity from oppression. It is an unholy alliance of power and self to dominate and to enslave the human race."

It appears that the "New World Order" is really just the "Old World Order" packaged with modern advertising slickness—new names, logos, and slogans. What once was traditional American conservatism has been molded into fascist forms, beginning with the infusion of National Socialism ideals into the military-industrial complex, which then spread into science, corporate life, the mass media, and even political parties.

This change has been engineered by the globalist elite who hold monopolies over basic resources, energy, pharmaceuticals, transportation, and telecommunications, including the news media. As detailed throughout this work, the same men, families, and companies that first supported communism in Russia funded and supported National Socialism in pre-war Germany. With the defeat of the Germans, they simply shifted their attention to the United States. They were abetted by Nazis financed by the stolen wealth of Europe—perhaps including Solomon's treasure—and utilizing a vast network of worldwide corporations. Thousands of Nazis escaped to both North and South America, their way facilitated by supporters in Wall Street, the Bank of England, and the Vatican.

Using German advances in the study of the human mind, behavior, and propaganda, these self-styled globalists are now attempting to subdue the American population through a maze of government policies, drugs, a dumbed-down education system, and a controlled corporate mass media. Political and corporate leadership continually swap roles, creating a merger of the state and industry—the very definition of fascism. Mergers and leveraged takeovers have concentrated corporate power into fewer and fewer hands, many of those directly connected through banking and corporate ties to prewar support for the Nazis. Law enforcement personnel increasingly no longer wear the blue uniforms of police sworn "to serve and protect," but black body armor with the German-style military helmets, initially dubbed the "Fritz" by the soldiers. Even the fields of religion, education, and entertainment are being used to transform whole generations of formerly free Americans into cowed and subservient members of an increasingly National Socialist system.

Is the new American Empire, as it is described in numerous books and articles, in danger of becoming an empire of the wealthy—a fascist Fourth Reich? Hitler's Thousand-year Reich collapsed after a mere twelve years. How long before the end of the New World Order's Fourth Reich in America?

An account of the fall of the Fourth Reich has not yet been written, for it has yet to happen. If, and how, this is to be accomplished, is up to you, dear reader.

SOURCES

If there is no citation, the information may be found in conventional histories and encyclopedias.

Introduction

A "spectacle presented by the Jews": William L. Shirer, *The Rise and Fall of the Third Reich* (New York: Simon and Schuster, 1960), p. 1131

Hitler's Testament: Louis L. Snyder, *Encyclopedia of the Third Reich* (New York: McGraw-Hill, 1976), pp. 165–166

Hitler's faraway look: Shirer, p. 1132

Remains never found: Ibid., p. 1134

Hitler's doubles: Pauline Koehler, *The Woman Who Lived in Hitler's House* (New York: Sheridan House, 1940), p. 47

The Escape of Adolf Hitler

Hitler cried "Get out!": Snyder, p. 125

Fegelein knew secret: Glenn B. Infield, *Hitler's Secret Life: The Mysteries of the Eagle's Nest* (New York: Stein and Day, 1979), p. 284

Linge to wait ten minutes: Ibid.

A Definition of Terms

Corporatism: Jasper Ridley, *Mussolini* (New York: St. Martin's Press, 1997), pp. 226–227

Fascism spells government: Thomas J. DiLorenzo, "Economic Fascism," *The Free-*

man, June 1994; http://www.banned-books.com/truth-seeker/1994archive/
121_3/ts213l.html

Communism versus National Socialism

Wall Street willing to finance revolution: William T. Still, *New World Order: The Ancient Plan of Secret Societies* (Lafayette, LA: Huntington House Publishers, 1990), p. 139

Jacob Schiff's $20 million: Gary Allen, *None Dare Call It Conspiracy* (Seal Beach, CA: Concord Press, 1971), p. 69

Elihu Root's special mission to Russia: http://nobelprize.org/nobel_prizes/peace/laureates/1912/root-bio.html

Arsene de Goulevitch: Allen, p. 72

American International Corporation: http://www.modernhistoryproject.org/mhp/EntityDisplay.php?Entity=AmerIntCorp

Trotsky to stop the war with Germany: G. Edward Griffin, *The Creature from Jekyll Island* (Westlake Village, CA: American Media, 1994), p. 265

Trotsky's release: Ibid., pp. 76, 266; Still, p. 140

Lenin and sealed train: Still, p. 140

Max Warburg and German High Command: Allen, p. 68

Power elite: A. K. Chesterton, *The New Unhappy Lords: An Exposure of Power Politics* (Hawthorne, CA: Christian Boys Club of America, 1969); http://www.watch.pair.com/heritage.html

Lenin speaks of another force: Still, p. 142

Rich back Bolshevik Revolution: Allen, p. 73

William Huntington Russell brought society to America: Antony C. Sutton, *America's Secret Establishment: An Introduction to the Order of Skull & Bones* (Walterville, OR: Trine Day, 2002), p. 6

Man at the heart of American ruling class: Ron Rosenbaum, "The Last Secrets of Skull and Bones," *Esquire,* September 1977

Russell in Germany: Sutton, *Skull & Bones,* p. 223

Links between Bones and Illuminati: Rosenbaum, pp. 87–88

"Man is not bad": A. Ralph Epperson, *The Unseen Hand: An Introduction to the Conspiratorial View of History* (Tucson, AZ: Publius Press, 1985), p. 79

End justifies the means: Ibid., p. 81

Secret is in concealment: Still, p. 73

Strict Observance under a different name: Lynn Picknett and Clive Prince, *The Templar Revelation* (New York: Touchstone, 1997), p. 142

Existence of name Illuminati irrelevant: Still, p. 81

Social programs grow: Dennis Cauchon, "Federal Aid Programs Expand at Record Rate," *USA Today,* March 13, 2005

PART ONE:
THE HIDDEN HISTORY OF
THE THIRD REICH

1. A New Reich Begins

The Thule Society: Jim Marrs, *Rule by Secrecy* (New York: HarperCollins Publishers, 2000), pp. 155–157

Hitler received orders: Adolf Hitler, *Mein Kampf* (New York: Houghton Mifflin, 1940), p. 291

Lower walks of life: Ibid., p. 292

Eckart as spiritual founder of National Socialism: http://www.jewishworldreview .com/cols/fields020606.asp

Party paper purchased from Thule Society: Editors, *The Twisted Dream* (Alexandria, VA: Time-Life Books, 1990), p. 156

Hitler manipulated and cast aside: Joseph P. Farrell, *The SS Brotherhood of the Bell* (Kempton, IL: Adventures Unlimited Press, 2006), p. 419

Dulles brothers at Schroeders: Eustace Mullins, *The Secrets of the Federal Reserve* (Staunton, VA: Bankers Research Institute, 1983), p. 75

Danger to the state and nation: Snyder, p. 287

Douglas Reed's comments: Ibid.

Poland's false flag attack: Shirer, p. 599

Build-up for European war: Antony C. Sutton, *Wall Street and the Rise of Hitler* (Seal Beach, CA: '76 Press, 1976), p. 21

Synthetic oil: Ibid., p. 22

J. P. Morgan production: Ibid., p. 24

Fight against Young Plan: Fritz Thyssen, *I Paid Hitler* (New York: Farrar & Rinehart, 1941), p. 88

Apex of the system: Sutton, *Wall Street,* p. 27

$30 million from City National Bank: Ibid., p. 34

Walter Teagle and tetraethyl: Charles Higham, *Trading with the Enemy: An Expose of the Nazi-American Money Plot—1933–1949* (New York: Delacorte Press, 1983), pp. 34–35

Teagle replaced by James V. Forrestal: Ibid., p. 135

Win working class from communism: James Pool, *Who Financed Hitler* (New York: Pocket Books, 1997), p. 122

Publicist Ivy Lee: Sutton, *Wall Street,* p. 43

American funds for Nazi propaganda: Ibid.

Max Warburg and Prescott Bush: http://www.georgewalkerbush.net/bushfamily history.htm

American I. G. Chemical Company as source of intelligence: Paul Manning, *Martin Bormann: Nazi in Exile* (Secaucus, NJ: Lyle Stuart, 1981), p. 57

Key German records destroyed: Sutton, *Wall Street,* p. 35

Petition from business leaders: John Toland, *Adolf Hitler* (New York: Doubleday, 1976), p. 277

Schroeder bank as Germany's agent: Mullins, *Federal Reserve,* p. 77

John Foster Dulles handled Schroeder loans: Sutton, *Wall Street,* p. 82

"No foolish economic experiments": Konrad Heiden, *Der Fuehrer* (Boston: Houghton Mifflin, 1944), p. 643

Schacht and Bank of England: Toland, pp. 185–186

Montague Norman and Bank of England: Howard S. Katz, *The Warmongers* (New York: Books in Focus, 1979), pp. 78–79

Schacht and Norman visits: Curt Riess, *The Nazis Go Underground* (Garden City, NY: Doubleday, Doran and Company, 1944), p. 34

Influence over political apparatus: Sutton, *Wall Street,* p. 27

BIS as money funnel: Higham, p. 2

Czech gold sent to Nazis: Ibid., p. 5

ITT German Chairman Westrick: Ibid., pp. 93–95

Banking connections to the Gestapo: Ibid., p. 20

50,000 artillery fuses: Ibid., p. 99

Interlocking directorships with I. G. Farben: William Bramley, *The Gods of Eden* (San Jose, CA: Dahlin Family Press, 1990), p. 415

Ford as "one great man": Hitler, p. 930

Ford's medal: Pool, p. 95

Ford built backbone trucks: Higham, p. 156

Synthetic fuel technology from GM: http://www.geocities.com/~virtualtruth/multi.htm

That company was International Business Machines: Edwin Black, *IBM and the Holocaust* (New York: Crown Publishers, 2001), p. 7.

Railroad cars: Edwin Black, "Final Solutions: How IBM Helped Automate the Nazi Death Machine in Poland," *The Village Voice* (March 27–April 2, 2002)

Watson's medal: Black (2001), p. 131

IBM obstruction of authors: Ibid. (2002)

Joseph Kennedy in Hoover note: Higham, p. 181

The coup of 1934: Ibid., pp. 163–164; also see Jules Archer, *The Plot to Seize the White House* (New York: Hawthorne Books, 1973)

Ambassador William E. Dodd: Ibid., p. 167

German bank support: Shirer, p. 144

Booster of Rome-Berlin Axis: Higham, p. 22

Loans to BMW and Mercedes: Manning, p. 67

2. The Strange Case of Rudolf Hess

Lunatic benevolence: Snyder, p. 144

Haushofer in Vril Society: William Bramley, *The Gods of Eden* (San Jose, CA: Dahlin Family Press, 1990), p. 409

General Karl Haushofer: Peter Levenda, *Unholy Alliance: A History of Nazi Involvement with the Occult* (New York: Avon Books, 1995), pp. 87–88

Influence only through Hess: Lynn Picknett, Clive Prince, and Stephen Prior, *Double Standards: The Rudolf Hess Cover-up* (London: Little, Brown and Company, 2001), p. 43

Hess dictated chapters: Ibid., p. 54

Terrible weapon will be in our hand: Ibid., p. 138

Authorities desperate to conceal a secret: Ibid., p. xxi

Cornerstone of German politics: Snyder, p. 139

Haushofer passed names: Jean-Michel Angebert, *The Occult and the Third Reich* (New York: Macmillan Publishing Company, 1974), p. 227

Six planets in Taurus: Levenda, *Unholy Alliance,* p. 235

Dream by supernatural forces: Albert Speer, *Inside the Third Reich* (New York: Macmillan Publishing Company, 1970), p. 211

Halifax and Sinclair kept quiet: Picknett, Prince, and Prior, p. 160

Willingness to Nazi occupation: Sophie Goodchild, "Queen Mum wanted peace with Hitler," *The Independent on Sunday,* March 5, 2000

Churchill quote: Emrys Hughes, *Winston Churchill in War and Peace* (Glasgow: Unity Publishing, 1950), p. 145

Churchill to Lord Robert Boothby: Sidney Rogerson, *Propaganda in the Next War* (foreword to the second edition, 2001), originally published in 1938; http://wontgetfooledagain.wikispaces.com/CentralBanking

Hess did not imagine a peace group: Picknett, Prince, and Prior, p. 262

England to cover one's back: Hitler, p. 183

Viktor Suvorov and Hitler's preemptive strike: http://www.ihr.org/jhr/v17/v17n4p30 _Michaels.html; Viktor Suvorov, *Icebreaker: Who Started the Second World War?* (London: Hamish Hamilton, 1990)

Joseph Bishop: http://www.ihr.org/jhr/v16/v16n6p22_Bishop.html#anchor283807

Hitler went down in history as the ultimate aggressor: Ibid.

Daniel W. Michaels: http://www.ihr.org/jhr/v18/v18n3p40_Michaels.html#anchor 1786488

Hitler said Europe would have been lost: Editors, "Hitler's Declaration of War Against the United States," *The Journal of Historical Review,* Winter 1988–89, vol. 8, no. 4, pp. 389–416

A small clique: Picknett, Prince, and Prior, p. 175

Sea Lion a sham: Ibid., pp. 120–121

Egbert Kieser: Ibid., p. 121

Mythologizing of Churchill: Ibid., p. 112

Hoover memo on Duke of Windsor: Higham, pp. 181–182

MI6, SOE conflicted: Picknett, Prince, and Prior, p. 262

There were two Hesses and the mind skids: Ibid., p. 436

Dulles dispatches Dr. Cameron: Ibid., p. 9

Dr. Cameron's background: John Marks, *The Search for the "Manchurian Candidate": The CIA and Mind Control* (New York, Times Books, 1979), p. 132

Prince Bernhard and Bilderberg: Marrs, *Rule by Secrecy,* pp. 39–40

3. Nazi Wonder Weapons

V-1 and V-2 damage in London: Snyder, p. 361

Tonne television guidance test: Joseph P. Farrell, *Reich of the Black Sun* (Kempton, IL: Adventures Unlimited Press, 2004), p. 184

Weaponry near completion: Brian Ford, *German Secret Weapons: Blueprint for Mars* (New York: Ballantine Books, 1969), pp. 6–7

Greatest technological leap: Igor Witkowski, *Truth About the Wunderwaffe* (Warsaw: European History Press, 2003), p. 10

Einsteinian physics as Jewish science: Nick Cook, *The Hunt for Zero Point* (London: Century, 2001), p. 194

Witkowski and Wright-Patterson records: Witkowski, p. 10

SS industrial concern: Albert Speer, *Infiltration: How Heinrich Himmler Schemed to Build an SS Industrial Empire* (New York: Macmillan Publishing Company, 1981), p. 3

Weapons plant concealed from Speer: Ibid., p. 179

Any quantum leap over state-of-the-art: Cook, p. 151

Feuerball description: Renato Vesco and David Hatcher Childress, *Man-Made UFOs 1944–1994: 50 Years of Suppression* (Stelle, IL: Adventures Unlimited Press, 1994), pp. 85–86

Kugelblitz: Ibid., p. 156–157

Rudolph Schriever's saucer: W. A. Harbinson, *Genesis* (New York: Dell Publishing Company, 1982), p. ix

Major Lusar's description: Die Deutschen Waffen und Geheimwaffen des 2 Weltkriegs und ihre Weiterentwicklung (Munich: J. F. Lehmanns Verlag, 1956), pp. 81–83

Schriever's death and plans: Harbinson, p. 590

Flying saucers in final stages of development: Ford, pp. 34–35

1954 CIA report: Nick Redfern, *The FBI Files: The FBI's UFO Top Secrets Exposed* (New York: Simon and Schuster, 1998), pp. 198–199

Viktor Schauberger: Vesco and Childress, pp. 243–245

General Twining's memo: Cook, p. 37

Barney Hill's description of German Nazi: Jim Marrs, *Alien Agenda* (New York: HarperCollins Publishers, 1997), pp. 198–199

Utter scientific nonsense: Farrell, *Reich of the Black Sun,* p. 23

Photochemical process: Ibid., p. 146

Karl Wirtz's comment: Ibid., p. 144

Developed only a uranium bomb: Ibid., p. 23

Allied engineers despaired: Ibid., pp. 34–35

Prima facie case for uranium bomb: Ibid., p. 154

Colossal air field in Norway: Ibid., p. 94

JU-390 flies to New York: Cook, p. 198

Blast effects on Manhattan Island map: Witkowski, p. 52

Mussolini's political testament: Farrell, *Reich of the Black Sun,* pp. 70–71

Nagasaki bomb given to Soviets: Ibid., p. 71

No viable critical mass: Carter Plymton Hydrick, *Critical Mass: How Nazi Germany Surrendered Enriched Uranium for the United States' Atomic Bomb* (Spring, TX: Whitehurst & Company, 2004), p. 133

Britain prepares for atomic attack: Farrell, *Reich of the Black Sun,* pp. 72–73

Berlin telephone service out: Ibid., pp. 75–76

Electromagnetic pulse explanation: Ibid., p. 77

Luigi Romersa: Farrell, *Reich of the Black Sun,* pp. 78–79, from his translation of Edgar Meyer and Thomas Mehner's *Hitler und die "Bombe"* (Rottenburg am Neckar, Germany: Kopp Verlag, 2002), pp. 62–66

A bomb which will surprise the whole world: Henry Stevens, *Hitler's Suppressed and Still-secret Weapons, Science and Technology* (Kempton, IL: Adventures Unlimited Press, 2007), p. 78

U.S. report on Hans Zissner: Witkowski, p. 218, actual document reproduced on p. 219

Japanese embassy report: Farrell, *Reich of the Black Sun,* pp. 43–45

Fuel-air bomb: Ibid., p. 48

Sevastopol explosion: Paul Carrell, *Hitler Moves East, 1941–1943* (New York: Ballantine Books, 1971), p. 503

Propaganda disaster: Farrell, *Reich of the Black Sun,* p. 69

A rational explanation: Ibid., p. 157

Bomb of German provenance: Edgar Meyer and Thomas Mehner, *Das Geheimnis der deutschen Atombombe* (Rottenburg am Neckar, Germany: Kopp Verlag, 2001), p. 122

Baron von Ardenne's laboratory: David Irving, *The German Atomic Bomb* (New York: Simon and Schuster, 1967), p. 290

Wilhelm Ohnesorge: Hydrick, p. 59

Reichspost awash with money: Farrell, *Reich of the Black Sun,* p. 40

Postal minister brought solution: Irving, p. 77

Not one pound of buna produced: Farrell, *Reich of the Black Sun,* p. 29

Buna plant and secrecy between top Nazis: Hydrick, pp. 60–61

U-234 and cargo: http://en.wikipedia.org/wiki/Unterseeboot_234

Dr. Velma Hunt: http://www.mikekemble.com/ww2/downfall.html

Officially no nuclear reactor in Germany: Cook, p. 180

Uranium highly enriched: Farrell, *Reich of the Black Sun,* p. 61

Crew anecdote: http://en.wikipedia.org/wiki/Unterseeboot_234

Wolfgang Hirschfeld: Ibid.

Sail only on fuehrer's orders: Hydrick, p. 152

Major John E. Vance: Ibid., p. 28

Document discrepancy: Stevens, pp. 7–8, with copies of archive letters and bills of lading

Luis Alvarez: Farrell, *Reich of the Black Sun,* p. 63

Senator Byrnes's memo: Farrell, *Reich of the Black Sun,* p. 57

Edward Hammel at Los Alamos: Hydrick, p. 29

U-234 bombs can scarcely be argued: Ibid., p. 143

German bomb on Indianapolis: Farrell, *Reich of the Black Sun,* p. 71

A present for the Americans: Stevens, pp. 76–77

Lieutenant Joseph Kennedy's flight: Witkowski, p. 44

The Germans did make atomic bombs: Stevens, p. 76

U-234 taken as part of Bormann's plan: Farrell, *Reich of the Black Sun,* p. 64

Covert plan for continuation of Nazi research: Hydrick, p. xv

Uranium and bombs courtesy of Nazi Party: Farrell, *Reich of the Black Sun,* p. 158

Something far more destructive: Ibid., p. 346

Himmler's letter: Speer, pp. 207–208

Kammler as commissioner general: Ibid., p. 209

Special projects office: Cook, p. 159

Dr. Konrad Zuse and computers: Stevens, p. 18

A whole new research and command structure: Ibid., pp. iii–vii

Development facilities moved: Ford, p. 22

Crucial that V-2s were spared: Mary Bennett and David S. Percy, *Dark Moon: Apollo and the Whistleblowers* (Kempton, IL: Adventures Unlimited Press, 2001), p. 173

A mounting body of evidence: Cook, p. 209

Walter Gerlach forbidden to talk: Ibid., p. 194

Rudolf Schuster: Ibid., p. 186

Whole area in deep valley: Witkowski, p. 263

Deeply secret: Cook, p. 196

No acknowledgment of Kommando: Cook, p. 187

A "fly trap": Witkowski, p. 265

Iodizing radiation and magnetic field of energy: Ibid., p. 234

All samples destroyed: Cook, p. 192

Scientists shot: Ibid., p. 184

Zero point energy: Colonel John B. Alexander, *Winning the War: Advanced Weapons, Strategies, and Concepts for the Post-9/11 World* (New York: Thomas Dunne Books, 2003), pp. 231–232

T. Townsend Brown: Editors, *Anti-Gravity & the Unified Field* (Kempton, IL: Adventures Unlimited Press, 2001), p. 93

Tesla on mastery of physical creation: Ibid., pp. 44–45

Interavia Aerospace Review: Cook, pp. 5–6

Ceramic superconductor and magnet: http://picturethis.pnl.gov/picturet.nsf/All/44CT2Y?opendocument

L. D. Bell, George S. Trimble, and William Lear: Ibid., p. 4

Bruce Cathie's worldwide energy grid: Bruce L. Cathie and Peter N. Temm, "The Anti-gravity Equation," *The Anti-Gravity Handbook* (Kempton, IL: Adventures Unlimited Press, 1993), p. 26

Distance is an illusion: Ibid.

Betty Cash, Vickie and Colby Landrum: Marrs, *Alien Agenda,* pp. 182–183

Material weighed less than pan: Laurence Gardner, *Lost Secrets of the Sacred Ark* (London: Element, 2003), p. 164

Philosopher's stone rediscovered: Ibid., p. 171

Dr. Hal Puthoff and exotic matter: Ibid., p. 168

Perpetual motion: Ibid., p. 166

Carl Sagan on time travel: http://www.pbs.org/wgbh/nova/time/sagan.html

Jenny Randles on time travel: Jenny Randles, *Breaking the Time Barrier: The Race to Build the First Time Machine* (New York: Paraview Pocket Books, 2005), p. 8

General Knerr letter to Spaatz: Witkowski, p. 10

Madcap dash to southern Germany and Prague: Farrell, *Reich of the Black Sun,* pp. 111–112

Czech-Austrian documents still secret: http://www.majesticdocuments.com/documents/1948-1959.php

Kammler had some of value to deal: Cook, p. 164

Deal cut with Dulles or Patton: Farrell, *Reich of the Black Sun,* p. 156

Plausible motivation for Patton's death: Ibid., p. 157

Kammler to bargain with Allies: Jean Michel, *Dora* (New York: Holt, Rinehart and Winston, 1980), p. 290

Other important developments: Speer, p. 243

Kammler disappeared without a trace: Michel, pp. 296–297

Kammler's good fortune: Witkowski, p. 271

Kammler's "deaths": Cook, pp. 180–181

Dozens not called to account: Ibid., p. 189

Collusion between United States and Nazi Germany: Hydrick, p. 144

Nazi science came with a virus: Cook, p. 270

4. A Treasure Trove

Otto Skorzeny background: Mark M. Bostner III, *Biographical Dictionary of World War II* (Novato, CA: Presidio Press, 1996), pp. 504–507

Otto Rahn: Otto Rahn, translated by Christopher Jones, *Crusade Against the Grail: The Struggle Between the Cathars, the Templars and the Church of Rome* (Rochester, VT: Inner Traditions, 2006), p. x

Much sorrow in my country: Rahn, p. xiii

Rosenberg on Rahn's death: Angebert, pp. 14–15

Solomon's treasure described: Colonel Howard Buechner, *Emerald Cup—Ark of Gold: The Quest of SS Lieutenant Otto Rahn of the Third Reich* (Metairie, LA: Thunderbird Press, 1991), pp. 39–42

Treasures went with Alaric: Ibid., p. 58

Casket part of treasure: Rahn, p. 95

Cathars influenced by Druids: Ibid., p. 74

Independent republics: Ibid., p. 54

Immigration of peoples: Ibid., p. 27

Abraham and the Table of Destiny: Laurence Gardner, *Genesis of the Grail Kings* (London: Bantam Press, 1999), pp. 100, 219–220

A shadowy prehistory: Alan F. Alford, *Gods of the New Millennium: Scientific Proof of Flesh & Blood Gods* (Walsall, England: Eridu Books, 1996), p. i

Payens as cousin to Count Champagne: Graham Hancock, *The Sign and the Seal* (New York: Touchstone, 1993), p. 93

Royal Engineers find Templar artifacts: Christopher Knight and Robert Lomas, *The Hiram Key* (New York: Barnes & Noble Books, 1998), p. 267

Templars acquired scrolls of knowledge: Ibid., pp. 267–269

French family and Templar connections: Michael Baigent, Richard Leigh, and Henry Lincoln, *Holy Blood, Holy Grail* (New York: Dell Publishing, 1983), p. 73

Catharism to Thule Society: Angebert, p. 53

Asmodeus as builder of Solomon's Temple: Baigent, Leigh, and Lincoln, p. 36

Speculation on Sauniere's discovery: Lynn Picknett and Clive Prince, *The Templar Revelation* (New York: Touchstone, 1997), p. 182

Rahn's letter to Weisthor: David Wood and Ian Campbell, *Geneset: Target Earth* (Middlesex, England: Bellevue Books, 1994), p. 243

1943 German expedition: Angebert, p. 48

Commandos at Montségur: Buechner, p. 190

Skorzeny finds treasure: Ibid., p. 192

Messages to and from Berlin: Ibid.

Skorzeny, pilgrims, and plane flight: Ibid., pp. 193–194

A Fieseler Storch: Angebert, p. 47

Pilgrims awestruck: Buechner, p. 194

Treasure's journey and intact for last time: Ibid., pp. 195, 203

Description of treasure: Ibid., pp. 202–203
Frau Bormann and gold coins: Ibid., p. 204

5. The Writing on the Wall
All the machinery of a well-organized Nazi state: Reiss, p. 2
Transfer apparatus back to Nazi Party: Ibid., p. 3
Himmler has private meeting with Bormann: Ibid., p. 4
Industrialists to cut ties with Nazi Party: Ibid., p. 17
Wish for world constantly on brink of war: Ibid., p. 18
No reason to believe Schnitzler: Ibid., p. 21
Thyssen proclaims his loyalty to Germany: Snyder, p. 348
Thyssen's neighbors on the Riviera: Reiss, p. 24
Thyssen's flight not genuine: Ibid., pp. 22–23
Schlacht Plan: Ibid., p. 31
Schlacht's lack of punishment: Snyder, p. 308
Hitler's amphetamine use: Associated Press, "Drug Tied to Hitler Behavior," *Fort Worth Star-Telegram,* June 14, 1979
Machiavelli of the office desk: Snyder, p. 36
Hitler's comment on Bormann: Ladislas Farago, *Aftermath: Martin Bormann and the Fourth Reich* (New York: Simon and Schuster, 1974), p. 105
Economic resurgence of Germany: Manning, p. 24
A profit-seeking track: Ibid.
Less conspicuous Nazis in factories: Ibid., p. 27
750 front corporations: Ibid., p. 136
Historic file shipped to Bormann archives: Ibid., p. 144
Bormann followed strategies by Schmitz: Ibid., p. 136
Interhandel: Ibid., p. 159
Demand deposits in U.S. banks: Ibid., p. 139
American firms effective in protecting Nazi interests: Ibid., p. 25
Orvis A. Schmidt: Ibid., p. 146
I. G. Farben breakthroughs: Ibid., p. 55
Farben's interest in seven hundred companies: Ibid., p. 153
Farben as banking hub: Ibid., p. 58
German combines as spearheads: Ibid., p. 148
Bormann takes reins of finance: Ibid., p. 87
Deutsche Bank leads Big Three: Ibid., p. 69
Gold to Turkey: http://www.deutsche-bank.de/csr/en/history/7636.html
Description of European loot: Farago, p. 201
Greatest bank robbery in history: http://legalminds.lp.findlaw.com/list/cyberjournal/msg00227.html
German gold sold to Swiss: Manning, p. 136

Swiss banks and "concept deficit": Adam LeBor, *Hitler's Secret Bankers: The Myth of Swiss Neutrality During the Holocaust* (Secaucus, NJ: Birch Lane Press, 1997), p. xiv

"Max Heiliger": Farago, p. 200

Abs headed consortium: Manning, pp. 278–279

John J. McCloy's chairmanships: http://en.wikipedia.org/wiki/John_J._McCloy

Abs exempts U.S. banks: Manning, p. 72

Bush's Nazi connections: John Buchanan, "Bush-Nazi Link Confirmed," *New Hampshire Gazette,* October 10, 2003

Thyssen transfers ownership: John Loftus, "How the Bush Family Made Its Fortune from the Nazis: The Dutch Connection," http://www.tetrahedron.org/articles/new_world_order/bush_nazis.html

Edward Boswell comment: http://www.georgewalkerbush.net/bushnazidealings continueduntil1951.htm

Oil to Spain: Higham, p. 59

Changed registry/close associates: Ibid., p. 35

Swedish firm as front for Hamburg-Amerika shipping line: Manning, pp. 132–133

Bush family complicity: John Loftus, "The Dutch Connection"; Robert Lederman, http://baltech.org/lederman/bush-nazi-fortune-2-09-02.html

Home school curriculum: Reiss, p. 77

Sheep-dipped SS: Ibid., p. 87

Functionaries provided anti-Nazi cover: Ibid., p. 97

Basis for card files: Ibid., p. 141

Nazis want World War III: Reiss, p. 189

Reiss's warning: Reiss, p. 201

PART TWO:
THE REICH CONSOLIDATES

6. The Ratlines

Heinrich Mueller's tombstone: Manning, p. 179

Bishop Johannes Neuhausler: Ibid., p. 15

Dr. Hugo Blaschke and Fritz Echtmann: Ibid., p. 16

SS general Heinrich Mueller: Ibid., p. 17

Simon Wiesenthal: Snyder, p. 37

Walter Buch's dying declaration: Manning, p. 45

Bormann's escape route: Ibid., pp. 198–200

San Domingo records destroyed: Ibid., p. 202

ODESSA as consortium of freelancers: Farago, p. 167

ODESSA as vast clandestine Nazi travel organization: Snyder, p. 259

Hans Rudel's 2,530 combat missions: http://www.achtungpanzer.com/gen9.htm

Refugee bureau of the Vatican: Farago, p. 167

Almost a billion in gold paid to the Vatican: Ridley, p. 198

Hitler studied at Catholic monastery: Snyder, p. 151

Hitler in accordance with Creator: http://thinkexist.com/quotation/i_believe_today
_that_my_conduct_is_in_accordance/182334.html

Hitler on positive Christianity: Shirer, p. 234

"With Burning Sorrow": Ibid., p. 235

Spiritually, we are all Semites: http://en.wikipedia.org/wiki/Pope_Pius_XI

Cardinal Eugene Tisserant: Ibid.

Cardinal at wedding and secret staircase: Ridley, p. 289

Colonel Rudel on church aid: Christopher Simpson, *Blowback: America's Recruitment of Nazis and Its Effects on the Cold War* (New York: Collier-Macmillan, 1988), p. 178

Bishop Alois Hudal and Lanz von Liebensfels: Levenda, *Unholy Alliance,* pp. 253–254

Wider indictment: John Cornwell, *Hitler's Pope* (New York: Viking, 1999), p. viii

Pius XII's little scope of action: http://wapedia.mobi/en/Hitler's_Pope

Ante Pavelic: http://en.wikipedia.org/wiki/Ante_Paveli%C4%87

Argentina as Nazi Gau: Reiss, p. 143

Rudel brings Luftwaffe staff: Linda Hunt, *Secret Agenda: The United States Government, Nazi Scientists, and Project Paperclip* (New York: St. Martin's Press, 1991), pp. 147–148

Kurt Tank and smuggled designs: Ibid., p. 149

General Wilhelm von Faupel: Reiss, p. 150

At least $100 million from Bormann: Manning, p. 203

Evita's deep resentments: Georg Hodel, "Evita, the Swiss and the Nazis," *iF Magazine,* January/February 1999, http://www.thirdworldtraveler.com/Global_Secrets_Lies/EvitaNazis.htm

Evita meets with Skorzeny and Alberto Dodero: Ibid.

Evita coordinates Nazi networks: Ibid.

Bormann's banking files: Manning, p. 205

Mueller still wields power: Manning, p. 204

Death Squad operations: Hodel

Barbie into drug trade: Levenda, *Unholy Alliance,* p. 293

Barbie's team and torture: Hodel

Passau robbery and murders: Author's interviews with William H. Spector, 1979–1980

Jewish buyers entered the marketplace: Manning, p. 194

Rothschilds create Israel: Marrs, *Rule by Secrecy,* pp. 82–83

Rockefeller to monopolize raw materials: John Loftus and Mark Aarons, *The Secret War Against the Jews: How Western Espionage Betrayed the Jewish People* (New York: St. Martin's Griffin, 1994), p. 165

Everything Germans wanted they got: Ibid.

John Foster Dulles a fellow conspirator: Ibid., p. 166

Dossier blackmailed Nelson Rockefeller: Ibid., pp. 167–169

Vengeance or a country: Ibid.

John Pehle: Manning, p. 151

Sixty thousand Nazis escaped: Ibid., p. 200

Well-planned communications system: Ibid., p. 193

Bormann's FBI file: Ibid., p. 205–207

Argentine police report: Farago, pp. 236–237

New York Times story on Bormann: Ibid., p. 35

Kammler's records cleaned up: Cook, p. 180

Secrecy even to families: Heinz Hoehne and Hermann Zolling, *The General Was a Spy* (New York: Coward, McCann & Ceoghegan, 1972), pp. 72–73

Hitler calls Gehlen a fool: Ibid., p. 45

Americans hold more objectivity: Reinhard Gehlen, translated by David Irving, *The Service: The Memoirs of General Reinhard Gehlen* (New York: Popular Library, 1972), pp. 70–73

Approach American military forces: Ibid., pp. 108–109

Gehlen as unrepentant Nazi: Gehlen, p. 73

All part of the plan: Ibid., pp. 115–116

Dulles's idea: Hoehne and Zolling, p. 56

Gentlemen's agreement: Gehlen, p. 121

Gentlemen's agreement detailed: Ibid., p. 122

Organization without authority from Washington: Ibid., p. 123

Gehlen tight with Allen Dulles: Carl Oglesby, *The Yankee and Cowboy War: Conspiracies from Dallas to Watergate and Beyond* (New York: Berkley Medallion Books, 1977), p. 41

Dulles provided aggregate of $200 million: Mae Brussell, "The Nazi Connection to the John F. Kennedy Assassination," *The Rebel,* November 22, 1983

Questionable intelligence: Loftus, p. 59

Gehlen promised immunity to Radislaw Ostrowsky: Loftus, p. 57

CIA admits Gehlen connection: Maria Alvarez, "CIA Admits Long Relationship with WWII German General Reinhard Gehlen," *New York Post,* September 24, 2000

J. Edgar Hoover urges Interpol membership: Vaughn Young, "The Men from Interpol," David Wallechinsky and Irving Wallace, editors, *The People's Almanac* (Garden City, NY: Doubleday, 1975), pp. 37–39

Paul Dickopf: Ibid.

Hoover acquires Interpol files: Spector
Best-connected cults: Levenda, *Unholy Alliance,* p. 282
Colonia Dignidad as torture center: Ibid., p. 305.

7. Project Paperclip and the Space Race

Truman assured: Linda Hunt, *Secret Agenda: The United States Government, Nazi Scientists, and Project Paperclip, 1945–1990* (New York: St. Martin's Press, 1991), p. 40
Paperclip documents missing: Ibid., p. x
Project continued nonstop until 1973: Ibid., p. 1
Dulles brothers as Republican replacements: Webster Griffin Tarpley and Anton Chaitkin, *George Bush: The Unauthorized Biography* (Washington, DC: Executive Intelligence Review, 1992), p. 76.
Dulles brothers attorneys for Averell Harriman: Ibid., p. 73
Dulles' empire of information: LeBor, p. 97
Henry Kissinger as translator for Allen Dulles: http://en.wikipedia.org/wiki/Allen_Dulles; http://groups.google.com/group/alt.assassination.jfk/browse_thread/thread/ba8247a7b4362766/78ca5ca61add29ec
Dulles as lifelong friend to Nixon: Peter Levenda, *Sinister Forces* (Walterville, OR: TrineDay, 2006), p. 39
Colonel Robert Schow: Hunt, p. 115
Ardent Nazi designation changed: Ibid., p. 119
Dossiers rewritten: Ibid., p. 121
Paperclip's dark secrets safely hidden: Hunt, p. 249
Von Braun no security threat: Ibid., p. 120
Magnus von Braun: Ibid., p. 45
Crimes never investigated: Ibid., p. 121
Activity unknown to the public: Ibid., p. 112
National Interest, prevailing myth and universities: Ibid., pp. 127–128
Yale receives Paperclip Nazis: Ibid., p. 203
Nazis obtain jobs in aircraft industry: Ibid., p. 176
Project 63 and violent reaction envisioned by McCloy: Ibid., p. 182
Ambassador James Conant: Ibid., p. 194
Project accelerated and records shredded or pulled: Ibid., pp. 199–200
Unspeakable evil transplanted to America: Ibid., p. 21
William Henry Whalen question: Ibid., p. 205
Whalen sold Americans down the river: Ibid., p. 216
No surveillance of dependents: Ibid., p. 49
Coded messages and expensive cars: Ibid., p. 50
Germans working for French suspected of working for new Reich: Ibid., p. 52
Von Braun caught sending/concealing information: Ibid.

Soviets use German zone as cover: Ibid., p. 55

GE manager complains to FBI: Ibid., pp. 53–54

Americans ordered to leave rockets to Russians: Lieutenant Colonel (Ret.) William E. Winterstein Sr., *Gestapo USA: When Justice Was Blindfolded* (San Francisco: Robert D. Reed Publishers, 2002), pp. 10–11

James Webb complained: Hunt, p. 219

Germans dominated the rocket program: Ibid., p. 218

Arthur Rudolph: Winterstein, pp. 232–233

OSI coordinated with KGB: Ibid., p. 236

Two space programs: Farrell, *Reich of the Black Sun,* p. 317

Kennedy finally released von Braun team: Winterstein, dedication

Continuance of their research: Bennett and Percy, p. 166

Criminal accusations against von Braun: Hunt, p. 109; Richard C. Hoagland and Mike Bara, *Dark Mission: The Secret History of NASA* (Los Angeles: Feral House, 2007), p. 237

Scientists spoke directly only to Korolev: Bennett and Percy, p. 192

A collaboration of two superpowers: Ibid., p. 201

Some entity or agency: Farrell, *Brotherhood of the Bell,* p. 133

Lieutenant Walter Jessel reports on trust: Hunt, pp. 43–44

Dr. Wilhelm Voss attempts to give material to the Americans: Cook, p. 177

Soviet moles in high places: Thomas Fleming, *The New Dealers' War: F.D.R. and the War Within World War II* (New York: Basic Books, 2001), p. 319

British Interplanetary Society: Bennett and Percy, pp. 182–183

5412 Committee and name changes: Committee on Foreign Relations, U.S. Senate, *The U.S. Government and the Vietnam War, Executive and Legislative Roles and Relationships* (Washington, DC, U.S. Government Printing Office, 1984), p. 310

Manual and Dr. Robert M. Wood: Ryan S. Wood, *MAJIC Eyes Only* (Broomfield, CO: Wood Enterprises, 2005), pp. 264–295; www.majesticdocuments.com/index.php

MJ-12 briefing papers: Copies in author's files; also see http://www.majestic documents.com

Hillenkoetter quote: Timothy Good, *Above Top Secret: The Worldwide UFO Cover-up* (New York: William Morrow, 1988), p. 387

Vannevar Bush and Averell Harriman: http://lcweb2.loc.gov/service/mss/eadxmlmss/eadpdfmss/1998/ms998004.pdf

Bush at Carnegie Institute: www.carnegieinstitution.org/about.html

Dillon, Read made largest profits from German loans: Sutton, *Wall Street,* p. 29

William Draper as brigadier general: Ibid., pp. 155–156

Defense photos shipped on Hamburg-Amerika steamship line: Higham, p. 134

Prince Bernhard worked for I. G. Farben intelligence: www.deepblacklies.co.uk/princes_of_plunder.htm

Twining trip canceled and William Moore: Good, p. 260

Twining's letter of September 23, 1947: Cook, p. 37

Vandenberg destroys Project Sign report: Marrs, *Alien Agenda,* p. 109

Detlev Bronk's connection to the Rockefellers: Eustace Mullins, *Murder by Injection: The Story of the Medical Conspiracy Against America* (Staunton, VA: The National Council for Medical Research, 1988), p. 343

Souers connections: www.trumanlibrary.org/hstpaper/souerss.htm

Kissinger as consultant to Gray and Rockefellers: www.sourcewatch.org/index.php ?title=Talk:Gordon_Gray; www.trumanlibrary.org/hstpaper/gray.htm

Gray directed UFO study: Good, p. 259

Menzel on poor observers: Howard Blum, *Out There* (New York: Simon and Schuster, 1990), p. 247

Menzel's double life: Good, pp. 249–250

Menzel, Howard Hughes, and the Rockefeller Foundation: www.ucar.edu/communications/ staffnotes/0010/bogdan.html

Berkner's AUI associations: http://www.aui.edu/history.php

Lyndon Johnson on masters of infinity: Bennett and Percy, p. 165

Weaponization of space: http://www.space.com/news/050617_space_warfare.html

Hoagland and astrological alignments: http://www.enterprisemission.com/ken2b .html; http://www.enterprisemission.com/table_of_coincidence.htm

Apollo lands on Tranquillity: http://www.enterprisemission.com/ken2j.html

Mary Ann Weaver: http://www.angelfire.com/ca3/citystars/

Baikonur Cosmodrome: http://www.enterprisemission.com/zarya.htm

JFK instructs James Webb on joint space exploration: National Security Action Memorandum No. 271, November 12, 1963 (source: Kennedy Library); copy in author's files

Sergei Khruschev: Frank Sietzen, "Soviets Planned to Accept JFK's Joint Lunar Mission Offer," *Space Daily,* October 2, 1997

Nick Rockefeller's connections: http://www.nicholasrockefeller.net/

Nick Rockefeller quote: http://www.jonesreport.com/articles/210207_rockefeller_ friendship.html

8. Nazi Mind Control

Churchill wanted no occultism revealed: Trevor Ravenscroft, *The Spear of Destiny* (York Beach, MA: Samuel Weiser, 1973), p. xiii

Airey Neave: Michael Baignet, Richard Leigh, and Henry Lincoln, *The Messianic Legacy* (New York: Dell Publishing, 1986), p. 161

Hitler's quote: Joachim C. Fest, *Hitler* (New York: Harcourt Brace Jovanovich, 1974), p. 555

Occult groups in incestuous embrace: Levenda, *Unholy Alliance,* p. 18

Awareness of nonhuman intelligences: Ravenscroft, p. 161

Hitler's union with dark powers: John Toland, *Adolf Hitler* (New York: Doubleday, 1976), p. 64

The legend of Thule: Ravenscroft, pp. 169–170

Discarnate entity or magical elite: Nicholas Goodrick-Clarke, *The Occult Roots of Nazism: Secret Aryan Cults and Their Influence on Nazi Ideology* (New York: New York University Press, 1992), p. 218

Remote viewing: For details, see Jim Marrs's book *Psi Spies: The True Story of America's Psychic Warfare Program* (Franklin Lakes, NJ: New Page Books, 2007)

Doktor Gruenbaum: Author's interview with Lyn Buchanan, April 2007

Central piece of the puzzle: Dr. Thomas Roeder, Volker Kubillus, and Anthony Burwell, *Psychiatrists—the Men Behind Hitler* (Los Angeles: Freedom Publishing, 1995), p. 8

Darwin and Malthus' strange marriage: Ibid., p. 14

Major theme of Hitler's major speeches: Snyder, p. 323

Being French as mental illness: Roeder, Kubillus, and Burwell, p. 23

Kaufmann therapy: Ibid., p. 26

Psychiatrist as judge of illness: Ibid., p. 28

Publicly funded German Research Institute for Psychiatry: Ibid., p. 150

Eugenics Records Office funded by Harriman and Rockefeller: Ibid., p. 247.

Kraepelin as conservative nationalist: Ibid., pp. 150–151

Kraepelin funded by James Loeb: http://www.hup.harvard.edu/loeb/founder.html

Leading advocates for a solution: John Cornwell, *Hitler's Scientists: Science, War, and the Devil's Pact* (New York: Viking, 2003), p. 89

Hereditary health courts: Robert Jay Lifton, *The Nazi Doctors: Medical Killing and the Psychology of Genocide* (New York: Basic Books, 1986), p. 25

Hitler's secret decrees: Roeder, Kubillus, and Burwell, p. 53

410,000 only preliminary: Lifton, p. 25

Fritz Lenz: Ibid., p. 26

Theodor Eicke: Roeder, Kubillus, and Burwell, p. 48; Mark M. Boatner III, *The Biographical Dictionary of World War II* (Novato, CA: Presidio Press, 1996), pp. 150–151

The recommended use of carbon monoxide: Lifton, p. 71

Program had achieved original goal of victims: Roeder, Kubillus, and Burwell, p. 63

An exact replica of the T4 program: Ibid., p. 65

Dr. Ernst Rudin: Tarpley and Chaitkin, p. 49

Rudin continued as psychiatric leader: Roeder, Kubillus, and Burwell, p. 95

Bayer and narcotics: Alfred McCoy, *The Politics of Heroin in Southeast Asia* (New York: Harper & Row, 1972), pp. 5–6

Rock Oil: Mullins (1988), pp. 320–321

Nujol and Flit: Ibid., pp. 322–323

Sloan-Kettering Institute: http://en.wikipedia.org/wiki/Memorial_Sloan-Kettering
_Cancer_Center

Rohm and Haas: http://www.rohmhaas.com/company/company_overview.html

The American College of Surgeons: Mullins, *Murder by Injection,* p. 342

The Rockefeller Sanitary Commission: Ibid., p. 343; http://archive.rockefeller.edu/
collections/rockorgs/hookwormadd.php

Dr. Olin West: http://www.mc.vanderbilt.edu/sc_diglib/archColl/86.html

Rockefeller Archives and GEB: http://archive.rockefeller.edu/collections/rockorgs/
geb.php

$100 million spent: Mullins, *Murder by Injection,* p. 343

Ailments traced to drug monopoly: Ibid., p. 348

Sir Oliver S. Franks: Ibid., p. 345

Alzheimer's as seventh leading cause of death: http://alz.org/alzheimers_disease_what
_is_alzheimers.asp

California increases fluoridation: Gig Conaughton, "Southern California Water Supplies to Be Fluoridated Starting in October," *North County Times,* August 1, 2007

Children will exceed safe dose: http://www.ewg.org/node/22445

Bill Walker and fluoride studies: Ibid.

Aluminum accumulates in brain cells: http://www.alzheimers.org.uk/Facts_about_
dementia/Risk_factors/info_aluminium.htm

Charles Eliot Perkins letter: Mullins, *Murder by Injection,* pp. 353–354

Christian Science Monitor survey: http://www.battery-rechargeable-charger.com/
water-filter-fluoride-poisoning-info.html

21 million prescriptions: http://www.drugtopics.com/drugtopics/data/articlestandard/
drugtopics/092007/407652/article.pdf

Oscar Ewing: http://www.trumanlibrary.org/oralhist/ewing3.htm

Congressman A. L. Miller's quote: Mullins, *Murder by Injection,* pp. 153–154

NYC leaflet: http://www.trumanlibrary.org/oralhist/ewing3.htm

West Germany, Sweden, and the Netherlands ban fluoride: Mullins, *Murder by Injection,* p. 158

MKULTRA approved by Helms and Allen Dulles: Marks, pp. 56–57

CIA officer details MKULTRA projects: Walter H. Bowart, *Operation Mind Control:
Our Secret Government's War Against Its Own People* (New York: Dell Publishing,
1978), p. 106

Psychiatrists naturally interested in LSD: Edward M. Brecher and the editors of *Consumer Reports, Licit & Illicit Drugs: The Consumer Union Report on Narcotics,
Stimulants, Depressants, Inhalants. Hallucinogens, and Marijuana—Including Caffeine, Nicotine, and Alcohol* (Mount Vernon, NY: Consumers Union, 1972), p. 349

LSD experimentation at Lexington: http://www.botany.hawaii.edu/faculty/wong/
BOT135/LECT13.HTM

Kaiser Family Foundation: http://www.kff.org/about/index2.cfm

Experience contradicts adult propaganda: Brecher and editors of *Consumer Reports,* p. 332

Frank Olsen's death: Marks, pp. 73–82

Senator Ted Kennedy's remarks: http://www.druglibrary.org/schaffer/history/e1950/mkultra/Hearing01.htm

Carol Rutz: http://members.aol.com/smartnews/CR05.htm

George Hunter White: http://www.everything2.com/index.pl?node_id=481520

Soldiers tested at Edgewood Arsenal: Bowart, p. 91

Carol Rutz's letter from soldier: http://members.aol.com/smartnews/CR05.htm

Paperclip scientists at Edgewood kept secret: Hunt, pp. 158–159

Kurt Rahr: Ibid., p. 159

Tabun and sarin poisons used: Ibid., pp. 160–161

Dr. L. Wilson Greene: Ibid., p. 162

James Moore quote: Hunt., p. 166

Brigadier General Walter Schreiber: Ibid., p.169

Schreiber with Gottlieb and Mengele: Gordon Thomas, *Mindfield: The Untold Story Behind CIA Experiments with MKULTRA & Germ Warfare* (http://www.gordonthomas.ie/mindfield.htm), pp. 126–127

Chemical warfare consultants: Ibid., p. 168

John K. Vance: http://www.washingtonpost.com/wp-dyn/content/article/2005/06/15/AR2005061502685.html

Kathleen Ann Sullivan: http://www.gordonthomas.ie/mindfield.htm

Experiments stemmed from Nazi science: Hunt, p. 234

9. Business as Usual

John J. McCloy pardons Nazis: Farrell, *Brotherhood of the Bell,* p. 79

McCloy commutes Malmedy convictions: Shirer, pp. 1095–1096

OPC answers to Forrestal and Dulles faction: Loftus and Aarons, p. 213

Emanuel Jasiuk and entanglements: John Loftus, *The Belarus Secret* (New York: Alfred A. Knopf, 1982), pp. 146–147

Belarus Brigade and Einsatzgruppen: http://www.geocities.com/dudar2000/Bcc.htm

OPC an arm of the State Department: Loftus, p. 69

The historical record speaks: James Perloff, *The Shadows of Power: The Council on Foreign Relations and the American Decline* (Appleton, WS: Western Islands, 1988), p. 7

Sociologist G. William Domhoff: Jonathan Vankin, *Conspiracies, Cover-ups and Crimes* (New York: Paragon House Publishers, 1992), p. 209

Laurie Strand: David Wallechinsky and Irving Wallace, editors, *The People's Almanac #3* (New York: Bantam Books, 1981), p. 87

Galbraith comment on CFR: David Halberstam, *The Best and the Brightest* (New York: Random House, 1972), p. 60

Dulles supports Skorzeny: Levenda, *Unholy Alliance,* pp. 281–282

Hassan al-Banna as admirer of Hitler: John Loftus, "The Muslim Brotherhood, Nazis and Al-Qaeda," *Jewish Community News,* October 4, 2004; http://www.front-pagemag.com/Articles/ReadArticle.asp?ID=15344

Arab Nazis: Marc Erikson, "Islamism, fascism and terrorism," *Asia Times,* December 4, 2002; http://www.atimes.com/atimes/Middle_East/DL04Ak01.html

Arab Nazis fight Israel: Loftus, "The Muslim Brotherhood"

Osama bin Laden and Saudi Arabia: Ibid.

Al-Qaeda as CIA database: http://www.globalresearch.ca/index.php?context=view-Article&code=BUN20051120&articleId=1291

Secrets have to come out: Loftus, "The Muslim Brotherhood"

Dulles brothers dynasty: Bowart, p. 139

Deutsche Bank regroups: http://en.wikipedia.org/wiki/Deutsche_Bank

Biggest postwar story: Manning, pp. 233–234

Three Wall Street houses financed German cartels: Sutton, *Wall Street,* p. 163

Russell A. Nixon: Higham, pp. 212–213

Three Wall Street houses: Sutton, *Wall Street,* p. 29

Freudenberg like Ford: Higham, p. 215

Nixon to subcommittee: Manning, p. 156

James Stewart Martin: Higham, pp. 215–216

Martin's quote: Ibid., p. 217

William H. Draper Jr.: Tarpley and Chaitkin, pp. 54–55

McCloy in Hitler's Olympic box: Ibid.

750 new corporations: Manning, pp. 134–135

Conspiracy of silence: Ibid., p. 152

Powerful friends: Ibid., p. 156

Marshall Plan opponents: Mullins, *Murder by Injection,* p. 339

Economic booster of the Rome-Berlin Axis: Higham, p. 22

Interlocking directorships with I. G. Farben: William Bramley, *The Gods of Eden* (San Jose, CA: Dahlin Family Press, 1990), p. 415

Schmitz's wealth: Manning, p. 280

The legal heir to Hitler: Ibid.

Nazis did not die: Jim Keith, *Casebook on Alternative 3: UFOs, Secret Societies and World Control* (Lilburn, GA: IllumiNet Press, 1994), p. 148

1942 Common Market: Manning, p. 95

Common market nurtured at Bilderberg meetings: Robert Eringer, *The Global Manipulators* (Bristol, England: Pentacle Books, 1980), p. 26

Nixon as mouthpiece for Allen Dulles: Loftus and Aarons, p. 222

Faustian bargain: Levenda, *Sinister Forces,* p. 45

Such interrelationships very common: Loftus and Aarons, p. 223

Nicolae Malaxa: Brussell, p. 27

Malaxa seamless tube business: Loftus and Aarons, pp. 223–224

Nixon's $2 million fee: Spector

Elmer Bobst as father figure: http://www.nixoncenter.org/publications/CLINTON
.html

Bobst and von Bolschwing: Brussell, p. 24

GOP's Ethnic Division: Loftus and Aarons, p. 222

No intention to roll back communism: http://www.nytimes.com/2006/10/29/books/
Heilbrunn.t.html?ei=5070&en=3910f0dadd13428d&ex=1177041600&page-
wanted=all

10. Kennedy and the Nazis

John F. Kennedy warns of secret societies: http://www.jfklibrary.org/Historical+Resources/
Archives/Reference+Desk/Speeches/JFK/003POF03NewspaperPublishers0427
1961.htm

Kennedy and Inga Arvad: http://www.answers.com/topic/inga-arvad

Kennedy's autonomous influence: Jim Marrs, *Crossfire: The Plot That Killed Kennedy*
(New York: Carroll & Graf, 1989); also see John Newman, *JFK and Vietnam:
Deception, Intrigue and the Struggle for Power* (New York: Warner Books, 1992);
Donald Gibson, *Battling Wall Street: The Kennedy Presidency* (New York: Sheri-
dan Square Press, 1994)

CIA involvement in the assassination: Marrs, *Crossfire*, pp. 187–202

George DeMohrenschildt as Nazi spy: Author's interviews with Jeanne DeMohrenschildt, 1978–1979

J. Walter Moore: House Select Committee on Assassinations, vol. XII, pp. 53–54

Ten separate reports from DeMohrenschildt: Associated Press, "Oswald Friend Labeled
CIA Informant in Memo," *Dallas Times Herald,* July 27, 1978

Dornberger as Michael Paine's boss: Brussell, p. 31

Oswald as CIA employee: Marrs, *Crossfire*, pp. 189–190; John Newman, *Oswald and
the CIA* (New York: Carroll & Graf, 1995)

David Copeland as William Torbitt: Author's interviews, 1979

Torbitt Document: http://www.newsmakingnews.com/torbitt.htm; original document in author's files

Warren Hinkle's phone call from Garrison: http://mcadams.posc.mu.edu/jimloon1
.htm

Garrison's rebuffed by NASA: Jim Garrison, *On the Trial of the Assassins* (New York:
Sheridan Square Press, 1988), pp. 133–135

Oswald expected NASA work: Anthony Summers, *Conspiracy* (New York: Paragon
House, 1989), p. 284

Louis M. Bloomfield: http://www.jfkmontreal.com/intro_to_bloomfield_gallery .htm

Permindex: Garrison, pp. 87–90; copies of Clay Shaw's biography in *Who's Who in the South and Southwest* for 1963 and 1964 in author's files

Bloomfield recruited into OSS: David Goldman and Jeffrey Steinberg, *Dope, Inc.: Britain's Opium War Against the U.S.* (New York: The New Benjamin Franklin House Publishing, 1978) p. 301

Karl Wolff, Goering, and Melvin Belli: Brussell, p. 24

A whole new form of government: Marrs, *Crossfire*, p. 429

Jack Ruby's letter: Copy in author's file

Helmet Streikher's comments: Brussell, p. 33

McCloy and Dulles as Establishment men: Donald Gibson, *The Kennedy Assassination Cover-up* (Commack, NY: Kroshka Books, 2000), pp. 227–228

Senate committee study and interlocking directorships: Donald Gibson, *Battling Wall Street*, pp. 130–131

Kennedy attacked: Ibid., p. 97

Mean-minded individual: Associated Press, "Hoover Called Oswald 'a Nut,' FBI Files Show," *Fort Worth Star-Telegram*, December 7, 1977

Hoover's concern: Marrs, *Crossfire*, p, 459

Strain of madness and violence: James Reston in *New York Times* as cited by Gibson, *The Kennedy Assassination Cover-up*, p. 225

Settling dust with same conclusion: Ibid.

LBJ transformed from opponent to proponent: Ibid., p. 70

Essentially an Establishment cover-up: Ibid., pp. 135–136

PART THREE:
THE REICH ASCENDANT

11. Rebuilding the Reich, American-style

LBJ's "wise men": http://www.pbs.org/wgbh/amex/presidents/36_l_johnson/filmmore/filmscript.html

CFR, Morgan, and Rockefeller a single entity: Gibson, *Battling Wall Street*, p. 72

United States overthrows foreign governments: http://en.wikipedia.org/wiki/William _Blum

Foreign adventurisms listed: Editors, *The New Encyclopaedia Britannica* (Chicago: Encyclopaedia Britannica, 15th edition, 1991)

G. Gordon Liddy and ODESSA: G. Gordon Liddy, *Will: The Autobiography of G. Gordon Liddy* (New York: St. Martin's Press, 1980), pp. 147–148

Nazi propaganda films: http://www.edwardjayepstein.com/agency/chap15.htm

Bebe Rebozo processed millions: Manning, p. 275

Brzezinski sees national sovereignty no longer viable: Zbigniew Brzezinski, *Between Two Ages: America's Role in the Technetronic Era* (New York: Viking, 1970), p. 296

The Trilateral Commission founders: Marrs, *Rule by Secrecy,* p. 24

The unsettling thing: William Greider, "The Trilateralists Are Coming! The Trilateralists Are Coming!" *Dallas Times Herald,* February 3, 1977

Trying to calculate the odds: Antony C. Sutton and Patrick M. Wood, *Trilaterals Over Washington* (Scottsdale, AZ: The August Corporation, 1979), p. 1

Lee Harvey and Osvaldo: Dennis A. Williams, "Was There a Plot to Kill Carter?" *Newsweek,* May 21, 1979

Carter claimed loss of control: http://tinwiki.org/wiki/carter_assassination_attempt

Reagan accepts George Bush: Ronald Reagan's remarks to the Republican National Convention, *Congressional Quarterly Weekly Reports, July–September 1980,* vol. 38, no. 29, July 19, 1980

Reagan's transition team: Epperson, p. 247

Jury foreman Mark Kristoff: David Armstrong and Alex Constantine, "The Verdict Is Treason," *Z Magazine,* July/August, 1990

John Judge interview: www.geocities.com/prohibition_us/dui.html

Bush's network of friends in high places: Robert Parry, *Secrecy & Privilege: Rise of the Bush Dynasty from Watergate to Iraq* (Arlington, VA: The Media Consortium, 2004), p. 247

Reagan at Bitburg: Mary Hladky, "Reagan, Kohl Take 'Painful Walk into the Past'" *Stars & Stripes,* European edition, May 7, 1985

Laszlo Pasztor and Paul Weyrich: http://www.onlinejournal.com/archive/01-28-00_Binion.pdf

"Bush will do anything" quote: Loftus and Aarons, p. 369

Bush must have known about ethnics: Ibid., p. 370

Nazis in Republican Party: Russ Bellant, *Old Nazis, the New Right and the Republican Party: Domestic Fascists Networks and Their Effect on U.S. Cold War Politics* (Boston: South End Press, 1991); http://books.google.com/books?id=ZWAHm-LuZeIoC&dq=russ+bellant+old+nazis&printsec=frontcover&source=web&ots=rdu8IAWRbq&sig=YD_MDenI-qGiHkpo4srBwqZ-RSk#PPP1,M1

Radi Slavoff as chairman of Heritage Council: www.reagan.utexas.edu/archives/speeches/1985/51785a.htm

Austin App quote: Russ Bellant, "G. H. W. Bush Used Nazi-Collaborators to Get Elected," *Press for Conversion!* August 2004, pp. 38–41; http://coat.ncf.ca/our_magazine/links/54/54_38-41.pdf

Florian Galdau, Nicholas Nazarenko, Method Balco, Walter Melianovich and Bohdan Fedorak: Ibid.

Bush backers fired: David Lee Preston, "Fired Bush Backer One of Several with Possible Nazi Links," *Philadelphia Inquirer,* September 10, 1988

One thing is certain: http://www.onlinejournal.com/archive/01-28-00_Binion.pdf

Number of CIA Nazi assets: P&O File 311.5 TS (Sections I, II, III), 1948 Top Secret Decimal File, Records of Army General Staff, RG 319, National Archives

A comfortable environment for the elite: Levenda, *Sinister Forces,* pp. 312–313

Continuity of government plan: James Mann, "The Armageddon Plan," *Atlantic Monthly,* March 2004, pp. 71–74

Hidden national security apparatus: Ibid.

George Lincoln Rockwell: http://en.wikipedia.org/wiki/George_Lincoln_Rockwell

Pro-Nazi supporters of Bilderberg: Wallechinsky and Wallace, eds. *The People's Almanac #3,* p. 82

Spencer Oliver on money: Parry, p. 248

Clinton at Bilderberg: Editors, "Clinton to Attend Meeting in Germany," *Arkansas Democrat Gazette,* June 4, 1991

Hillary Clinton at Bilderberg: James P. Tucker Jr., "Bilderberg Tracked to Scotland," *The Spotlight,* May 18, 1998

Robert Rubin: Gibson, *Battling Wall Street,* pp. 148–149

Wilson quote: Woodrow Wilson, *The New Freedom* (New York: Doubleday, 1914), pp. 13–14

Financial element owns government: Elliot Roosevelt, ed., *F.D.R. His Personal Letters 1928–1945,* vol. I (New York: Duell, Sloan & Pearce, 1950), pp. 371–373

The Secret Team: L. Fletcher Prouty, Colonel, USAF (Ret.), *The Secret Team: The CIA and Its Allies in Control of the United States and the World* (Englewood Cliffs, NJ: Prentice-Hall, 1973), pp. 2–3

Men not incompetent or stupid: Epperson, p. 305

A "unipolar moment": Patrick J. Buchanan, "Doesn't Putin Have a Point?" http://www.vdare.com/buchanan/070212_putin.htm

An aggressive strategy: Mike Whitney, "Putin's Censored Press Conference: The Transcript You Weren't Supposed to See," http://www.informationclearinghouse.info/article17856.htm

Thomas C. Schelling: http://www.lse.ac.uk/collections/CPNSS/events/Abstracts/HistoryofPoswarScience/sent_schelling.pdf

Republican neoconservatives: John W. Dean, *Broken Government: How Republican Rule Destroyed the Legislative, Executive and Judicial Branches* (New York: Viking, 2007), pp. xi–xii

Arnold Schwarzenegger and quotes: Bob Fitrakis and Harvey Wasserman, "Sieg Heil: The Bush-Rove-Schwarzenegger Nazi Nexus and the Destabilization of California," *The Free Press,* October 6, 2003; http://www.buzzflash.com/contributors/03/09/17_arnold.html.

12. Guns, Drugs, and Eugenics

Barrier between military and basic research a vain hope: Cornwell, *Hitler's Scientists,* pp. 447–448.

Hitler on disarmed subject races: Dr. Henry Picker, ed., translated by Norman Cameron and R. H. Stevens, *Hitler's Table-Talk at the Fuehrer's Headquarters 1941–1942* (Bonn: Athenaum-Verlag, 1951), second edition (1973), pp. 425–426

JPFO statement and inherited gun owners lists: www.jpfo.org/GCA_68.htm

Dodd supplied Nazi Weapons Act: Ibid.; http://en.wikipedia.org/wiki/Gun_Control_Act

Master manipulators of popular emotion: www.guncite.com/gun_control_gcnazimyth.html

Rockefeller "Medical Monopoly": Mullins, *Murder by Injection,* p. 342

Committee on the Costs of Medical Care: http://www.innominatesociety.com/Articles/The%20Committee%20%20On%20The%20Costs%20Of%20Medical%20Care.htm

Dr. Smith and health care crisis: Ibid.

Dr. Detlev Bronk and the Rockefeller Institute: Mullins, *Murder by Injection,* p. 343

Rockefeller Education Board and medical schools: Ibid.

Burroughs Wellcome and Lord Oliver Franks: Ibid., p. 345

Big Bill Rockefeller as carnival medicine-show barker: Marrs, *Rule by Secrecy,* pp. 44–45

I. G. Farben, Sandoz, and Ciba-Geigy: Mullins, *Murder by Injection,* p. 337

Rat studies irrelevant: Louis J. Elsas II, MD, "Nutrasweet: Health and Safety Concerns," testimony before the U.S. Senate Committee of Labor and Human Resources, November 3, 1987

Dr. Betty Martini on aspartame release: http://www.newswithviews.com/NWVexclusive/exclusive15.htm

Patty Wood Allott: http://www.soundandfury.tv/pages/rumsfeld.html

Dr. Janet Starr Hull and academics: http://www.janethull.com/about/index.php. http://www.janethull.com/newsletter/0206/bella_italia_the_soffritti_aspartame_study.php

Dr. Morando Soffritti study: Daniel J. DeNoon, reviewed by Dr. Michael W. Smith, "Study Links Aspartame to Cancer," http://www.cbsnews.com/stories/2005/07/28/health/webmd/main712605.shtml. See study at: http://www.ehponline.org/members/2007/10271/10271.pdf

Barcelona study: http://www.presidiotex.com/barcelona/

No link between cancer and aspartame: http://www.cbsnews.com/stories/2006/04/05/health/webmd/main1473654.shtml

Michael F. Jacobson: Ibid.

$182 million through June 2006: http://www.publicintegrity.org/rx/report.aspx?aid =823

Drug lobby second to none: http://www.publicintegrity.org/rx/report.aspx?aid=723

Direct-to-consumer advertising: Ibid.

Dr. Marcia Angell and rise of Big Pharm: http://www.nybooks.com/articles/17244

Disease-mongering: Dr. Michael Wilkes, "Inside Medicine: Some 'Diseases' Invented for Profit," *Sacramento Bee,* May 26, 2007

Eight-hundred-pound gorilla: http://www.nybooks.com/articles/17244

Professor Peter Piper: http://news.independent.co.uk/health/article2586652.ece

Big Pharm contributions: http://www.opensecrets.org/pres08/select.asp?Ind=H04

Nazis recognized asbestos-cancer link: Robert N. Proctor, *The Nazi War on Cancer* (Princeton, NJ: Princeton University Press, 1999), p. 111

Consensus not obtained for two decades: Ibid., p. 113

German recognition of tobacco addiction and cancer: Ibid., p. 173

Fritz Lickint: Ibid., p. 186

Hitler's comment on the Red Man: Ibid., p. 219

Antitobacco war criminals: Ibid., p. 241

Differences of Holfelder and Fischer: Ibid., p. 251

Otto Warburg: Ibid., p. 37

Oxygen replaced by fermentation of sugar: Mullins, *Murder by Injection,* p. 351

Frank Howard and "Dusty" Rhoads: Ibid., p. 331

Coca-Cola unsuitable for children: Proctor, p. 147

Natural foods and drugs: Ibid., p. 265

The future will be vegetarian: Picker, Section 66

83,000 vegetarians: http://www.geocities.com/hitlerwasavegetarian/

Hitler not a poacher: Picker, Section 308

Appreciating the complexities: Proctor, p. 278

Justice Oliver Wendell Holmes: Ibid., p. 21

Eugenics Records Office: Ibid., p. 23

Dr. Ernst Rudin: Tarpley and Chaitkin, p. 49

Rudin honored in 1992: Roeder, Kubillus, and Burwell, p. 95

General William Draper: Tarpley and Chaitkin, p. 54–56

Eugenics and sterilizations: Jonathan Vankin and John Whalen, *Fifty Greatest Conspiracies of All Time: History's Biggest Mysteries, Cover-ups, and Cabals* (New York: Carol Publishing Group, 1995), p. 22

Planned Parenthood prevents unwanted pregnancies: http://www.plannedparenthood .org/about-us/who-we-are.htm

Planned Parenthood contributions: http://www.plannedparenthood.org/files/PPFA/ Annual_report.pdf

Plutocrats in league with scientists: Vankin and Whalen, p. 24

Maxwell Taylor quote: Editors, "Maxwell Taylor: 'Write Off a Billion,'" *Executive Intelligence Review,* September 22, 1981

The desire to remake humanity: Steve Sailer, "Q&A: Steven Pinker of 'Blank State,'" *United Press International,* October 30, 2002.

13. Religion

National Socialism was a religion: George L. Mosse, *Nazi Culture* (New York: Grosset & Dunlap, 1966), pp. xxxi–xxxii

Evangelical Christians: Ibid.

Faith as the sole basis of a moral life: Hitler, p. 365

Fanatical preaching: Ibid., p. 487

Hitler kept Party aloof from religion: Hitler's Table Talk (London: Weidenfeld and Nicolson, 1953), pp. 58–62

Nazi, Christian concepts incompatible: Mosse, p. 244

Bishop Ludwig Mueller: Erwin W. Lutzer, *Hitler's Cross* (Chicago: Moody Press, 1995), p. 130

Fuehrer bequeathed by the Lord: Mosse, p. 241

Nazi era shouts its lessons: Lutzer, p. 13

Jim Wallis and Sojourners: http://www.sojo.net/index.cfm?action=about_us.mission

Wallis as leftist political operative: www.lewrockwell.com/anderson/anderson107.html

Traditional Values Coalition attack on Jim Wallis: http://www.traditionalvalues.org/modules.php?sid=2664

Separation of church and state given sinister twist: Lutzer, p. 19

Pastor Mark Holick and the IRS: Bob Unruh, "IRS to Church: Shut Up—Church to IRS: No Way," http//wnd.com/news/article.asp?ARTICLE_ID=55979

Marsha West: http://www.newswithviews.com/West/marsha46.htm

Bill Keller: Liveprayer.com

Karl W. B. Schwarz: http://www.yourchristianpresident.com/Assets/Index%20Commentary/Beware%20the%20GOP%20has%20become%20a%20fascist%20cult.htm

Pat Robertson on assassination: Richard Cizik, "Robertson Apologizes for Chavez Assassination Remarks," *Washington Post,* August 25, 2005

Erik Prince and Blackwater: Jeremy Scahill, *Blackwater: The Rise of the World's Most Powerful Mercenary Army* (New York: Avalon Books, 2007) p. 339.

Some alien flag: Lutzer, p. 191

Easier to obey than accept dangers of freedom: Gerald Suster, *Hitler: The Occult Messiah* (New York: St. Martin's Press, 1981), p. 135

14. Education

Education is crucial: Mosse, p. xxxiv

The Lincoln School: http://www.britannica.com/eb/article-9001067/New-Lincoln-School

American Baptist Education Society: http://archive.rockefeller.edu/publications/res-rep/rose1.pdf

Senator William Benton and Encyclopaedia Britannica: http://bioguide.congress.gov/scripts/biodisplay.pl?index=B000399

Paolo Lionni: http://www.sntp.net/education/leipzig_connection_6.htm

Tremendous control for one group: Ibid.

Norman Dodd: Epperson, p. 209; http://video.google.com/videoplay?docid=-7373201783240489827

CIA on campuses: David N. Gibbs, "The CIA Is Back on Campus: Spying, Secrecy and the University," http://www.counterpunch.org/gibbs04072003.html

German-American Youth: http://www.longwood.k12.ny.us/history/yaphank/german_american_bund.htm

Rockefeller shaping a new industrial social order: William H. Watkins, *The White Architects of Black Education: Ideology and Power in America, 1865–1954* (New York: Teachers College Press, 2001), pp. 133–134

Pat Buchanan and NEA criticism: http://en.wikipedia.org/wiki/National_Education_Association#_ref-Ontheissues-Ed_0

Ex-Gay voice: http://www.exgaywatch.com/wp/2004/02/nea-ex-gay-educ/

Kevin Jennings: George Archibald, "Changing Minds: Former Gays Meet Resistance at NEA Convention," *Washington Times,* July 27, 2004

Connecticut House vote: http://www.namiscc.org/newsletters/Sept01/Alternative.htm#connecticut

University of Wisconsin study: http://omnihealthcaregroup.com/ADD.htm

Gretchen LeFever's reports prompt firing: http://www.pbs.org/newshour/bb/health/jan-june00/adhd_2-24.html; http://www.ahrp.org/infomail/05/04/10.php

Failure to find real disease: P. R. Breggin, *Toxic Psychiatry* (New York: St. Martin's Press, 1991), chapters 12 and 13

Alan Larson: Bruce Wiseman, *Psychiatry: The Ultimate Betrayal* (Los Angeles, CA: Freedom Publishing, 1995), p. 287

German trained psychoanalysts: Cornwell, *Hitler's Scientists,* p. 161

Psychiatry developed along lines of Nazi founders: Roeder, Kubillus, and Burwell, p. 141

DSM disorders grow: Wiseman, p. 355

Psychiatry may still be suspect: Ibid., p. 31

Prozac, Zoloft, Effexor, and Paxil: Roeder, Kubillus, and Burwell, p. 287

Dr. Helmut Remschmidt: Ibid., pp. 136–137, 142–143

Human screening deemed indefensible: Ibid.

Growth of school psychologists: Wiseman, p. 287

105 adverse reactions to Ritalin: Ibid., p. 285

WHO compares Ritalin to cocaine: Kelly Patricia O'Meara, "New Research Indicts Ritalin," *Insight on the News,* October 1, 2001

Cho's prescription drugs: http://www.nytimes.com/2007/04/18/us/18gunman.html?_r=1&pagewanted=2&hp&oref=slogin

Dr. Breggin and Luvox at Columbine: Bob Unruh, "Are Meds to Blame for Cho's Rampage?" http://www.wnd.com/news/article.asp?ARTICLE_ID=55310

Statistic rarely mentioned in news reports: http://www.teenscreentruth.com/index.html

Massacres have drugs in common: Dr. Julian Whitaker, MD, "Prescription Drugs—the Reason Behind the Madness," *Health and Healing,* November, 1999

TeenScreenTruth quote and list: http://www.teenscreentruth.com/psychiatry_drugs_suicide.html

Big Pharma as one of largest advertisers: www.publicintegrity.org/rx/report.aspx?aid=723

Alan Mathios: http://www.onthemedia.org/transcripts/2007/08/10/03

Purpose of NCLB: http://www.ed.gov/policy/elsec/leg/esea02/pg1.html#sec1001

Funds for migratory children: Ibid.

Utah rejects NCLB: http://select.nytimes.com/gst/abstract.html?res=F70B15FC3D550C738EDDAD0894DD404482

CEP report and Jack Jennings: Alex Kingsbury, "Do Schools Pass the Test?" *U.S. News & World Report,* June 18, 2007

Obese kids data: Helyn Trickey, "No Child Left Out of the Dodgeball Game?" http://www.cnn.com/2006/health/08/20/pe.nclb/index.html

Garrett Lydic: Ibid.

Too destructive to salvage: Alfie Kohn, "NCLB: 'Too Destructive to Salvage,'" *USA Today,* May 31, 2007

Paul Weyrich and Heritage Foundation: http://www.watch.pair.com/heritage.html

Hitler quote: Witkowski, p. 7

John D. Rockefeller on wanting workers: http://www.thenation.com/blogs/edcut?pid=173207

Hitler requires instincts: Mosse, p. xxviii

Bush's instincts: Ron Suskind: "Without a Doubt," *New York Times Magazine,* October 17, 2004; http://www.ronsuskind.com/articles/000106.html

Margaret Spellings and Siemens: http://www.girlscouts.org/news/news_releases/2006/doe.asp

Art Ryan, Elizabeth Weiss Green, and John Dewey: Elizabeth Weiss Green, "Grade School Goes Corporate," *U.S. News & World Report,* May 7, 2007

Corporate support for schools debate: http://www.nea.org/neatoday/0610/debate.html

Advertisers expelled: Gary Ruskin, "Much of What Advertisers Are Doing Is An Invasion of Privacy," *Advertising Age* (April 26, 2004)

Commercialism in schools: http://www.consumersunion.org/other/captivekids/problem.htm

Young set against old: Mosse, p. xxxiii

Fetish of youthfulness: Ibid.

Ten Nazi friends: Milton Mayer, *They Thought They Were Free: The Germans, 1933–45* (Chicago: The University of Chicago Press, 1966), pp. xviii-xix

Pastor Martin Niemoeller: http://www.jewishvirtuallibrary.org/jsource/Holocaust/Niemoller_quote.html

No one noticed widening gap: Ibid., pp. 166–173

15. Psychology and Public Control

Jack Heyman and ILWU: Lee Sustar, "Taft-Hartley, Bush and the Dockworkers," *CounterPunch,* October 11, 2002

The Rockefeller Syndicate not independent: Mullins, *Murder by Injection,* p. 312

Rockefeller Oil Trust became the military-industrial complex: Ibid., p. 318

David Rockefeller as courier: Ibid., p. 313

The Real ID Act: http://www.dhs.gov/xprevprot/laws/gc_1172767635686.shtm

The slow encirclement of law-abiding U.S. citizens: Steven Yates, " 'Your Papers, Please': National ID, 2002," http://www.lewrockwell.com/yates/yates64.html

Representative Jane Harman and "smart card": Dee Ann Davis and Nicholas M. Horrock, "Ridge Eyes New Driver's Licenses," *Washington Times,* May 2, 2002; http://www.newdem.org/press/newsreleases/2002-04-08.304.phtml

VeriChip: Press releases and Web site, Applied Digital Solutions, Inc. http://www.adsx.com/prodservpart/verichip.html

Chips cause tumors and Tommy Thompson: Todd Lewan, "Chip Implants Linked to Animal Tumors," *Washington Post,* September 8, 2007

Echelon like vacuum cleaner: Ned Stafford, "Newspaper: Echelon Gave Authorities Warning of Attacks," *Newsbytes.com,* September 13, 2001; http://www.globalresearch.ca/articles/STA205B.html

Healing frequency codes isolated: Dr. Nick Begich, *Controlling the Human Mind: The Technologies of Political Control or Tools for Peak Performance* (Anchorage, AK: Earthpulse Press, 2006), p. 19

Dr. Thomas Rivers: Barry Lynes, *The Cancer Cure That Worked: Fifty Years of Suppression* (Queensville, Ontario: Marcus Books, 9th printing 2001), p. 42; www.rense.com/general31/rife.htm

Four hundred tests: Lynes, p. 51

Terminal cancer patients cured: Ibid., p. 60

Rife's description of his machine's operation: Ibid., pp. 60–61

Broken agreements: Ibid., pp. 78–79

Thinking manipulated by external means: Begich, p. 28

Neurophone: http://www.toolsforwellness.com/neurophone.html

Psychotronic generators: Begich, pp. 37–38

Brain decoding and interaction: Ibid., p. 56

U.S. Patent: Ibid., p. 119

Levels ignored by the West: Ibid., p. 74

Complaint letter to Putin: Ibid., p. 80

Enhancement of human potentials: Ibid., p. 151

Dr. Anna Leiter: Roeder, Kubillus, and Burwell, p. 129

The Fixated Threat Assessment Centre: Joanna Bale, "VIP 'Stalker' Squad Set Up by Government," *The Times,* May 27, 2007; http://www.timesonline.co.uk/tol/news/uk/crime/article1847697.ece

Nazi special courts: Roeder, Kubillus, and Burwell, p. 81

Authority of command: Snyder, p. 104

The decider: Kenneth T. Walsh, "History's Verdict," *U.S. News & World Report,* January 29, 2007

Kathryn Dunn Tenpas: http://www.brookings.edu/papers/2006/07governance_tenpas.aspx

Bush's obligation to keep presidency robust: http://www.whitehouse.gov/news/releases/2002/03/20020313-8.html

Kucinich on the PATRIOT Act: Eli Pariser, editor, "Can Democracy Survive an Endless 'War'?" *MoveOn Bulletin,* July 18, 2002; www.moveon.org/moveonbulletin

Representative Ron Paul: Kelly Patricia O'Meara, "Police State," *Insight Magazine,* Nov. 9, 2001

Representative Bernie Sanders: Ibid.

U.S. Senate's Special Committee on Termination of National Emergency: http://www.constitution.org/mil/lawnanti.htm; also see http://www.fas.org/sgp/crs/natsec/98-505.pdf

State of Emergency: http://www.whitehouse.gov/news/releases/2001/09/20010914-4.html

Presidential directives: http://www.whitehouse.gov/news/releases/2007/05/20070509-12.html

Sharon Bradford Franklin: Charlie Savage, "White House Revises Post-disaster Protocol," *Boston Globe,* June 2, 2007; http://www.boston.com/news/nation/washington/articles/2007/06/02/white_house_revises_post_disaster_protocol/

Dick Cheney on NBC: http://writ.news.findlaw.com/dean/20020524.html

Bush on presidency: Ibid.

John W. Dean: Ibid.

Phyllis Schlafly and Bruce Fein: Ibid.

Statement can bypass law: Jennifer Van Bergen, "Why the Bush Doctrine Violates the Constitution: The Unitary Executive," http://www.counterpunch.org/vanbergen01122006.html

Bush quietly files signing statements: Charlie Savage, "Bush Challenges Hundreds of laws," *Boston Globe,* April 30, 2006

Jennifer Van Bergen: http://writ.news.findlaw.com/commentary/20060109_bergen.html

Traudl Junge: "Hitler's Final Witness," *BBC News,* February 4, 2002; http://news.bbc.co.uk/1/hi/world/europe/1800287.stm

German Justice Minister Herta Daeubler-Gmelin: http://www.time.com/time/magazine/article/0,9171,1003368,00.html

16. Propaganda

Media tell us what to think about: Michael Parenti, *Inventing Reality: The Politics of the Mass Media* (New York: St. Martin's Press, 1986), p. 23

Pew Research Center poll and Editor & Publisher: John Leo, "Elephant in the Living Room," *U.S. News & World Report,* April 20, 1998

The uncritical transmission of official opinions: Parenti, p. 51

Britt Hume: Ibid., p. 52

People fall more easily for a great lie: Hitler, p. 313

Goering's quote: G. M. Gilbert, *Nuremberg Diary* (New York: Signet Books, 1947); http://www.snopes.com/quotes/goering.htm

Hitler's closeness to Goebbels no coincidence: Mosse, p. xl

AOL Time Warner Anywhere: Frank Gibney Jr., "Score One for AOLTW," *Time,* December 25, 2000; http://www.time.com/time/magazine/article/0,9171,998845,00.html

Viacom and Senator John McCain: http://www.newint.org/features/2001/04/01/facts/

Rupert Murdoch's remark: Ward Harkavy, "World War Free," *Village Voice,* September 13–19, 2000; http://www.villagevoice.com/news/0037,harkavy,18140,5.html

Bertelsmann owned by two families: http://www.mediawiredaily.com/labels/BERTELSMANN.html

Bertelsmann Foundation: http://www.answers.com/topic/bertelsmann-foundation

IHC Chairman Saul Friedlaender: http://news.bbc.co.uk/2/hi/business/2308415.stm

Thielen and Arnold's statements: Ibid.

$486 million to influence elections and legislation: http://www.publicintegrity.org/telecom/report.aspx?aid=778

Charles Lewis on deregulation: http://www.icij.org/telecom/report.aspx?aid=96

A drop in straight news: http://www.journalism.org/node/442

Arthur Ochs Sulzberger quote: Project for Excellence in Journalism, "The State of the News Media 2007," annual report, p. 4; http://www.stateofthenewsmedia.com/ 2007/execsummary.pdf

A growing pattern of offering solutions: Ibid., p. 7

Jim Garrison and America as empire: http://www.worldforum.org/home/AmAs Empire.htm

Morley Safer: Editors, "Disgust within the Ranks," *Quill* (a publication of the Society of Professional Journalists), May 1998

Mark Crispin Miller: http://www.co-bw.com/General%20Reference%20and%20 Search%20CBAW.htm

History's actors: Ron Suskind, "Without a Doubt," *New York Times Magazine,* October 17, 2004

Sameness of popular taste: Mosse, p. xxix

Poverty increase: http://www.census.gov/Press-Release/www/releases/archives/ income_wealth/010583.html; David Leonhardt, "U.S. Poverty Rate Was Up Last Year," *New York Times,* August 31, 2005

Middle class decreasing: Janny Scott, "Cities Shed Middle Class, and Are Richer and Poorer for It," *New York Times,* July 23, 2006

Universal economic ruin: Roeder, Kubillus, and Burwell, p. 42

FBI has no hard evidence on Osama bin Laden: http://www.teamliberty.net/id267. html

2007 Zogby International poll: http://www.zogby.com/news/ReadNews.dbm?ID= 1355

2004 Zogby International poll: www.911truth.org/article.php?story= 20040830120349841

MSNBC National Poll: http://www.contramotion.com/pictures/issues/mediareal-ity/msnbc/

Sergeant Lauro Chavez: Author's interview, September 28, 2006

Emergency teams in NYC: http://www.abodia.com/911/Link/Topic%20Pages/War %20Games%20On%209-11.htm

Bin Laden family members allowed to fly: Jane Mayer, "The House of Bin Laden," *The New Yorker,* November 12, 2001; Craig Unger, "Saving the Saudis," *Vanity Fair,* October 2003

Pentagon photos show no large hole: http://www.911studies.com/911photostudies1 .htm

Mayor Giuliani warned: www.prisonplanet.com/eye_witness_account_from_new_ york.html; http://physics911.org/net/modules/wfsection/article.php?articleid=15

FEMA probe a half-baked farce: Bill Manning, "Selling Out the Investigation," *Fire Engineering,* January 2002

Low probability of occurrence: Federal Emergency Management Agency, *World Trade*

Center Building Performance Study (Washington: U.S. Government Printing Office, 2002), pp. 7–8

Short-selling of stock and links to CIA: http://www.onlinejournal.com/artman/publish/article_842.shtml

9/11 warnings from other nations: Jim Marrs, *The Terror Conspiracy: Deception, 9/11, and the Loss of Liberty* (New York: The Disinformation Company, 2006), pp. 70–84

Close ties of Bushes and bin Ladens: Mike Ward, "Bin Laden Relatives Have Ties to Texas," *Austin American-Statesman,* November 9, 2001

Epilogue

Blackwater as new Nazi Brown Shirts: Jeremy Scahill, *Blackwater: The Rise of the World's Most Powerful Mercenary Army* (New York: Nation Books, 2007), p. xxv

Laurence W. Britt and principles of fascism: Laurence W. Britt, "Fascism Anyone? The Fourteen Identifying Characteristics of Fascism," *Free Inquiry Magazine,* vol. 22, no. 2, July 15, 2003; http://www.fascistic.net/

2007 military budget: http://www.whitehouse.gov/omb/budget/fy2007/defense.html

Military spending: http://www.warresisters.org/piechart.htm

Congress cannot ignore media control: http://www.commondreams.org/views02/0615-03.htm

Corporate protection: http://www.federalismproject.org/preemption/

Federal agencies shield industries: Alan C. Miller and Myron Levin, "Industries Get Quiet Protection from Lawsuits," *Los Angeles Times,* February 18, 2006

Academics fired or reprimanded: http://www2.nea.org/he/heta05/images/2005pg119.pdf

Bush presidency compared to U. S. Grant: http://hnn.us/articles/5019.html

$2.3 trillion missing: http://www.cbsnews.com/stories/2002/01/29/eveningnews/main325985.shtml

Nepotism in the Bush administration: Dana Milbank, "In Appointments, Administration Leaves No Family Behind," *Washington Post,* March 12, 2002

Politicization of science: http://www.ucsusa.org/scientific_integrity/interference/prominent-statement-signatories.html

Culture of cronyism and corruption: http://www.buzzflash.com/alerts/05/09/ale05149.html

GOP prevented Kerry's election: Robert F. Kennedy Jr., "Was the 2004 Election Stolen?" *Rolling Stone,* June 1, 2006

Eschew "My country, right or wrong": Lutzer, p. 204

Democracy or democratic dissent: http://www.thirdworldtraveler.com/Parenti/Exposing_Terrorism_Trap.html

Political correctness defined: http://www.merriam-webster.com/dictionary/political
+correctness

Walter Schultze: Mosse, pp. 314–316

A fool's errand: Jonathan Rauch, "In Defense of Prejudice: Why Incendiary Speech Must Be Protected," *Harper's Magazine,* May 1995

A network of rival leaders: Mosse, p. xxxvi

Basic attitudes still with us: Ibid., p. xli

Despicable forces: Farago, p. 327

Fascism on the march again: Martin Lee, *The Beast Awakens: Fascism's Resurgence from Hitler's Spymasters to Today's Neo-Nazi Groups and Right-Wing Extremists* (London: Little, Brown and Company, 1997), p. 389

President Roosevelt's 1940 words: http://www.americanrhetoric.com/speeches/fdrarsenal ofdemocracy.html

INDEX

National Archives, U.S., 141–42, 251
National City Bank of Cleveland, 323
National City Bank of New York, 25, 30, 112, 141, 210, 214
National Defense Research Council, 166
National Economic Council, 256
National Education Association, 303
National Emergencies Act (1976), 337–38
National Institutes of Health (NIH), 271, 274
National Interest program, 153–54, 155
nationalism, 15, 362
National Program Office, 252–53
National Research Council, 169
National Security Act (1947), 163
National Security Agency, 63, 171, 180, 329, 338
National Security Council (NSC), 163–64, 170, 219, 252
National Security Presidential Directive/NSPD-51, 338
National Socialist German Workers Party, *see* Nazis, Nazism
Nation Betrayed, A (Rutz), 199–200
NATO, 145, 169, 258, 300
Nazarenko, Nicholas, 249
Nazi Culture (Mosse), 296, 317, 372
Nazi Doctors, The (Lifton), 185
Nazis, Nazism, 6, 21, 40, 107, 110, 121, 184, 254
  anti-Semitism of, 30–31, 254, 279, 302, 354
  Aryan myth of, 353, 354
  Bormann's control of, 110
  Catholic Church and, 128–31, 138, 141
  communism vs., 11–12, 26, 27
  "ends justifies means" ideology of, 15, 254
  eugenics programs of, 183–84, 282
  euthanasia program of, 185–87, 279, 333

foreign front corporations of, 111, 115, 117–18
  Fuehrerprinzip of, 335
  GHW Bush supported by, 249–51
  as globalist project, 5, 11–12, 162, 229, 231, 296
  industrialists and, 108–9
  Kennedy assassination and, 222, 223, 228
  Muslim Brotherhood and, 207–8
  occultism and, 93–94, 101–2, 173, 178–81, 287
  postwar plans of, 12, 107, 120–22, 135, 215
  propaganda used by, 26–27, 343–45, 354
  psychic experiments of, 180–81
  ratlines of, 127–42, 205, 323
  religion and, 286–88, 294
  Republican Party and, 218–19, 248–51, 292
  sterilization programs of, 183, 184–85
  in U.S., 216–19, 231, 236, 248–51, 257, 323; *see also* Paperclip, Project
  youth culture of, 317–18
  see also Germany, Nazi
Nazis Go Underground, The (Reiss), 107
Nazi technology, 4–5, 15, 38, 49, 50–91, 162
  aircraft, 53, 54, 60
  anti-gravity, 74, 76
  atomic bomb, *see* atomic bomb, Nazi
  computer, 74, 263
  conventional weapons, 51–52, 72
  flying saucers, 54–57, 165
  rocketry, 50, 53, 72, 74–76, 89–90, 157–58, 159, 161, 163, 263
  SS control of, 53–54, 73–79
  television, 51, 263
  U.S. acquisition of, *see* Paperclip, Project